ST. MARY'S COLLEGE OF MARYLAND LIBRARY
ST. MARY'S CITY, MARYLAND

Benchmark Papers in Behavior

Series Editors: Martin W. Schein
West Virginia University
and
Stephen W. Porges
University of Illinois
at Urbana-Champaign

Volume

1. HORMONES AND SEXUAL BEHAVIOR / *Carol Sue Carter*
2. TERRITORY / *Allen W. Stokes*
3. SOCIAL HIERARCHY AND DOMINANCE / *Martin W. Schein*
4. EXTERNAL CONSTRUCTION BY ANIMALS / *Nicholas E. Collias and Elsie C. Collias*
5. IMPRINTING / *Eckhard H. Hess and Slobodan B. Petrovich*
6. PSYCHOPHYSIOLOGY / *Stephen W. Porges and Michael G. H. Coles*
7. SOUND RECEPTION IN FISHES / *William N. Tavolga*
8. VERTEBRATE SOCIAL ORGANIZATION / *Edwin M. Banks*
9. SOUND PRODUCTION IN FISHES / *William N. Tavolga*
10. EVOLUTION OF PLAY BEHAVIOR / *Dietland Müller-Schwarze*
11. PARENTAL BEHAVIOR IN BIRDS / *Rae Silver*
12. CRITICAL PERIODS / *John Paul Scott*
13. THERMOREGULATION / *Evelyn Satinoff*

Related Titles in BENCHMARK PAPERS IN ECOLOGY Series

Volume

1. CYCLES OF ESSENTIAL ELEMENTS / *Lawrence R. Pomeroy*
2. BEHAVIOR AS AN ECOLOGICAL FACTOR / *David E. Davis*
3. NICHE: Theory and Application / *Robert H. Whittaker and Simon A. Levin*
4. ECOLOGICAL ENERGETICS / *Richard G. Wiegert*
5. ECOLOGICAL SUCCESSION / *Frank B. Golley*
6. PHYTOSOCIOLOGY / *Robert P. McIntosh*
7. POPULATION REGULATION / *Robert H. Tamarin*
8. PATTERNS OF PRIMARY PRODUCTION IN THE BIOSPHERE / *Helmut H. Lieth*
9. SYSTEMS ECOLOGY / *H. H. Shugart and R. V. O'Neill*

Benchmark Papers in Behavior / 13

A BENCHMARK® Books Series

THERMOREGULATION

Edited by
EVELYN SATINOFF
University of Illinois at Urbana-Champaign

STROUDSBURG, PENNSYLVANIA

Copyright © 1980 by **Dowden, Hutchinson & Ross, Inc.**
Benchmark Papers in Behavior, Volume 13
Library of Congress Catalog Card Number: 79-10267
ISBN: 0-87933-349-9

All rights reserved. No part of this book covered by the copyrights hereon may be reproduced or transmitted in any form or by any means—graphic, electronic, or mechanical, including photocopying, recording, taping, or information storage and retrieval systems—without written permission of the publisher.

82 81 80 1 2 3 4 5
Manufactured in the United States of America.

LIBRARY OF CONGRESS CATALOGING IN PUBLICATION DATA
Main entry under title:
Thermoregulation.
   (Benchmark papers in behavior; 13)
   Includes bibliographical references and indexes.
   1. Body temperature—Regulation—Addresses, essays, lectures.
I. Satinoff, Evelyn, 1937–   II. Series
QP135.T48   591.1'916   79-10267
ISBN 0-87933-349-9

Distributed world wide by Academic Press,
a subsidiary of Harcourt Brace Jovanovich,
Publishers.

SERIES EDITOR'S FOREWORD

It was not too many years ago that virtually all research publications dealing with animal behavior could be housed within the covers of a very few hard-bound volumes that were easily accessible to the few workers in the field. Times have changed! The present-day students of behavior have all they can do to keep abreast of developments within their own area of special interest, let alone in the field as a whole; and of course we have long since given up attempts to maintain more than a superficial awareness of what is happening "in biology," "in psychology," "in sociology," or in any of the broad fields touching upon or encompassing the behavioral sciences.

It was even fewer years ago that those who taught animal behavior courses could easily choose a suitable textbook from among the very few that were available; all "covered" the field, according to the bias of the author. Students working on a special project used *the* text and *the* journal as reference sources, and for the most part successfully covered their assigned topics. Times have changed! The present-day teacher of animal behavior is confronted with a bewildering array of books to choose among, some purported to be all-encompassing, others confessing to strictly delimited coverage, and still others being simply collections of recent and profound writings.

In response to the problem of the steadily increasing and overwhelming volume of informatin in the area, the Benchmark Papers in Behavior was launched as a series of single-topic volumes designed to be some things to some people. Each volume contains a collection of what an expert considers to be *the* significant research papers in a given topic area. Each volume, then, serves several purposes. To teachers, a Benchmark volume serves as a supplement to other written materials assigned to students; it permits in-depth consideration of a particular topic while at the same time confronting students (often for the first time) with original research papers of outstanding quality. To researchers, a Benchmark volume serves to save countless hours digging through the various journals to find *the* basic articles in their area of interest; often the journals are not easily available. To students, a Benchmark volume provides a readily accessible set of original papers on the topic, a set that forms the core of the more extensive bibliography that they are likely to compile;

Series Editor's Foreword

it also permits them to see at first hand what an "expert" thinks is important in the area, and to react accordingly. Finally, to librarians, a Benchmark volume represents a collection of important papers from many diverse sources, thus making readily available materials that might otherwise not be economically possible to obtain or physically possible to keep in stock.

The choice of topics to be covered in this series is no small matter. Each of us could come up with a long list of possible topics and then search for potential volume editors. Alternatively, we could draw up long lists of recognized and prominent scholars and try to persuade them to do a volume on a topic of their choice. For the most part, we have followed a mix of both approaches: match a distinguished researcher with a desired topic, and the results should be outstanding. So it is with the present volume.

Dr. Evelyn Satinoff is no newcomer to the problem of temperature regulation in warm-blooded animals; she has been publishing in this area for a number of years. More significantly, her recent outstanding reviews (1974, 1978) demonstrate an overall grasp and approach to the topic that is of interest to students and colleagues. Hence, it was singularly appropriate to invite her to serve as editor for this Benchmark volume. We are very pleased that she accepted.

MARTIN W. SCHEIN
STEPHEN W. PORGES

PREFACE

When I was asked to put together a set of papers that were critical in the field of temperature regulation, I was at first delighted. There had never been such a collection of original works in one place, and I had often wished for one to give my students so that they could not only learn about the field but also develop an appreciation of what excellent science is and how one goes about looking for critical problems to work on. Upon reflection, however, I realized that the task would be extremely difficult. Papers on heat and cold adaptation, clinical applications of hypothermia, pharmacological aspects of thermoregulation, hibernation and estivation, bioenergetics, control theory analysis of thermoregulation, and problems in peripheral and central thermodetection could all legitimately be included in a book entitled, "Thermoregulation." However, given the confines of no more than 400 pages, to include papers in all these areas would give the reader only a fragmentary feeling for what is central in the field. Happily, several of these subjects are discussed in Professor Theodor Benzinger's two-volume collection in the Benchmark series in human physiology.*

For me, the question that holds the field together is, "How do animals maintain a relatively constant body temperature in the face of a notoriously inconstant environment?" This question, as it is answered in many of the papers in this volume, may be restated as, "How does the central nervous system integrate an animal's available thermoregulatory responses so as to maintain thermal homeostasis?" The first set of papers defines the concept of thermal homeostasis and illustrates how animals maintain it by using the reflexive and behavioral means available to them.

The second group of papers concerns one of the most exciting stories in neurophysiology—the search for the thermostat. A thermostat implies a master temperature regulating "center" in the brain, and indeed, for many years the concept of centers dominated not only thermoregulation but also work on feeding, drinking, sexual behavior, sleep and wakefulness, and indeed, almost all behaviors that were known to be

---

*Benzinger, T. H. 1977. *Temperature, Part I: Arts and Concepts; Part II: Thermal Homeostasis*. Benchmark Papers in Human Physiology, vols. 9 and 10. Stroudsburg, Pa.: Dowden, Hutchinson & Ross.

Preface

under neural control. The concept is simple and parsimonious, and is therefore a highly attractive way to categorize brain-behavior interactions, but unfortunately, it is almost certainly inadequate. I have included several papers that demonstrate why it is inadequate and the new theoretical direction in which I believe the field is going.

Third, because I believe that the pathology of any system can be tremendously useful in understanding how the normal system operates, I have included several fundamental papers on fever, the most ubiquitous and striking pathology in the thermoregulation (although, as we shall see, it may be inaccurate to describe fever as a pathological condition even though the infection that causes it may be so described). Finally, since the evolution of endothermy, or warm-bloodedness, is a source of endless and fascinating speculation, and must be considered in any theory of thermoregulatory function, I have included two papers on this subject that are highly provocative and may even be true.

This is an idiosyncratic collection. Some of the papers included here are significant because they directed the general course of thermoregulatory research for many years. Others beautifully illustrate important theoretical or experimental points or striking, unexpected findings. Others had great influence over me in the course of my intellectual development. A further analysis of many of the phenomena described in these papers will be extremely fruitful in understanding the basic principles of homeostasis and motivated behavior in general, as well as thermoregulation. The papers have been collected in the hope that some of the people who read this book will be stimulated to investigate the subject further.

EVELYN SATINOFF

CONTENTS

Series Editor's Foreword	v
Preface	vii
Contents by Author	xiii
Introduction	1

PART I: THE CONCEPT OF THERMAL HOMEOSTASIS

Editor's Comments on Papers 1 Through 4   8

1   **BERNARD, C.:** Constant or Free Life   10
Lectures on the Phenomena of Life Common to Animals and Plants, H. E. Hoff, R. Guillemin, and L. Guillemin, trans., Springfield, Ill.: Charles C Thomas, 1974, pp. 83–91

2   **CANNON, W. B.:** The Constancy of Body Temperature   19
The Wisdom of the Body, N. Y.: W. W. Norton & Co., 1963, pp. 177–201

3   **RICHTER, C. P.:** Total Self Regulatory Functions in Animals and Human Beings   44
Harvey Lectures **38**:63–64, 68–71, 100–101 (1943)

4   **BARTHOLOMEW, G. A.:** The Roles of Physiology and Behaviour in the Maintenance of Homeostasis in the Desert Environment   50
Soc. Exp. Biol. Symp. **18**:7–29 (1964)

PART II: THE SEARCH FOR THE THERMOSTAT

Editor's Comments on Papers 5 Through 8   76

5   **OTT, I.:** The Relation of the Nervous System to the Temperature of the Body   80
J. Nerv. Ment. Dis. **11**:141–147 (1884)

6   **BAZETT, H. C., and W. G. PENFIELD:** A Study of the Sherrington Decerebrate Animal in the Chronic as Well as the Acute Condition   87
Brain **45**:185–194, 205–214, 237–239, 251–252, 257–262 (1922)

7   **RANSON, S. W.:** Regulation of Body Temperature   119
Assoc. Res. Nerv. Ment. Dis. Proc. **20**:342–343, 350–379, 390–396, 398–399 (1940)

8	SATINOFF, E.: Behavioral Thermoregulation in Response to Local Cooling of the Rat Brain *Am. J. Physiol.* **206**:1389–1394 (1964)	158

PART III: EVIDENCE FOR MANY THERMOSTATS

Editor's Comments on Papers 9 Through 14		166
9	KELLER, A. D.: Observations on the Localization in the Brainstem of Mechanisms Controlling Body Temperature *Am. J. Med. Sci.* **185**:746–748 (1933)	170
10	SIMON, E., W. RAUTENBERG, and C. JESSEN: Initiation of Shivering in Unanaesthetized Dogs by Local Cooling Within the Vertebral Canal *Experientia* **21**:476–477 (1965)	173
11	CHAMBERS, W. W., M. S. SEIGEL, J. C. LIU, and C. N. LIU: Thermoregulatory Responses of Decerebrate and Spinal Cats *Exp. Neurol.* **42**:282–299 (1974)	177
12	ROBERTS, W. W., and R. D. MOONEY: Brain Areas Controlling Thermoregulatory Grooming, Prone Extension, Locomotion, and Tail Vasodilation in Rats *J. Comp. Physiol. Psychol.* **86**:470–480 (1974)	195
13	CARLISLE, H. J., and D. L. INGRAM: The Effects of Heating and Cooling the Spinal Cord and Hypothalamus on Thermoregulatory Behaviour in the Pig *J. Physiol.* **231**:353–364 (1973)	206
14	CABANAC, M., P. RAMEL, R. DUCLAUX, and M. JOLI: Indifférence a la douleur et confort thermique (2 cas) *La Presse Med.* **77**:2053–2054 (1969)	218
	English translation: Indifference to Pain and Thermal Comfort	220

PART IV: FEVER

Editor's Comments on Papers 15 Through 18		224
15	BERNARD, C.: Lectures 20 and 21 Translated from *Leçons sur la chaleur animale: sur les effets de la chaleur et sur la fièvre*, Paris: Bailliere, 1876, pp. 405–445	227
16	KING, M. K., and W. B. WOOD, Jr.: Studies on the Pathogenesis of Fever. IV. The Site of Action of Leucocytic and Circulating Endogenous Pyrogen *J. Exp. Med.* **107**:291–303 (1958)	244
17	VAUGHN, L. K., H. A. BERNHEIM, and M. J. KLUGER: Fever in the Lizard *Dipsosaurus dorsalis* *Nature* **252**:473–474 (1974)	257
18	SATINOFF, E., G. N. McEWEN, Jr., and B. A. WILLIAMS: Behavioral Fever in Newborn Rabbits *Science* **193**:1139–1140 (1976)	259

PART V: THERMOREGULATORY BEHAVIOR

Editor's Comments on Papers 19 Through 24   262

19   NEWPORT, G.: On the Temperature of Insects, and Its Connexion with the Functions of Respiration and Circulation in This Class of Invertebrated Animals   264
Philos. Trans. Royal Soc. London, Part II, **127**:259–263, 294–299, 324–328 (1837)

20   ROZIN, P. N., and J. MAYER: Thermal Reinforcement and Thermoregulatory Behavior in the Goldfish, *Carassius auratus*   280
Science **134**:942–943 (1961)

21   REGAL, P. J.: Voluntary Hypothermia in Reptiles   282
Science **155**:1551–1553 (1967)

22   KINDER, E. F.: A Study of the Nest-building Activity of the Albino Rat   285
J. Exp. Zool. **47**:117–125, 139–161 (1927)

23   WEISS, B.: Thermal Behavior of the Subnourished and Pantothenic-acid-deprived Rat   316
J. Comp. Physiol. Psychol. **50**:481–485 (1957)

24   STRICKER, E. M., and F. R. HAINSWORTH: Evaporative Cooling in the Rat: Interaction with Heat Loss from the Tail   321
Q. J. Exp. Physiol. **56**:231–241 (1971)

PART VI: EVOLUTION

Editor's Comments on Papers 25 and 26   334

25   HEATH, J. E.: The Origins of Thermoregulation   336
Evolution and Environment, E. T. Drake, ed., New Haven, Conn., and London: Yale University Press, 1968, pp. 259–278

26   COWLES, R. B.: Possible Origin of Dermal Temperature Regulation   356
Evolution **12**:347–357 (1958)

Author Citation Index   367
Subject Index   373
About the Editor   377

CONTENTS BY AUTHOR

Bartholomew, G. A., 50
Bazett, H. C., 87
Bernard, C., 10, 227
Bernheim, H. A., 257
Cabanac, M., 218, 220
Cannon, W. B., 19
Carlisle, H. J., 206
Chambers, W. W., 177
Cowles, R. B., 356
Duclaux, R., 218, 220
Hainsworth, F. R., 321
Heath, J. E., 336
Ingram, D. L., 206
Jessen, C., 173
Joli, M., 218, 220
Keller, A. D., 170
Kinder, E. F., 285
King, M. K., 244
Kluger, M. J., 257
Liu, C. N., 177
Liu, J. C., 177

Mayer, J., 280
McEwen, G. N., Jr., 259
Mooney, R. D., 195
Newport, G., 264
Ott, I., 80
Penfield, W. G., 87
Ramel, P., 218, 220
Ranson, S. W., 119
Rautenberg, W., 173
Regal, P. J., 282
Richter, C. P., 44
Roberts, W. W., 195
Rozin, P. N., 280
Satinoff, E., 158, 259
Seigel, M. S., 177
Simon, E., 173
Stricker, E. M., 321
Vaughn, L. K., 257
Weiss, B., 316
Williams, B. A., 259
Wood, W. B., Jr., 244

THERMOREGULATION

INTRODUCTION

If an animal is to stay alive and healthy, certain basic physiological functions must be held fairly constant. The tissues of the body must receive an adequate but not excessive supply of oxygen. The blood must not become too acidic or too alkaline. Blood glucose levels and body temperature must be maintained within very narrow limits. This is no problem for an organism that lives in a congenial, unvarying environment, like a tapeworm or a fetus. But the rest of us must depend on our own resources to achieve such stability, and the degree to which our resources are automatic determines the time we have to devote to such activities as mating, defending a territory, or reading a book. For instance, if the liver did not store glucose and release it at a controlled rate, we would have to spend most of our time preparing and eating small quantities of food to give us energy to be active. Because of the liver's automatic functions, we are able instead to eat only two or three times a day and spend the rest of our time doing other things. Our bodies do not store and release heat, but we have very efficient behavioral and reflexive mechanisms for maintaining a constant body temperature and so we are largely free from worries about climatic changes. The great French physiologist Claude Bernard (1813-1878) was the first to enunciate this principle. In 1868 he wrote, "The constancy of the internal environment is the condition of a free and independent life" (see Paper 1, this volume). This is the first statement of the principle Walter B. Cannon was later to call "homeostasis" (see Paper 2). *Homeostasis* refers to the state of equilibrium in the body with respect to various functions and chemical composition of fluids and tissues and also to the processes through which such equilibrium is maintained.

Temperature regulation is an outstanding example of a homeostatic process. To understand how this regulation is accomplished, three major questions must be answered: (1) What is the source of the heat that allows endothermic mammals and birds to be warmer than their environment? (2) What responses does an animal use to increase heat production and conservation and to decrease heat loss? (3) How does the central nervous system integrate these responses so that a constant body temperature is maintained? Most of the papers in this collection concern answers to the second and third questions. However, these questions could begin to be studied only after the first problem was resolved. Therefore, for reasons of historical interest and also because I believe that an understanding of the answer to the first question will make it easier to integrate the other two, I will begin with a short history of the inquiry into the source of animal heat. (For much of this history, I am indebted to Goodfield, 1960.)

WHAT IS THE SOURCE OF ANIMAL HEAT?

One of the most obvious differences between living animals and dead ones is that living beings are warm and dead ones are cold. Inquiry into the source of heat in living organisms has been one of the central problems in physiology. The ancient Greeks thought that the heart generated the heat, which was then distributed to the rest of the body through the arteries. The function of the lungs and respiration was primarily to cool the blood. By the end of the eighteenth century, we find several different theories. Some were mechanical, for example, the heat was generated by friction between the blood and the blood vessels. However, such theories were very shortlived. The problem with them was that the blood was fluid and the surfaces of the vessels were so smooth that not enough friction could possibly be generated to provide adequate heat.

Other theories were chemical, for example, that the heat was produced by some sort of fermentation or chemical mixture. In fact, one of these ideas was partially correct in assuming that the decomposition of food eaten was responsible for animal heat. A different, combustion, theory put forward by the Scottish chemist Joseph Black (1728-1799), asserted that animal heat was generated in the lungs which, through respiration, emitted carbon dioxide (or "fixed air"), which was accompanied by heat. (Black did not conceive of heat as a form of energy but rather as a substance that combined with the thing being heated—a sort of physical analog of phlogiston.) Black's theory was correct in equating respiration with combustion, but it was in error in focusing all attention on the lungs. At some point, attention had to be directed to the individual cells in the body.

Introduction

It remained for the great German organic chemist, Justus von Liebig (1803-1873) to provide an adequate account of the chemical processes involved in the generation of animal heat. He was the first to realize that the carbon dioxide and water breathed out came from the oxidation of the complex foods taken in, and he was able to account for the chemical transformations that took place in the body during this conversion.

> The mutual action between the elements of the food and the oxygen conveyed by the circulation of the blood to every part of the body is the source of animal heat. (von Liebig, quoted by Goodfield, 1960.)

The picture of animal heat we have today is basically unchanged since von Liebig. Oxygen, carried by the lungs through the bloodstream to each cell in the body, combines with the nutrients from the food eaten. This process liberates energy, which is dissipated mainly in the form of heat.

Although von Liebig gave a precise and accurate account of the physicochemical nature of animal heat, he still could not understand how a constant body temperature was maintained—how an animal in contact with the varying external would appeared to be unaffected by it. He invoked the existence of a "vital force," some unique characteristic of living organisms that enabled them to resist the influence of external agents. Here Claude Bernard's great concept of the "milieu interieur" enters, removing the last traces of vitalism from explanations of animal functions, explaining the constancy of body temperature, and establishing physiology as a science in its own right, not subservient to physics and chemistry.

Claude Bernard was a determinist; he believed that all biological phenomena were determined by physical and chemical conditions. But the conditions to which he referred were those of the internal environment of the body, not those of the external world. He was the first to insist that one must look at the whole complex organism—not simply at an isolated organ or biochemical process—and the way in which it interacts with its environment.

> As the organism becomes more and more complex, so this environment becomes more and more isolated from the outside world, and the differences between simpler and more complex creatures are simply differences in this degree of "isolation and protection." (Bernard, quoted by Goodfield, 1960.)

When vital phenomena are looked at in this light, we can see that they are surrounded by their own environment, which has its own regulatory mechanisms. To understand the constancy of body temperature, then, we must understand the regulatory mechanisms of the internal

Introduction

environment, and when they are understood, Bernard said, it will be found that they follow the laws of physics and chemistry.

Thus the animal is not out of touch with or unaffected by the outside world. On the contrary, it is in intimate contact with it, and it is the immediate and successful operation of the animal's buffering systems in response to changes in the external environment that allows for that internal constancy we call *homeostasis.*

HOW IS BODY TEMPERATURE REGULATED?

Body temperature can be regulated only if rates of heat production and heat loss are controlled. This control is exerted by the central nervous system, and it was only after the development of neurophysiological techniques that the nervous control of body temperature could be experimentally investigated. Isaac Ott (1847-1916) (Paper 5) in Philadelphia and Charles Richet (1850-1935) in France first demonstrated that temperature regulation could be deranged by puncturing an area of the brain between the corpus striatum and the optic thalamus. This localization was further supported by observations that, as successive levels of the brainstem were transected, animals were essentially unable to regulate their internal temperature at all unless the hypothalamus remained connected to the rest of the brain below (Bazett and Penfield 1922, Paper 6).

Besides lesions and transections, the technique of thermal stimulation was used to define more accurately the neural area concerned with thermoregulation. Barbour (1912) first showed that heating and cooling the tissue around the corpus striatum in rabbits altered deep body temperature. Magoun, Harrison, Brobeck, and Ranson (1938) then localized the areas more precisely by demonstrating that radio-frequency heating of discrete hypothalamic regions in anesthetized cats evoked the heat-loss responses of increased respiration and panting. These reactions were most marked when the heating electrodes were in the anterior hypothalamus and the preoptic area just rostral to it, but they could still be elicited by more caudal stimulation. These results were confirmed in monkeys (Beaton et al, 1941) and interpreted as identifying a reactive region in the anterior hypothalamus containing elements excited by rising temperature, which in turn activated heat-loss mechanisms.

All these data implied that the hypothalamus is sensitive to its own temperature and to temperature signals from the rest of the brain and body. In 1963 temperature-sensitive units, that is, cells that alter their firing rate in response to changes in their temperature, were identified by Nakayama, Hammel, Hardy, and Eisenman. Since that time many other areas of the brain have been shown to contain temperature-sensitive units (Bligh 1973).

For many years, however, the hypothalamus was considered to be the only area of the brain controlling thermoregulatory responses. It was divided, after a theory of Hans Meyer (1913), into two antagonistic but interrelated centers, the anterior portion responsible for heat loss and the posterior portion responsible for heat production (see Satinoff 1974 for review). Nevertheless, all through this period, Keller (Paper 9) in the United States was pointing out that heat-loss responses could be obtained in decerebrate dogs, with no hypothalamus attached, and Thauer (1935) in Germany was arguing that the spinal cord contained thermosensitive structures. The evidence is now incontrovertible that many areas of the brain and spinal cord are indeed capable of inducing appropriate heat-loss and heat-production responses, even in the absence of the hypothalamus.

What then is the function of the hypothalamus in temperature regulation? Within a Jacksonian framework (Jackson 1958), it serves to coordinate and adjust the activity of thermoregulatory systems located at several lower levels of the neuraxis. Selective facilitation and inhibition from the hypothalamus to lower levels would ensure that appropriate thermoregulatory reflexes are activated and inappropriate ones suppressed. Facilitatory influences from the hypothalamus would also ensure that thermoregulatory responses are initiated promptly in the presence of an appropriate stimulus (Satinoff 1978).

All the neurophysiological work discussed here was devoted to understanding the neural control of reflexive thermoregulatory responses such as panting, sweating, and vasodilation in the heat and shivering, nonshivering thermogenesis and vasoconstriction in the cold. Very little attention was paid to behavior, possibly a more important means of achieving a stable internal temperature. Zoologists (for example, Cowles and Bogert 1944) had long been aware of the importance of behavior in reptiles. Indeed, insects, fish, amphibia, and reptiles have highly developed behavioral methods of thermoregulating whereas automatic mechanisms are either nonexistent or few and inefficient. Some psychologists (for example, Richter, Paper 3; Kinder, Paper 22) had been studying nest building in the cold in rats, but in general, thermoregulatory behavior in mammals was ignored, probably because there was no convenient method available for quantifying it. Then in the late 1950s, Carlton and Marx (1958) and Weiss (Paper 23), using the operant techniques developed by B. F. Skinner, demonstrated that behavior is precisely attuned to regulating body temperature. As several papers in this volume attest, behavior alone is extremely successful in dealing with thermal stresses.

REFERENCES

Barbour, H. G., 1912. Die Wirkung unmittelbarer Erwärmung und Abkühlung der Wärmezentra auf die Körpertemperatur. *Arch. Exp. Pathol. Physiol. Pharmakol.* **70**:1-26.

Beaton, L. E.; McKinley, W. A.; Berry, C. M.; and Ranson, S. W., 1941. Localization of cerebral center activating heat-loss mechanisms in monkeys. *J. Neurophysiol.* **4**:478-485.

Bligh, J., 1973. *Temperature Regulation in Mammals and Other Vertebrates.* North-Holland, Amsterdam.

Carlton, P. L., and Marks, R. A., 1958. Cold exposure and heat reinforced operant behavior. *Science* **128**:1344.

Cowles, R. B., and Bogert, C. M., 1944. A preliminary study of the thermal requirements of desert reptiles. *Bull. Am. Mus. Nat. Hist.* **83**:265-296.

Goodfield, G. J., 1960. *The Growth of Scientific Physiology.* Hutchinson, London.

Jackson, J. H., 1958. *Selected Writings of John Hughlings Jackson.* J. Taylor, ed. Basic Books, New York.

Magoun, H. W.; Harrison, F.; Brobeck, J. R.; and Ranson, S. W., 1938. Activation of heat loss mechanisms by local heating of the brain. *J. Neurophysiol.* **1**:101-114.

Meyer, H. H., 1913. Theorie des Fiebers und seiner Behandlung. *Verh. Dtsch. Bes. Inn. Med.* **30**:15.

Nakayama, T.; Hammel, H. T.; Hardy, J. D.; and Eisenman, J. S., 1963. Thermal stimulation of electrical activity of single units of the preoptic region. *Am. J. Physiol.*, **204**:1122-1126.

Satinoff, E., 1974. Neural integration of thermoregulatory responses. *In Limbic and Autonomic Nervous System: Advances in Research*, L. V. Dicara, ed. Plenum Press, New York.

Satinoff, E. 1978. Neural organization and evolution of thermal regulation in mammals. *Science* **201**:16-22.

Thauer, R., 1935. Wärmeregulation und Fieberfähigkeit nach operativen Eingriffen am Nervensystem homoiothermer Säugetiere. *Pflug. Arch. ges. Physiol.* **236**: 102-147.

Part I

THE CONCEPT OF
THERMAL HOMEOSTASIS

Editor's Comments on Papers 1 Through 4

1 **BERNARD**
   Constant or Free Life

2 **CANNON**
   The Constancy of Body Temperature

3 **RICHTER**
   Excerpts from *Total Self-Regulatory Functions in Animals and Human Beings*

4 **BARTHOLOMEW**
   The Roles of Physiology and Behaviour in the Maintenance of Homeostasis in the Desert Environment

The four selections in this section begin with the concept of thermal homeostasis and then go on to describe the various reflexive and behavioral adjustments that make a constant body temperature possible.

Paper 1 presents Claude Bernard's written statement of the constancy of the internal environment, one of the cornerstones of modern physiology. Bernard distinguishes three forms of life: (1) *latent life*, in which organisms are *potentially* alive but temporarily "have fallen into the state of chemical indifference," that is, when the external physicochemical conditions, especially heat, humidity, and oxygen, are inadequate, vital motion is suspended. Seeds, yeast, and some simple animal forms are examples of latent life.

(2) *Oscillating life,* in which vital signs can vary within wide limits depending on the external environment. All plants that are dormant in winter exhibit oscillating life. All invertebrates and all ectothermic vertebrates, whose body temperature, and therefore activity, depends on the environment, also exhibit oscillating life. "Cold makes them dormant and if they cannot be removed from its influence during the winter, life becomes attenuated, respiration slows down, digestion is suspended, movements become feeble or disappear."

(3) *Constant life,* the subject of the present excerpt. It also depends on water, oxygen, heat, and the chemical reserves that the animal has

available to it. In the section on heat Bernard mentions the vasomotor nerves, which by causing vasoconstriction, prevent heat from being lost to the periphery and by causing vasodilation, allow animals to lose heat. The discovery of the vasomotor nerves was one of Bernard's greatest contributions to physiology. (The book from which this excerpt is taken, *Lectures on the Phenomena of Life Common to Animals and Plants,* has been translated into English and is well worth reading not only for a more extended understanding of the principle but also for Bernard's arguments for determinism in physiology.)

In Paper 2, Walter B. Cannon discusses the homeostatic mechanisms by which one critical aspect of the internal environment, body temperature, is maintained. Cannon is concerned mainly with reflexive mechanisms in mammals and discusses physical methods of heat production and heat loss such as shivering and vasoconstriction and chemical means of heat production such as increased thyroid and adrenal output. He shows in his experiments that these responses are delicately balanced and concludes that this implies a regulatory thermostat somewhere in the brain.

In Paper 3, Curt Richter extends the mechanisms maintain homeostasis to behavior. Richter did a long series of experiments on specific hungers in animals and humans, demonstrating that if homeostatic balance could not be maintained physiologically, for instance, if salt balance were disturbed because of extirpation of the adrenal glands, the animals would attempt to maintain a constant internal environment through behavioral means such as ingesting larger than normal amounts of a highly salty diet. Similarly, if heat production were disturbed through extirpation of various endocrine glands, rats would build much larger than average nests and thereby maintain a normal body temperature.

In Paper 4, George Bartholomew demonstrates how reflexive and behavioral mechanisms interact to maintain homeostasis in mammals in the harsh desert environment and further, how reptiles are able to maintain highly stable internal temperatures using behavior almost entirely.

Copyright © 1974 by Charles C Thomas, Publisher

Reprinted from pages 83-91 of *Lectures on the Phenomena of Life Common to Animals and Plants*, H. E. Hoff, R. Guillemin, and L. Guillemin, trans., Charles C Thomas, 1974, 288pp.

CONSTANT OR FREE LIFE

Claude Bernard

[*Editor's Note:* In the original, material precedes and follows this excerpt.]

Constant, or free, life is the third form of life; it belongs to the most highly organized animals. In it, life is not suspended in any circumstance, it unrolls along a constant course, apparently indifferent to the variations in the cosmic environment, or to the changes in the material conditions that surround the animal. Organs, apparatus, and tissues function in an apparently uniform manner, without their activity undergoing those considerable variations exhibited by animals with an oscillating life. This is because in reality the *internal environment* that envelops the organs, the tissues, and the elements of the tissues does not change; the variations in the atmosphere stop there, so that it is true to say that *physical conditions of the environment* are constant in the higher animals; it is enveloped in an invariable medium, which acts as an atmosphere of its own in the constantly changing cosmic environment. It is an organism that has placed itself in a hothouse. Thus the perpetual changes in the cosmic environment do not touch it; it is not chained to them, it is free and independent.

I believe I was the first to insist upon this idea that there are really two environments for the animal, an *external environment* in which the organism is placed, and an *internal environment* in which the elements of the tissues live. Life does not run its course within the external environment, atmospheric air for the air breathing creatures, fresh or salt water for the aquatic animals, but within the *fluid internal environment* formed by the circulating organic liquid that surrounds and bathes all of the anatomical elements of the tissues; this is the lymph or plasma, the liquid portion of the blood which in the higher animals perfuses the tissues and constitutes the ensemble of all the interstitial fluids, is an expression of all the local nutritions, and is the source and confluence of all the elementary exchanges. A complex organism must be considered as an association of *simple beings,* which are the anatomical elements, and which live in the fluid internal environment.

The constancy of the internal environment is the condition for free and independent life: the mechanism that makes it possible is that which assured the maintenance, within the *internal environment,* of all the conditions necessary for the life of the elements. This enables us to understand that there could be no free and independent life for the simple beings whose constituent elements are in direct contact with the cosmic environment, but that this form of life is on the contrary the exclusive attribute of beings that have arrived at the summit of complication or organic differentiation.

The constancy of the environment presupposes a perfection of the organism such that external variations are at every instant compensated and brought into balance. In consequence, far from being indifferent to the external world, the higher animal is on the contrary in a close and wise relation with it, so that its equilibrium results from a continuous and delicate compensation established as if by the most sensitive of balances.

The conditions necessary for the life of the elements which must be brought together and maintained constant in the internal environment for the exercise of free life, are those that we know already: water, oxygen, heat, and chemical substances or reserves.

These are the same conditions as those which are necessary for life in the simple beings, except that in the more perfect animals with independent life, the nervous system is called upon to regulate the harmony among all these conditions.

1. Water

This is an indispensable element, qualitatively and quantitatively, in the constitution of the environment in which the living elements function and evolve. In free-living animals there must exist an ensemble of dispositions regulating output and intake so as to maintain the necessary quantity of water within the internal environment. In the lower beings the quantitative variations in water compatible with life are more extensive, but the creature is on the other hand without means of regulating them. This is why it is chained to the vicissitudes of the climate, dormant in latent life during dry weather, revived in wet weather.

The highest organism is inaccessible to hygrometric variations, thanks to artifices of construction and to physiological functions which tend to maintain the relative constancy of the quantity of water.

In man, especially, and in general in the higher animals, loss of water occurs in all the secretions, in the urine and the sweat especially, and secondarily in respiration, which carries off a notable quantity of water vapor, and finally by cutaneous perspiration.

As to the intake, this is accomplished by the ingestion of fluids or of foods that include water, or even in some animals by absorption through the skin. In all events, it is most likely that the whole quantity of the water in the organism comes from the outside by the one or the other two routes. It has not been possible to demonstrate that the animal organism really produces water; the contrary opinion appears to be nearly certain.

It is the nervous system, we have said, that provides the mechanism for compensation between intake and output. The sensation of thirst, which is under the control of this system, makes itself felt whenever the proportion of fluid diminishes within the body as the result of some condition such as hemorrhage or abundant sweating; the animal thus finds itself induced to drink in order to restore the losses it has undergone. But even this ingestion is regulated in the sense that it cannot increase the quantity of water present in the blood beyond a certain level; urinary and other excretions eliminate the surplus as a sort of overflow. The mechanisms that vary the quantity of water and reestablish it are thus most numerous; they set in motion a host of mechanisms of secretion, exhalation, ingestion, and circulation which transport the ingested and absorbed fluid. These mechanisms are varied, but cooperate to the same end: the presence of water in effectively fixed proportions within the internal environment, the condition for free life.

These compensatory mechanisms exist not only for water; they are observed also for most of the mineral and organic substances contained in solution in the blood. It is known that the blood cannot take on a considerable load of sodium chloride, for example; above a certain limit the excess is eliminated in the urine. As I have established, it is the same for sugar, which normally present in the blood, is, above a certain quantity, eliminated in the urine.

2. Heat

We know that for each organism, elementary or complex, limits of external temperature exist between which its activity is possible, with a midpoint which corresponds to the maximum of vital energy. This is true not only for beings that have arrived at the adult state but also for the egg or embryo. All these creatures are subject to oscillating life, but for the higher animals, the so-called warm-blooded animals, the temperature compatible with the manifestations of life is closely fixed. This fixed temperature is maintained within the internal environment despite extreme climatic variations, and assures the continuity and the independence of life. In a word, there exists in animals with constant and free life a function of calorification which does not exist at all in animals with an oscillating life.

For this function there exists an ensemble of mechanisms governed by the nervous system. There are *thermic* nerves and *vasomotor* nerves to which I have called attention, whose activity produces sometimes an elevation and sometimes a fall in temperature, according to the circumstances.

The production of heat is due, in the living world as in the inorganic world, to chemical phenomena; such is the great law whose understanding we owe to Lavoisier and Laplace. It is in the chemical activity of the tissues that the higher organism finds the source of the heat it conserves within its internal environment, at a nearly constant level, from 38 to 40 degrees for mammals, and 45 to 47 degrees for birds. The regulation of heat takes place, as

I have said, by means of two kinds of nerves; the nerves that I call *thermic,* which belong to the sympathetic system and serve as brakes, so to speak, on the chemicothermic activities taking place within the living tissues. When these nerves act, they diminish the interstitial combustions and lower the temperature; when their influence is weakened by suppression of their action or by the antagonism of other nervous influences, then combustions are increased, and the temperature of the internal environment rises considerably. The *vasomotor* nerves, by accelerating circulation in the periphery of the body, or in the central organs, intervene also in the mechanisms for the equilibration of animal heat.

I will add only this last fact. When the action of the cerebrospinal system is considerably attenuated, while that of the sympathetic *(thermic nerve)* is permitted to remain intact, temperature is seen to fall considerably, and the warm-blooded animal is so to speak converted into a cold-blooded animal. I have carried out this experiment on rabbits, cutting the spinal cord between the seventh cervical vertebra and the first dorsal. When, on the contrary, the sympathetic is destroyed, leaving the cerebrospinal system intact, the temperature is noted to rise, at first locally and then generally; this is the experiment I carried out in horses by cutting the sympathetic trunk, especially when they were weakened beforehand. A true fever then follows. I have elsewhere developed the history of all these mechanisms at length (see Leçons sur la Chaleur Animale, 1873); I recall them here only to establish that the calorific function characteristic of warm-blooded animals results from the perfecting of the nervous mechanism which, by an incessant compensation, maintains an apparently fixed temperature within the *internal environment,* within which there live the organic elements to which ultimately we must always attribute all the vital manifestations.

3. Oxygen

The manifestations of life require for their production the intervention of air, or better, its active portion, oxygen, in a dissolved form and in an appropriate state for it to reach the elementary organism. It is moreover necessary for this oxygen to be in proportions that are to a certain degree constant within the internal environment; too small a quantity or too great a quantity are equally incompatible with the vital functions.

Thus in animals with a constant life appropriate mechanisms are required to regulate the quantity of this gas which is assigned to the internal environment, and to keep it more or less constant. In the highly organized animals the penetration of oxygen into the blood is dependent upon the respiratory movements and the quantity of this gas present in the ambient environment. Moreover, the quantity of oxygen that is found in the air depends, as physics teaches it, on the percentage composition of the atmosphere and its pressure. Thus it can be understood that an animal could live in an atmosphere less rich in oxygen if an increase in pressure compensated for this decrease, and inversely, that the same animal could live in an environment richer in oxygen than ordinary air if a diminution in pressure compensated for the increase. This is an important general proposition, resulting from the work of Paul Bert. It can be seen in this case that the variations in the environment compensate and balance each other, without the intervention of the animal. If the percentage composition diminishes or increases in the opposite direction, when the pressure rises or falls, the animal ultimately finds the same quantity of oxygen in the environment and its life goes on under the same conditions.

But there can be mechanisms within the animal itself that establish this compensation when it is not accomplished on the outside, and which insure the penetration into the internal environment of the quantity of oxygen required by the vital functions; we would mention the different variations that can take place in the quantity of hemoglobin, the active absorbing material for oxygen, variations that are still little known but which certainly also intervene for their own part.

All these mechanisms, like the preceding, are without effect except within rather restricted limits; they are perverted and become powerless in extreme conditions. They are regulated by the nervous system. When the air becomes rarefied for some reason, such as during ascension in a balloon or on mountains, the respiratory movements become deeper and more frequent, and compensation is established. Nevertheless, mammals and man cannot sustain this struggle for compensation very long when rarification is extreme, as when for instance they are transported to altitudes above 5000 meters.

We cannot enter here into the particular details that the question deserves. It suffices for us to propose it. We call attention only to an example related by Campana. It is relative to the high-flying birds, such as the birds of prey and particularly the condor, which rises to heights of 7000 to 8000 meters. They remain there, moving around for long periods of time, although in an atmosphere that would be fatal to a mammal. The principles set forth above permit the prediction that the internal respiratory environment of these animals ought to escape, by some appropriate mechanism, from the depression of the external environment; in other terms, that the oxygen contained in their arterial blood ought not to vary at these great heights. In fact there are in the birds of prey enormous pneumatic sacs, connected to the wings, which do not operate except when these move. When the wings lift, they are filled with external air, when they fall, they pump the air into the pulmonary parenchyma. So that, as the air is rarified, the work of the bird's wing which supports it is necessarily increased and consequently the supplementary volume of air passing through the lungs is also increased. The compensation for the rarification of the external air by an increase in the quantity inspired is thereby assured, and with it the constancy of the respiratory environment characteristic of the bird.

These examples, which we could multiply, demonstrate to us that all the vital mechanisms, however varied they might be, always have one purpose, that of maintaining the integrity of the conditions for life within the internal environment.

4. Reserves

Finally, it is necessary for the maintenance of life that the animal have reserves that assure the constancy of the constitution of its internal environment. Highly organized beings draw the materials for their internal environment from their food, but as they cannot be subjected to an identical and exclusive kind of diet, they must have within themselves mechanisms that derive similar materials from these varied diets and regulate the proportion of them that must enter the blood.

I have demonstrated, and we shall see later, that nutrition is not *direct* according to the teaching of accepted chemical theories, but that on the contrary, it is *indirect* and carried out by means of reserves. This fundamental law is a consequence of the variety of the diet as compared with the constancy of the environment. In a word, *one does not live by his present food, but by that which he has eaten previously,* modified, and in some way created by assimilation. It is the same with respiratory combustion; nowhere is it *direct,* as we shall demonstrate later.

Thus there are reserves, prepared from the food, and consumed at each moment in greater or lesser proportions. The vital manifestations thus destroy the provisions which no doubt have their primary origin from the outside, but which have been elaborated within the tissues of the organism, and which, added to the blood, insure the constancy of its chemicophysical constitution.

When the mechanisms of nutrition are disturbed, and when the animal finds it impossible to prepare these reserves, when it only consumes those that it had accumulated beforehand, it is on its way to ruin, that can end only in the impossibility of life, in death. It would then be of no use for it to eat; it would not be nourished, it would not assimilate, it would waste away.

Something of the kind takes place when the animal is in a state of fever; it uses without restoring, and this state becomes fatal if it persists to the complete exhaustion of the materials accumulated through previous nutrition.

Thus, the nutritive substances that enter an organism, whether animal or plant, do not participate in nutrition directly or immediately. The nutritive phenomenon takes place in two stages and these two stages are always separated from one another by a longer or shorter period, whose duration is a function of a host of circumstances. Nutrition is preceded by a particular elaboration that is terminated by a *storage of reserves* in the animal as well as in the plant. This fact permits one to understand how a being can continue to live, sometimes for a long time, without taking food; it lives on its reserves, accumulated within its own substance; it consumes itself.

These reserves are of variable importance depending upon the creatures concerned, and the various substances, in different animals and plants, and in annual or biennial plants, etc. This is not the place to analyze such a vast subject; we have wanted to show that the formation of reserves is not only the general law of all forms of life, but that it constitutes also an active and indispensable mechanism for the maintenance of a constant and free life, independent of variations in the ambient cosmic environment.

Conclusion

We have examined in succession the three general forms in which life appears: *latent* life, *oscillating* life, and *constant* life, in order to see whether in any of them we might find an internal vital principle capable of producing its manifestations, independently of external physicochemical conditions. The conclusion to which we find ourselves led is easy to draw. We see that in latent life the being is dominated by external physicochemical conditions, to the point that all vital manifestations can be arrested. In oscillating life, if the living being is not as absolutely subject to these conditions, it nevertheless remains so chained to them that it is subject to all their variations. In constant life the living being seems to be free, and vital manifestations appear to be produced and directed by an inner vital principle free from external physicochemical conditions; this appearance is an illusion. On the contrary, it is particularly in the mechanism of constant or free life that these close relations exhibit themselves in their full clarity.

We cannot therefore admit the presence of a free vital principle within living beings, in conflict with physical conditions. It is the opposite fact that is demonstrated, and thus all the contrary concepts of the vitalists are overthrown.

2

Copyright © 1932 by Walter B. Cannon; copyright renewed 1960 by Cornelia J. Cannon. Revised edition copyright © 1939 by Walter B. Cannon; renewed 1966, 1967 by Cornelia J. Cannon

Reprinted from pages 177-201 of *The Wisdom of the Body*, W. W. Norton & Co., 1963, 333pp., by permission of W. W. Norton & Company.

THE CONSTANCY OF BODY TEMPERATURE

Walter B. Cannon

[*Editor's Note:* In the original, material precedes and follows this excerpt.]

I

One of the most striking and easily observed constants of the internal environment is that of the temperature of "warm-blooded" animals. Although in normal human beings there is a daily swing from a low point about 4 A. M., when the thermometer readings average 36.3° C. (97.3° Fahrenheit) to a high point about 4 P. M., when they average 37.3° C. (99.1° F.), it does not vary much beyond this range. The constancy is so reliable that the thermometer makers can stamp "98.6°" on the Fahrenheit scale with assurance that it will mark closely the mean temperature of the healthy person everywhere.

The uniformity of our body temperature is not always maintained. Alcohol and anesthetics may abolish the regulatory processes, and then, on exposure to cold, heat is rapidly lost. Thus during alcoholic intoxication a man has suffered a fall of temperature to 24° C. (75° F.) and later has recovered a normal state. On the other hand, in the course of infectious diseases the fever may rise to 40° C. (104° F.) or higher without causing disability. But these variations bring dangers and limitations. For example, if the temperature rises to about 42°–43° C. (107°–

109° F.), as it may in sunstroke, and remains there for some hours, it produces profound disturbances in nerve cells of the brain. Also, 24° C. (75° F.) is much lower than is compatible with activity. As Britton has noted, the heart beats very slowly, respirations are shallow and infrequent, and deep lethargy prevails. There is a good reason, therefore, in avoiding the extreme variations.

There is good reason also in preserving the normal constancy of body temperature. Its value becomes clear as soon as we compare the influence of cold on ourselves and on lower animals, e.g., amphibia and reptiles, which have no heat regulatory apparatus. I have already mentioned the effect of low temperature on the frog. As the weather turns cold he becomes cold, too, and his actions are more and more sluggish. His heart beats rarely, and as he lies inert, deep in a frigid pool, he does not breathe at all. Thus he remains until he is warmed again. This behavior of the frog is chiefly due to the fact that many of the essential processes going on in organisms are chemical and that the speed of chemical processes varies with the temperature, an increase of 10° C. practically doubling the rate. The "cold-blooded" animals, therefore, having the temperature of their surroundings, can act with alacrity only when the weather is warm; the warm-blooded, which maintain a fairly fixed high temperature in spite of external cold, can act quickly at all times. By the preservation of constancy of the internal environment they are freed from the influence of vicissitudes in the external environment.

II

To understand the regulation of temperature in our bodies we must realize, first, that heat is being continuously produced by every variety of activity in which our organs engage. All the energy of the powerfully contracting heart is ultimately turned to heat inside us, for the mechanical work which the heart performs in building up the head of arterial pressure is spent in overcoming frictional resistance in the blood vessels. About three-fourths of the energy of our muscular activity appears necessarily as heat. And the processes in the liver and other glands are all accompanied by heat production. We have learned that when an organ becomes specially active, the blood flows through it in larger volume. The heat developed by the activity of the organ diffuses from the warm cells to the cooler blood. Thus the cells are prevented from becoming overheated, and the heat produced in one part is made serviceable in other parts. The man who swings his arms and stamps his feet vigorously on a cold morning is making heat in his muscles that the circulating blood renders generally useful for the cold regions of his body. An important function which the moving part of the fluid matrix performs, therefore, is that of equalizing the temperature throughout the organism. As we shall see later, it plays also an essential rôle in the management of heat loss through the skin.

There is evidence that heat production in the body is under control, and that in normal persons, under standard conditions, it is remarkably uniform. The standard conditions are satisfied by a fast of about eighteen hours after taking a mixed diet (commonly over night) and then a rest in the reclining position for about twenty minutes before the test. With the subject awake and lying quietly on his back, in a room temperature of about 20° C. (68° F.), the intake of oxygen and the output of carbon dioxide are measured during a series of short periods. From the figures for O_2 consumed and CO_2 discharged, the heat produced by the burning can be readily calculated. It is usually expressed in calories per square meter of body surface per hour or per day, and measures the so-called "basal metabolism." As we shall see later, there is a gradual diminution of the rate of heat production by the body as one progresses towards the years of senescence. Within limited periods of the existence of the organism, however, constancy of the basal metabolism is the rule. A man studied for six years by Benedict and Carpenter, at the Carnegie Laboratory in Boston, had a variation as little as 3.7 per cent each year from the average of all the years. In a dog tested by Lusk during two years the basal metabolic rate, in 17 observations, differed only 2.9 per cent. Such uniformity is astonishing.

We know that the metabolic rate is affected by disturbances of several glands of internal secretion; for example, the pituitary at the base of the brain and the cortex of the adrenal gland. But the gland that is most markedly and most directly influential is the thyroid, in the neck. When the thyroid is over-active, as in exophthalmic goiter, the metabolic rate commonly rises 50 or 75 per cent above normal, and cases of hyperthyroidism are known in which the rate doubled, i.e., the processes of heat production under standard conditions were actually going on twice as fast as in the healthy person. When the surgeon, in treating this state, removes a large part of the gland the proper metabolic rate can be restored. On the other hand, when the thyroid gland is defective or deficient, as in cretinism and myxedema, the rate of combustion may be from 30 to 40 per cent below the normal level. The metabolism of such patients can easily be raised to the normal level by feeding them thyroid gland or giving them an extract of it in substitution for the thyroid secretion which they lack.

What keeps the thyroid gland constantly active is not known. In cats which were studied by a group in the Harvard Physiological Laboratory, after different stages in the removal of the sympathetic nervous system, there was a slight drop in the metabolic rate when the nerve strands to the neck region were extirpated, but the effect was so slight as to be, perhaps, insignificant. The glands could not have been kept active in these circumstances by secretion of adrenin from the adrenal glands, because in certain cases these too were denervated and there was no clear difference on that account. For the present the control of the internal secretion of the thyroid must be left for further research. All that we know definitely is that the basal metabolic rate is one of the constants of the organism, that its constancy seems to be directly dependent on the proper functioning of the thyroid gland, and that the reliable uniformity of the basal metabolism is a condition for other phases of homeostasis, especially that of body temperature.

III

The homeostasis of body temperature, like that of the oxygen supply to the tissues, is achieved by modifying the speed of a continuous process. As we have seen, heat is continuously being produced by organic activity. Constant temperature can be maintained by increasing or decreasing the rate of heat loss or by increasing or decreasing the rate of its production, according to need. We shall consider first the agencies concerned in heat loss.

Let us suppose that conditions favor a rise of body temperature because, for example, a large amount of heat has been produced by very vigorous muscular work. In these circumstances the vasomotor nerves governing the size of the surface arterioles relax their grip, the vessels dilate, and the blood, warmed by the active muscles, flows in much larger volume through these arterioles and through the capillaries to which they contribute. In consequence the skin becomes red. If the surrounding air is cool the extra heat brought to the skin will pass out to it by radiation and conduction, and a rise of body temperature will be prevented.

If the outer air is too warm to permit the heat to pass to it, however, another process is invoked. Heat is lost by evaporation. When water evaporates, as much heat is taken from neighboring objects as would be required to cause the water to evaporate. This is a fact well known to persons who live in hot dry climates and who use porous earthen jars or canvas bags to cool the drinking water. The greater delivery of warm blood to the skin can be combined with the pouring out of sweat on the skin surface. Just as the evaporation of moisture from the outside of the porous containers cools the water within, so likewise the evaporation of sweat cools the skin and consequently the blood flowing through its capillaries. If the air is dry, large amounts of heat may be lost in this way. It is by this means that high external temperatures are withstood by stokers and by foundry workers who are exposed to the intense heat of open furnaces. Occasionally, however, persons are found with defective sweat glands. A man with that affliction, when exposed to the sun for a short time on a summer day, soon had a body temperature of 41.5° C. (nearly 107° F.). When such a person has to work hard in hot weather his only resource in avoiding the development of a fever is to wet his garments repeatedly and let the vaporization of water from them replace the vaporization of sweat. The very uncomfortable experience which we have on a day which is not only hot but also muggy is due to the high vapor density or humidity of the air, which interferes with the change of sweat to water vapor and thus prevents cooling.

To a considerable degree we lose heat also by evaporation of fluid from the surfaces of the respiratory tract. On a morning in winter we "see the breath" because the moisture added to the inspired air is promptly condensed when that air is breathed out to the cold surrounding atmosphere. On a hot day a similar evaporation is continuously going on and can be augmented by rapid respiration. As human beings we do not ordinarily use this process for cooling purposes alone, though in one case on record a man who could not sweat because of an atrophied skin and who breathed 6.32 liters of air per minute when he was at rest and his body temperature was normal, breathed nearly three times that much when his temperature rose to 39.9° C. (103.8° F.), and at the rate of 90 breaths per minute! In some lower animals—in the dog, for example—the quick movement of air to and fro in the respiratory passages, while panting, is the chief means of losing heat if the temperature tends to rise. In man, also, during and after vigorous exercise, the fast, deep breathing caused by excess production of carbon dioxide, has the nice quadruple effect of preventing the accumulation of carbon dioxide in the lungs, assuring the presence of plenty of oxygen there, pumping onward the venous blood and helping to get rid of the extra heat which results from the muscular activity.

As we have already noted, there is a considerable amount of heat produced by the organism as the inevitable by-product of existence. The basal metabolism, by definition, is the lowest degree of chemical oxidation when the body is at rest. Only by complete idleness can heat production be reduced to a minimum. When the surrounding temperature is high, therefore, not much advantage can be gained by restricting the development of heat. The main reliance must be the increase of heat loss by the methods which we have been considering.

IV

If the body temperature tends to fall, an interesting series of adjustments occur, all directed towards preservation of the steady state. First, heat which is being lost through the skin is conserved. To that end perspiration is reduced to a minimum. The surface vessels are contracted so that the warm blood from the interior is not exposed to the cold surroundings. And in animals provided with hair or feathers these appendages of the skin are lifted to enclose in their meshes a thicker layer of air, which is a poor conductor of heat. Of this last protective reaction only futile "gooseflesh" remains in us, and the single little hair standing upright in each hummock of the gooseflesh signals its futility. In place of the efficient protection which fur would afford, mankind has to resort to extra clothing—often the fur of lower animals!—to prevent too great loss of heat.

In addition to the constriction of surface vessels and erection of hairs, it is interesting to learn that a rise in the level of blood sugar occurs when the body is chilled. Erect hairs, constricted arterioles and hyperglycemia are signs that the sympathetic division of the autonomic nervous system is active. The question naturally arises, is secretion of adrenin, which is admittedly under sympathetic control, augmented when cold causes a discharge of sympathetic impulses? The answer to that question has important bearings, because adrenin not only collaborates with the sympathetic impulses which are diminishing the caliber of the surface vessels, but it has the power, as shown by Aub and McIver and their associates in the Harvard Physiological Laboratory, to accelerate the processes of combustion in all parts of the organism. Adrenin would have effects like those caused by opening the dampers of a furnace; burning would go on more rapidly. If, therefore, the adrenal medulla is made to secrete by chilling the body, the extra adrenin discharged would effect a faster production of heat just when there would be special need for it.

Microscopical studies of the adrenal glands by Cramer and others indicated that exposing animals to cold reduces the substance from which adrenin is derived. This was interpreted as evidence that cold stimulates adrenal secretion. But perhaps cooling retards or diminishes the production of that substance—the result would be what was observed. More direct evidence was required in order to determine the fact. Such evidence was furnished by Hartman and his collaborators in Buffalo. They made use of the completely denervated iris in cats. When these animals were wet with cold water, or cooled after being wet with warm water, the pupil dilated if the adrenal glands were functional, but not if they had been rendered incapable of action. The natural aversion of cats for water on the skin caused a certain amount of excitement, however, and the observed effect could not be sharply differentiated from the effect of excitement itself. To avoid this possible error and also for other reasons Querido and Britton, Miss Bright and I decided to carry on the investigation further.

In our experiments we again made use of the denervated heart, in unanesthetized animals, as an indicator of increased adrenin in the blood stream. We employed several methods of exposing the animals to cold. At first we held the animal, warm and comfortable, on a cushion near a window; and after the heart rate had been counted or registered we merely opened the window to the outer cold. The great advantage of this method was that the element of excitement and emotion was wholly eliminated because the surroundings were familiar; the only change was the opening of the window. In figure 27 is presented a copy of the original records of the effects on the rate of the denervated heart of exposing an animal thus to the cold air. Both adrenal glands were intact. The room temperature was 16° C. (60.8° F.), the outdoor temperature was −4° C. (24.8° F.). Observe that the basal rate was 118 beats per minute. Four minutes after the window was opened the

CONSTANCY OF BODY TEMPERATURE 187

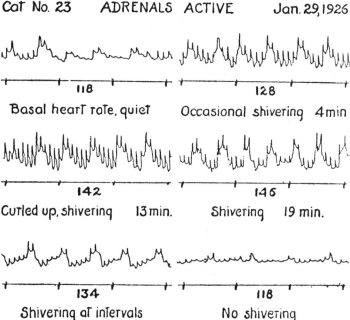

Fig. 27. Original records showing increases of rate of the denervated heart when cat 23 was exposed to air at –4°, January 29, 1926. Below the records taken while the window was open are written figures telling the time after the opening. Nine minutes after closing the window the rate had fallen from 146 to 134, and seven minutes later to 118 (the basal). Time in five-second intervals.

heart rate had increased 10 beats per minute. In thirteen minutes it had increased 24 beats. At that time the door was opened so that a strong draught of cold air rushed by. At the end of the next six minutes, that is, nineteen minutes after the start, the rate had reached 146 beats per minute— i.e., 28 beats above the basal, a rise of twenty-four per cent. The window was then closed. Nine minutes thereafter the rate had fallen from 146 to 134. It continued to fall until, sixteen minutes after the window was closed, it again reached the basal level. I would stress the fact that the animal throughout the test was resting quietly so that the reaction was not at all complicated by adrenal secretion stimulated through excitement or bodily movements. That the sympathetic system was active, however, was made evident by the erection of the hairs four minutes after the window was opened and by continuance of this state until about four minutes after the window was closed, when the animal was covered with a blanket.

In figure 28 is shown by a series of graphs the increase of heart rate above a basal level when animals with adrenals intact were exposed to cold air in the manner just described. The period of exposure is denoted by a thickening of the base line. The zig-zag or V-shaped marks represent shivering. Observe especially that the shivering is not a necessary condition for the faster heart rate, because the rate was increased long before the shivering began. There is clear proof that the faster pulse is due to adrenal secretion. In cases 46 and 49, with adrenal secretion excluded, the dash-lines show the changes of the heart rate under the usual conditions of exposure to cold. Note that the primary effect was a decrease of the rate instead of an increase.

The defect of the method just described lies in its limitation to periods of cold weather. In order to have a method which could be applied at any time we devised the procedure of introducing into the stomach a known amount of ice water. Of course, ice water can be given at any time and

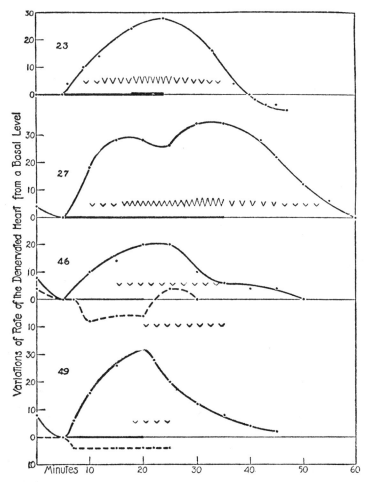

Fig. 28. Increases of rate of the denervated heart above a basal level when animals with adrenals intact (nos. 23, 27, 46 and 49) were exposed to cold air. The period of exposure is denoted by a thickening of the base line; the double thickening in the case of no. 23 marks the time of a cold draught through an open door. Shivering is indicated by separate v's when it was intermittent, and by connected v's when it was continuous. The size of the v's marks roughly the intensity. The dash lines in cases 46 and 49 show the changes of heart rate under similar conditions of exposure after exclusion of medulliadrenal secretion.

190   THE WISDOM OF THE BODY

in familiar surroundings. It readily permits registration of changes in the heart rate and it is quantitatively accurate. The last advantage is most important. The weight of the animal and its temperature and specific heat, the volume of the introduced water and its temperature and specific heat, can all be known. The heat of the animal passes into the cold water in the stomach and intestines. Indeed, the circulating blood makes the addition of cold water to the body the approximate equivalent of mixing the water with the bulk of fluid which forms the internal environment of the organism. Thus it is possible to calculate how much a certain amount of cold water would lower the body temperature if no extra heat were produced. The amount of extra heat which the organism must produce in order to maintain its normal temperature can thus be well estimated. We have called this amount the "heat debt." The main defect of the method is a slight disturbance of the animal at the start because the water must be introduced into the stomach through a tube. This disturbance is, however, only temporary.

The heat debt which we established in cats varied between approximately 1500 and 2000 small calories per kilogram (about two pounds) of body weight. In figure 29 are shown the original records of the pulsations of the denervated heart when cat 33 was given water at 1.0° C. (33.8° F.) in an amount which established a heat debt of 1850 small calories. Observe that there was an initial increase in rate of 42 beats, due, no doubt, to the disturbance of giving the water. This soon dropped 8 beats, and thereafter for more than a half-hour the rate continued at a high level. Indeed, a full hour after the heat debt was estab-

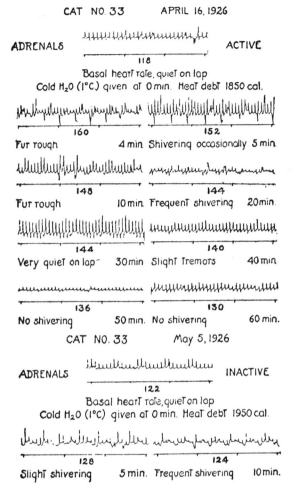

Fig. 29. Original records showing increases of rate of the denervated heart when cat 33 with active adrenals was given, April 16, 120 cc. of water at 1° (heat debt, 1850 small calories), and when, with adrenals inactivated, it was given, May 5, 116 cc. of water at 1° (heat debt, 1950 calories). Below each record are stated the prevailing conditions and the interval since the water was given. Time in five-second intervals.

lished the heart rate was still 12 beats above the initial level. Later, after the adrenal glands had been inactivated a heat debt of 1950 calories was established, and, as you note, within ten minutes the heart rate had practically returned to the basal level. In figure 30 similar results are

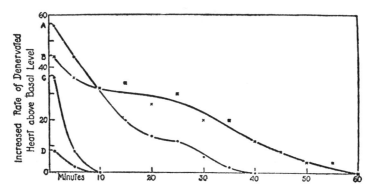

Fig. 30. Increases of rate of the denervated heart in cat 27, responding to heat debts produced by introducing water into the stomach under different conditions. A, B and C, with adrenals active. A, a heat debt of 2000 small calories when the water was given at 10°. B, a heat debt of 1820 calories, water at 1°. C, a heat debt of 250, water at 33°. D (with adrenals inactivated), a heat debt of 2000 calories, water at 1°.

shown graphically. In line A a heat debt of 2000 small calories was established by water given at 10° C. (50° F.), and in line B a heat debt of 1820 calories by water given at 1.0° C. The colder water had a considerably longer effect. In line C a small heat debt of 250 calories was established by giving water at 33° C. (91.4° F.). The main effect here was that of excitement; within ten minutes all of the disturbance had subsided.

From these and many other similar observations we drew

the conclusion that conditions which would naturally cause a reduction of body temperature induce an increased discharge of adrenin into the circulating blood.

V

Various investigators have found that adrenin has a heat-producing effect. Boothby and Sandiford have demonstrated that a milligram of adrenin injected into a man augments the basal heat production by fifty large calories. Since adrenal secretion is increased when the organism is exposed to the danger of a too rapid loss of heat, and since adrenin secreted in natural amounts is capable of accelerating the oxidative processes, it should be possible to demonstrate the service which this physiological reaction performs for the organism. We attempted in two ways to test the value of the reaction. It seemed possible that the heat produced by the automatic muscular contractions in shivering would be relied upon to a greater degree if the adrenal glands were rendered inactive, and that increased metabolism could be demonstrated without the assistance of the shivering mechanism. Accordingly we studied the effects of a certain heat debt on shivering when the adrenal glands are active or inactive, and we observed in man the influence of a heat debt on the metabolic rate in the absence of shivering.

Let us consider first the effect of a heat debt on shivering when the adrenal glands are present or absent. If the heat debt is large, i.e., if it amounts to 1000 small calories or more per kilogram, with the water at 1.0° C., and if the room temperature is about 20° C., it is commonly met by two calorigenic agencies: by an increased output of adrenin and by shivering. In figure 28 the fact is demonstrated that shivering coincides with the period of greatest discharge of adrenin. I emphasized the point, however, that shivering is not a necessary condition for the discharge of adrenin; indeed, shivering may be wholly absent although the heart

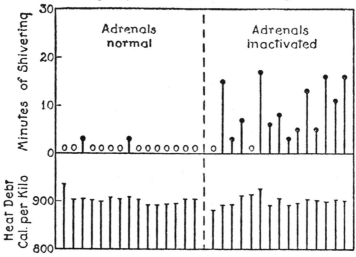

Fig. 31. Presence or absence of shivering when there was a heat debt of about 900 calories in a series of cats with normally innervated adrenal glands, and in another series with one adrenal gland removed and the other denervated.

rate is well accelerated. Now if the environmental temperature is about 20° C. and a heat debt of only 900 calories is to be paid, we found that shivering rarely occurs, or if it occurs it is of short duration. As shown in figure 31, in fifteen tests made in these circumstances shivering occurred in only two instances and lasted only three minutes. Note, on the other hand, what occurs if one adrenal has been removed and the other denervated. The same heat debt, established under the same conditions, resulted in shivering in all but two instances and it lasted as long as fifteen, sixteen and seventeen minutes. Thus when the heat-producing service of the adrenal medulla is lacking the shivering mechanism is resorted to.

The other observations, to test the effect of a heat debt on the metabolic rate in the absence of shivering, were made on human beings. We established a heat debt which averaged 449 small calories per kilogram. In twenty-two observations on eleven human subjects there was an average maximal increase in metabolism of a little more than 16 per cent, with variations above that average ranging as high as 38 per cent. These increases in the metabolic rate were not accompanied by shivering. The reader might suppose that the effect was due to the disturbance of taking the cold water and ice which established the heat debt. The highest point of the increase, however, averaged twenty-three minutes after the water and ice were taken and therefore occurred too long after the ingestion to be due to that. Furthermore, when warm water, equivalent in amount to the cold water, was drunk, the average increase in metabolism was only 3.1 per cent, and since the time of maximal increase occurred regularly in the first seven minutes of the experiment, it must have resulted chiefly from the disturbance of taking the water. This fact is brought out clearly in figure 32. A heat debt of 46 small calories per kilogram, due to the taking of 520 cubic centimeters of water at 30° C. (86° F.), caused in J. L. H. an immediate increase of 4 per cent in the metabolic rate. Later a heat debt of 409 calories per kilogram, due to the drinking of 354 cubic

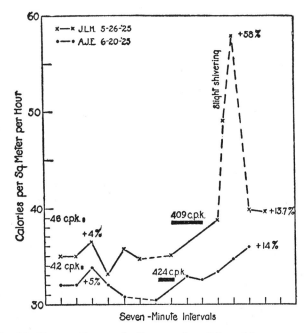

Fig. 32. Changes in metabolism produced by taking warm water and later by taking cold water and ice. A heat debt of 46 small calories (per k.) due to taking 520 cc. of water at 30° caused in J. L. H. an immediate increase of 4 per cent in the metabolic rate; a heat debt of 409 calories from taking 354 cc. of water at 1° and 130 g. of ice caused an increase of 13.7 per cent. Slight shivering was attended by a sudden and temporary increase of 58 per cent. A heat debt of 42 calories from 420 cc. of water at 30.2° caused in A. J. E. an immediate increase of 5 per cent in the metabolic rate; a heat debt of 424 calories, from 260 cc. of water at 1° and 139 g. of ice, caused a gradual increase of 14 per cent, without shivering. Each point breaking the line represents a metabolism record.

centimeters of water at 1.0° C. and swallowing 130 grams of ice, caused an increase of metabolic rate of 13.7 per cent. Note that a brief period of shivering was attended by a sudden and temporary increase of 58 per cent in the rate.

The observations on A. J. E. are quite free from the criticism that shivering might have affected the final result. In that case a heat debt of 42 calories from 420 cubic centimeters of water at 30.2° C. (86.4° F.) caused an immediate increase of 5 per cent in the metabolic rate. Later the taking of 260 cubic centimeters of water at 1.0° C. and 139 grams of ice established a heat debt of 424 calories. Observe that this caused a gradual increase in metabolism, amounting to 14 per cent, without any shivering whatever.

The foregoing experiments have shown that the same conditions that increase adrenal secretion in lower animals increase the metabolism in man and that they may do this without the accompaniment of shivering. It seems reasonable to conclude, therefore, that a disturbing heat loss evokes activity of the adrenal medulla in man as it does in the lower animals and that the extra output of adrenin in both organisms has the same effect of accelerating combustion.

It may be that the thyroid gland as well as the adrenal is involved in the processes of temperature regulation. We have seen that when it is overactive in disease, heat production in the body is greatly increased. Perhaps the thyroid is stimulated to action by cold and coöperates with the adrenal glands, but in a less quickly responsive manner, to accelerate oxidation. At least there are some observations which suggest that possibility. When cattle on our Western plains are first exposed to cold weather in early winter an enlargement of the thyroid gland has been noted as a characteristic change in them. And Loeb has reported that when part of the gland has been surgically removed the remnant grows more actively if the animal is placed in cold surroundings than if it is kept warm. But this is only suggestive evidence. Here again, more information is needed before we can draw definite conclusions.

As we survey the arrangements which check a shift of body temperature in one direction or the other it is interesting to observe that there are successive defenses which are set up against the shift. If dilation of the skin vessels does not stop the rise of body temperature, sweating and even panting supervene. If conservation of heat by constriction of the skin vessels does not prevent a fall of temperature, there is a chemical stimulation of more rapid burning in the body by means of secreted adrenin, and if that in turn is not adequate to protect the internal environment from cooling, greater heat production by shivering is resorted to. It is noteworthy that in all these functions except shivering the sympathico-adrenal mechanism is at work. As investigations by Dworkin in the Harvard Physiological Laboratory have shown, shivering itself has its most complete expression when that part of the brain, the diencephalon (see fig. 33, p. 233), which is the coördinating center for the sympathetic system, is intact.

We must recognize that among civilized people the physiological devices for the maintenance of constant temperature may have little opportunity to function. In wintry weather we spend our days in heated houses and offices and travel about in heated cars. Encased in warm clothing we carry with us everywhere a temperate climate. Thus only few occasions arise which demand either the conservation of the heat always being produced by the organism, or the development of extra heat by bodily activity. And in summer, likewise, mechanically operated fans, cold drinks, ice cream and refrigerated rooms lessen the use of the natural arrangements for keeping cool. It is not impossible that we lose important protective advantages by failing to exercise these physiological mechanisms, which were developed through myriads of generations of our less favored ancestors. The man who daily takes a cold bath and works until he sweats may be keeping "fit," because he is not permitting a very valuable part of his bodily organization to become weakened and inefficient by disuse.

VI

The delicate control of body temperature indicates that somewhere in the organism a sensitive thermostat exists which regulates the operations which we have been considering. Experiments on rabbits have shown that this part of the regulatory apparatus is located in the base of the brain, in the diencephalon (see fig. 33). The cerebral hemispheres and other parts anterior to that region can be removed, as Isenschmid has shown, and although the surrounding temperature may vary from 10° C. (50° F.) to 28° C. (82.4° F.), the normal temperature of the animal is preserved. If now the diencephalon is separated from the rest of the body, heat regulation is lost; the reactions to temperature changes in the environment are those of a cold-blooded animal. It is interesting to note that in the diencephalon are the central stations for the secretion of sweat, for shivering and probably also for panting—in short, for the automatic reactions which govern the production and loss of heat.

The thermostat in the diencephalon may be affected in

two ways, either by the temperature of the blood going to it or by nerve impulses from the surface of the body. Warming the blood in one of the large arteries in the neck that distributes to the brain, will cause dilation of the blood vessels of the skin and sweating. On the other hand, cooling that part of the circulation will cause shivering. And Leonard Hill found that sweating in a hot room could be stopped by placing his hand in cold water but not if the circulation through the arm was checked. Persistence of the local sensation of cold proved that nerve connections had not been disturbed. It is clear, therefore, that the flowing blood itself may affect directly the regulatory center. The evidence for reflex nervous influence on this center is also clear. A sudden dash of cold water on the skin causes not a fall of temperature but a rise, due to reflex constriction of the surface vessels and consequent interference with the normal loss of heat. Furthermore, the observation has been reported that if a person takes a bath in water at 29° C. (84.2° F.), he feels cold, shivers, and by the reaction keeps constant his body temperature. If he takes a bath at the same temperature in carbonated water, however, he does not feel cold, there is no reaction, and the body temperature falls. Although we have this evidence of a double control of the responses of the thermostat in the brain, the actual mode of control—for example, the influence on it of the nerves which give us sensations of heat and cold—is not yet well understood.

In spite of gross interference by civilized man with the physiological mechanisms for homeostasis of body temperature we know that these mechanisms exist and are always ready for action. If conditions are such that there is a tendency to tip the organism in one direction, a series of processes are at once set at work which oppose that tendency. And if an opposite tendency develops, another series of processes promptly oppose it. Thus quite automatically the remarkable uniformity of the temperature of the internal environment is preserved, in opposition to both internal and external disturbing conditions.

REFERENCES

Aub, Bright and Forman. Am. Journ. Physiol., 1922, lxi, 349.
Benedict and Carpenter. Publication No. 261, Carnegie Institution, Washington.
Britton. Quart. Journ. Exper. Physiol., 1928, xiii, 55.
Cannon, Querido, Britton and Bright. Am. Journ. Physiol., 1927, lxxix, 466.
Cannon, Newton, Bright, Menkin and Moore. Ibid., 1929, lxxxix, 84.
Cramer. Report, Imper. Cancer Research Fund, London, 1919, 1.
Dworkin. Am. Journ. Physiol., 1930, xciii, 227.
Hartman, McCordock and Loder. Ibid., 1923, lxiv, 1; cf. also lxv, 612.
Hill, Journ. Physiol., 1921, liv, p. cxxxvi.
Isenschmid. Hanbk. d. norm. u. path. Physiol., Berlin, 1926, xvii, 56.
Loeb. Journ. Med. Res., 1920, xlvii, 77.
Lusk. Journ. Physiol., 1924, lxix, 213.
McIver and Bright. Am. Journ. Physiol., 1924, lxviii, 622.

TOTAL SELF REGULATORY FUNCTIONS IN ANIMALS AND HUMAN BEINGS[1]

CURT P. RICHTER[2]

Associate Professor of Psycho-Biology, Johns Hopkins University School of Medicine

IN 1859 Claude Bernard (1) first described what he called the internal environment of the body, consisting largely of the body fluids, and showed that in mammals the properties of this internal environment ordinarily vary within fixed limits, variation outside of these ranges endangering life. He described many of the physiological mechanisms by means of which the body keeps these properties at fixed levels, and pointed out that it is by virtue of the existence of these mechanisms that mammals are able to live and thrive under widely varying external conditions.

Cannon (2), in a long series of remarkable experiments, collected in 1932 in his book "The Wisdom of the Body," not only confirmed Bernard's concept but greatly extended it. Largely through his efforts this concept has become almost an axiom of modern medicine. Cannon speaks of a constant state or homeostasis. Thus he states: "The constant conditions which are maintained in the body might be termed equilibria. That word, however, has come to have a fairly exact meaning as applied to relatively simple physico-chemical states, in closed systems, where known forces are balanced. The coordinated physiological processes which maintain most of the steady states in the organism are so complex and so peculiar to living beings—involving, as they may, the brain and nerves, the heart, lungs, kidney, and spleen, all working cooperatively—that I have suggested a special designation for these states, homeostasis. The word does not imply something set and immobile, a stagnation. It means a

[1] Lecture delivered November 19, 1942.

[2] This work was carried out with the following collaborators: Doctors Bruno Barelare, John E. Eckert, D. Clarence Hawkes, L. Emmett Holt, Elaine Kinder, Katherine Rice, Edward C. H. Schmidt, Jr., and Mr. John Birmingham, Miss Alice MacLean and Mrs. Kathryn H. (Campbell) Clisby.

condition—a condition which may vary, but which is relatively constant."

Both Bernard and Cannon concerned themselves almost entirely with the physiological and chemical regulators of the internal environment. They showed, for instance, that when an animal is placed in a cold external environment and is consequently threatened with a decrease in body temperature, loss of heat is minimized by decreased activity of the sweat glands and constriction of the peripheral blood vessels, and more heat is produced by increased burning of stored fat and by shivering. These are all physiological or chemical regulators.

The results of our own experiments have shown that behavior or total organism regulators also contribute to the maintenance of a constant internal environment. The existence of such behavior regulators was first established by the results of experiments in which it was found that after elimination of the physiological regulators the animals themselves made an effort to maintain a constant internal environment or homeostasis. I will give you a few examples which are taken mainly from experiments on the endocrine glands. Thus, operative removal of the adrenal glands from animals eliminates their physiological control of sodium metabolism, and as a result large amounts of sodium are excreted as salt in the urine and the internal environment is greatly disturbed (3). If given access only to a stock diet, such animals die in 8–15 days. However, if given access to salt in a container separate from their food they will take adequate amounts to keep themselves alive and free from symptoms of insufficiency.

[Editor's Note: Material has been omitted at this point.]

One more example should suffice for the present purpose. It is taken from the field of body temperature regulation and is concerned with the rat's effort to maintain a constant body temperature after the physiological heat regulating mechanisms have been seriously disturbed (11, 12). The individual cages used for these experiments were each equipped with a roll of soft paper ½ inch wide and 500 feet long, with the free end readily accessible to the rat within the cage. Figure 4 shows a cross-sectional view of one of these cages. By means of a cyclometer and a scale to compensate for the progressively decreasing diameter of the roll, the amount of paper used each day was measured, and interpreted as an effort made by the rat to conserve heat by covering itself.

FIG. 4. Side view of nest building cage, showing paper roll and cyclometer.

All used paper was removed each day at noon. It was found that normal male and female rats used approximately equal amounts of paper to build nests which varied in size with changing external temperatures, for example, a drop in room temperature from 80 to 45 degrees increased the amount of paper used daily from 500 to 6000 centimeters. With this method we were also able to show that hypophysectomized rats built much larger nests than normal animals, as a result of their inability to produce adequate amounts of heat, which consequently threatened them with a fatal reduction in body temperature. Figure 5 shows the effect produced on nest building activity of a rat by hypophysectomy. The length of paper used daily increased from 700 to 3500 centimeters. When nest building paper was no longer made available, the rat died after 35 days, with a body temperature more than 15 degrees below normal. Thyroidectomized rats, which likewise have lost their ability to produce adequate amounts of heat, also built very large nests in an effort to cover themselves and thus to conserve heat. Both thyroidectomized and normal rats treated

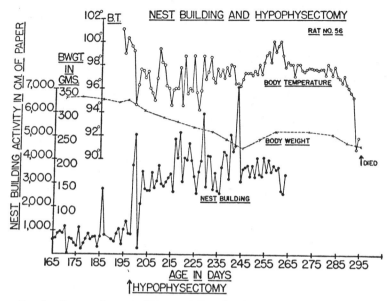

FIG. 5. Increased nest building activity after hypophysectomy, and fall in body temperature with consequent death, after removing nesting paper.

with large amounts of thyroid extract stopped building nests altogether. Some of the hypophysectomized and thyroidectomized rats used the entire roll of 15,000 centimeters (500 feet) of paper in 24 hours. Thus we have another instance in which, after removal of the physiological regulators, homeostasis was maintained by a total organism response.

On the basis of the results of these different experiments, it would seem very likely that in the normal, intact animal the maintenance of a constant internal environment depends not only

SELF REGULATORY FUNCTIONS

on the physiological or chemical regulators, but also on the behavior or total organism regulators. We do not yet know, however, the relative parts played by each: whether, for instance, the physiological responses take care of most of the regulation, or whether they function only when the behavior mechanisms have failed or broken down, or whether both are constantly and simultaneously in action.

[*Editor's Note:* Material has been omitted at this point.]

Thus, in summary, I have tried to show that the maintenance of a constant internal environment depends not only on the physiological or chemical regulators, but as well on behavior or total organism regulators. Proof of the existence of the behavior regulators was taken from experiments in the field of endocrinology and nutrition. It was shown that disturbances created in the internal environment by removal of one or the other of the endocrine glands were corrected by the animals themselves. It was demonstrated that the ability to select diets with relation to internal needs seems to depend more upon taste sensations than on experience, and it was pointed out that this knowledge of the ability of animals to make beneficial selections can be used to study a variety of problems in the fields of endocrinology and nutrition. Evidence was further presented for the existence and successful operation of similar behavior regulators in human beings. Thus, we believe that the results of our experiments indicate that in human beings and animals the effort to maintain a constant internal environment or homeostasis constitutes one of the most universal and powerful of all behavior urges or drives.

BIBLIOGRAPHY

1. Bernard, C., Leçons sur les propriétés physiologiques et les altérations pathologiques des liquides de l'organisme, Paris, Bailliers, 1859.
2. Cannon, W. B., The wisdom of the body, New York, Norton and Company, Inc., 1932.
3. Richter, C. P., *Am. Jour. Physiol.*, 1936, 115, 155.
11. Kinder, E. F., *Jour. Exp. Zool.*, 1927, 47, 117.
12. Richter, C. P., *Cold Spring Harbor Symposia on Quant. Biol.*, 1937, 5, 258.

4

Copyright © 1964 by the Society for Experimental Biology

Reprinted from *Soc. Exp. Biol. Symp.* **18**:7-29 (1964)

THE ROLES OF PHYSIOLOGY AND BEHAVIOUR IN THE MAINTENANCE OF HOMEOSTASIS IN THE DESERT ENVIRONMENT

By GEORGE A. BARTHOLOMEW

Department of Zoology, University of California, Los Angeles

There is an endless fascination in attempting to understand the ways in which animals are able to live and reproduce under physically difficult circumstances. However, in the temperate parts of the world where biologists have attained a high population density, the physiological capacities of animals are usually not of great ecological importance as determinants of the distributional limits. Other largely inferential factors such as competition, predation, food supply, or suitable habitats, restrict the distribution of animals to areas much smaller than those permitted by their physiological capacities *per se* (see Bartholomew (1958), Gordon (1962), for discussion and documentation).

There are, of course, parts of the world where the physical environment is so demanding and inhospitable that it directly limits animal distribution. One of the most widespread of these physically demanding environments exists in the low-latitude deserts which dominate the continental areas and oceanic islands in the horse latitudes of both hemispheres, and extend long tongues north and south on the lee sides of mountain ranges, particularly in the New World.

A POINT OF VIEW

It is my purpose in this paper to discuss and evaluate some of the ways in which vertebrates have come to terms with this difficult physical environment, but, to develop a point of view, I will first describe briefly a way of looking at organisms that is useful to the student of the ecologically relevant aspects of physiology, particularly in studying problems of homeostasis. The complexity of contemporary biology has inevitably led to extreme specialization and 'tunnel vision', and to poor communication between disciplines. In this situation, members of each specialized field begin to feel that their own work is fundamental and that the work of other groups, although sometimes technically ingenious, is either irrelevant or at best peripheral to the understanding of basic problems and issues. The biochemist is apt to feel that all important biological problems will be solved at the molecular or submolecular level, while the ecologist feels that the molecular biologist is preoccupied with details of machinery whose significance he does not appreciate. There is a familiar solution to this problem, widely recognized intellectually but sometimes difficult to accept emotionally. This is the idea that there are a number of levels of biological integration and that each level offers unique problems and insights, and further, that each level finds its explanations of mechanism in the levels below, and its significance in the levels above.

This elementary philosophical idea is relevant here because physiological ecology demands preoccupation with many levels of biological integration, continuously and usually simultaneously. Any attempt to attain an adequate understanding of the relation of an organism to its environment presents a problem of such enormous complexity that the biologist must, unfortunately, be reconciled from the beginning to obtaining an incomplete answer. The task, which all scientists face, of isolation and simplification of problems, is particularly acute for the student of physiological ecology. He cannot reduce the problem until only a single variable remains; he cannot restrict the data to a single level of biological integration, or, as is done in most other biological disciplines, even to several adjacent levels. Further, he cannot limit his data-gathering to the techniques of any one specialized field. As well as seeing the whole of the problem with sufficient clarity to enable him to ask the right questions, the student must have enough intellectual brutality not to be deterred by the knowledge that few men are able to handle these complexities. More than most students, he must recognize that biology is a continuum.

ORGANISM AND ENVIRONMENT

It is unavoidable that we biologists, because of our limitations, divide ourselves into categories of specialization and then pretend that these categories exist in the biological world. As everyone knows, organisms are functionally indivisible and cannot be split into the conventional compartments of morphology, physiology, behaviour and genetics. Each of these is only one aspect of the organism as a whole and since it is the organism which deals with the physical environment, where do we start? First, we must decide what an organism is. Obviously, it is not just a museum specimen, nor is it just an animal or plant caged in the laboratory, or observed in the field. It is useful to think of the organism as an interaction between a complex, self-sustaining physicochemical system and the substances and conditions which we usually think of as the environment. As Claude Bernard pointed out almost a century ago, organism and environment form an inseparable pair; one can be defined only in terms of the other. The separation of organism from environment seems natural to us, because we are organisms and we think of ourselves as separate from the environment. It is clear that the organism exists as a dynamic equilibrium and thus, as long as it is alive, it is the example *par excellence* of the phenomenon of homeostasis.

CLIMATE AND ECOCLIMATE

Although organism and environment form a single functional unit it is convenient to maintain a verbal distinction between them. Fortunately, we can do this if we never treat the two ideas separately and if we always remember that, when dealing with a population of a given species, it must not be related to the gross environment; instead, we must relate it to the specific and limited environment with which the organisms in question are maintaining a dynamic equilibrium. This point of view is particularly important when dealing with terrestrial animals, for on land, in contrast to the sea, there is an almost infinite series of physical situations available, and terrestrial animals can by their behaviour select from this array of environmental conditions in an intricate and precise way. Terrestrial animals, because of their mobility and capacity for complex behaviour, including acceptance and rejection, can actively seek out and utilize those specific and limited facets and aspects of the physical environment that allow their anatomical and physiological attributes to function adequately for survival and reproduction. Consequently, by their behaviour these animals can fit the environment to their functional capacities. From this point of view, it follows that there is no such thing as 'the environment', at least in the terrestrial situation, but rather an enormous series of environments. Indeed, it is probable that, in a given part of the world, as many terrestrial environments exist as there are species.

For a large mammal, such as a human being, the climate, in the usual meteorological sense of the word, would appear to be a reasonable approximation of the conditions of temperature, humidity, radiation, and air movement in which terrestrial vertebrates live. But, in fact, it would be difficult to find any other lay assumption about ecology and natural history which has less general validity. Ostriches, kangaroos, prong-horned antelope, and shepherds may live in the meteorologists' climate but few other vertebrates do.

Most vertebrates are much less than a hundredth of the size of man and his domestic animals, and the universe of these small creatures is one of cracks and crevices, holes in logs, dense underbrush, tunnels and nests—

a world where distances are measured in yards rather than miles and where the difference between sunshine and shadow may be the difference between life and death. Climate in the usual sense of the word is, therefore, little more than a crude index to the physical conditions in which most terrestrial animals live.

Each species selects from the variations in the local macroclimate and microclimates a particular combination of physical conditions—its ecoclimate—appropriate to its functions and capacities. Each species or other adaptive group in a given area has its own ecoclimate and, if one is to study the relations of an animal to its physical environment, interpretations must be made in terms of ecoclimate.

The ecoclimate of each species or organism must be measured separately and the measurements that are taken and the places where the measurements are made must be carefully selected on the basis of accurate knowledge of the natural history of the population concerned. It is obvious that, because of the great diversity of physical conditions which are available in the terrestrial environment, an individual animal can exploit its physiological capacities most fully by utilizing its behaviour to place itself in those situations with which it can cope by physiological regulation, or by restricting its period of exposure to intolerable physical conditions, so that its limits of physiological tolerance are not exceeded in spite of its inability to maintain an adequately steady state. The rest of this paper will consider a few especially instructive instances of the interplay of behaviour and physiology by which various desert-dwelling reptiles, birds and mammals either maintain a steady state, or restrict to acceptable rates and periods of time the changes in their physiological state in situations with which they cannot cope fully by physiological means.

Adequate general coverage of the adaptations of the different groups of vertebrate animals to conditions of desert life is available in the literature (Schmidt-Nielsen & Schmidt-Nielsen, 1952; Schmidt-Nielsen, 1964; Chew, 1961; Bartholomew & Cade, 1963); this paper will not attempt an exhaustive summary, but instead will discuss selected patterns of adjustment to the desert environment in relation to homeostasis.

THE DESERT ENVIRONMENT

The heat and aridity of low-latitude deserts pose an ecological challenge so severe that few kinds of vertebrates can occupy the deserts permanently. The few species which do occur in deserts are often present in some numbers, however, a situation similar to that found in other areas, such as the arctic, where the limiting factors are physical rather than biotic. The difficulty of desert life, of course, is the result, either directly or indirectly, of an inseparable pair of circumstances—high temperatures and water scarcity—which together present so great a challenge that few vertebrates can meet it by physiological regulation alone. However, although the macroclimate of the desert is an impossible one from the biological view, it contains an impressive array of biologically exploitable micro-environments. Moreover, the existence of daily and seasonal cycles of temperature and rainfall means that there is no region where animals must cope continuously with extremely high temperatures.

Patterns of adjustment to deserts

Any ecologically relevant understanding of physiology or any functional understanding of anatomy must depend on a knowledge of behaviour. The physiology and the anatomy of vertebrates change extremely slowly; consequently, all vertebrates are in a sort of phylogenetic trap as regards the kinds of physiological and anatomical adjustments they can make to temperature conditions. In contrast, behavioural adjustments to the environment can be drastic, rapid, precise, and of exquisite flexibility.

Since natural selection demands only adequacy, elegance of design is not relevant; any combination of behavioural adjustment, physiological regulation, or anatomical accommodation that allows survival and reproduction may be favoured by selection. Since all animals are caught in a phylogenetic trap by the nature of past evolutionary adjustments, it is to be expected that a given environmental challenge will be met in a variety of ways by different animals. The delineation of the patterns of the accommodations of diverse types of organisms to the environment contributes much of the fascination of ecologically relevant physiology.

From the standpoint of homeostasis, the particular interest of the environment of low-latitude deserts is that it poses environmental problems which carry two of the basic challenges of terrestrial life to their most extreme limits—for many hours of each day for many months of the year surface temperatures far exceed the upper limits within which active life can exist, and most of the time no free surface water is available. Such a situation is difficult for all vertebrates, but mammals and birds contend with an additional complication: not only must they avoid acquiring excessive quantities of heat from the environment, but they must contrive to lose to the environment the large quantities of heat produced by their own high rates of metabolism. The problem is further complicated because evaporation of water is the most readily available mechanism for heat loss when ambient temperature equals or exceeds body temperature, and in deserts water is in extremely short supply.

Vertebrates have two primary classes of solution to the desert-imposed problems of heavy heat load and water scarcity: (1) relaxation of the limits for physiological homeostasis with regard either to thermoregulation or osmoregulation (or both); and (2) avoidance of both problems by means of behavioural adjustments. As might be expected, both these patterns as well as many combinations of them are employed by one group or other of desert vertebrates.

Nocturnal rodents

The thermal environment of the desert surface regularly exceeds the limits which living systems can tolerate. Nevertheless, the desert supports large populations of small mammals, although of few species. Despite the almost overwhelming physiological demands of a severe low-latitude desert, an animal can escape the dilemma of too much heat and not enough water if it can defer activity until the sun goes down, or move a few centimetres underground. Thus, the most obvious way for a small animal to live in the desert is to be nocturnal or fossorial.

In the deserts of the New World, most desert mammals belong to genera that are widespread in more mesic environments, and at least some of the species which live in the desert have not yet been seen to show any clear-cut or striking physiological adjustments to it (Murie, 1961). Others, such as kangaroo rats (*Dipodomys*), kangaroo mice (*Microdipodops*), and pocket mice (*Perognathus*) show marked physiological adaptations to the aridity of their environment, but show no particular tolerance of high temperatures. All members of this fauna of small nocturnal desert mammals share a general behavioural adjustment to the desert. They remain underground during the day and confine their surface activity to the hours of darkness. Thus they avoid any special problems of thermoregulation and so avoid the necessity for squandering on evaporative cooling the limited supply of water obtained from their food and from their oxidative metabolism. The parameters of this mode of desert life for small mammals have been delineated by Schmidt-Nielsen & Schmidt-Nielsen (1952). The effectiveness of this pattern of heat avoidance and minimal water loss in excretion, defaecation and respiration, is witnessed by the large populations of rodents, and their dependent population of carnivores, which occupy all but the most severe deserts of the holarctic.

This general pattern of fossorial and nocturnal habits offers a classical demonstration of the way in which behavioural adjustments allow animals to cope only with those aspects of the physical environment with which they are physiologically equipped to contend. These adjustments are clearly adaptive to the hot desert summer, but pose thermoregulatory problems during the winter when, over large areas, particularly at altitudes above 2000 ft., surface temperatures at night are well below freezing. As a result, the primary thermoregulatory adjustment which many small desert rodents have to make is to low, rather than to high, temperatures.

Diurnal mammals

A very large mammal obviously cannot be primarily fossorial; therefore, if it lives in a desert region it cannot avoid thermal problems. The type-case of the large desert mammal is of course the dromedary camel; Schmidt-Nielsen, Schmidt-Nielsen, Jarnum & Houpt (1957) have shown that toleration of dehydration and hyperthermia allows this species to adjust to

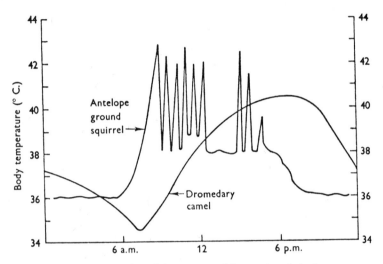

Fig. 1. A schematic representation of the patterns of fluctuations in body temperature of a large and a small desert mammal. (Data are based on Schmidt-Nielsen, Schmidt-Nielsen, Jarnum & Houpt, 1957; Bartholomew & Hudson, 1961.)

extremely demanding conditions. By storing heat during the day and unloading heat at night it can avoid excessive dependence on evaporative cooling (Fig. 1). Its mobility is such that it can seek out surface water when the limits of its tolerance of dehydration are approached. This solution, of heat storage during the daytime and periodic drinking, is of course possible only for a large animal.

The scarcity of small diurnal desert mammals may be considered as an indication of the difficulty of the problems they face. The rapidity of their metabolic turn-over, their relative lack of mobility and consequent inability to seek surface water, and their small size and resultant limited capacity for heat storage, have forced them to a precarious, behaviourally maintained balance between physiological capacity and environmental stress.

Among the diurnal desert mammals which weigh 100 g. or less only one, the antelope ground squirrel, *Citellus (Ammospermophilus) leucurus*, has been systematically examined from the standpoint of temperature regulation and water economy (Dawson, 1955; Bartholomew & Hudson, 1959; Hudson, 1962). *Citellus leucurus* neither hibernates nor aestivates and must therefore cope with the problems of desert life at all seasons.

If body temperature is to remain uniform, heat loss must at all times be at least equal to heat production. In most environmental situations birds and mammals are warmer than their environment and the dissipation of metabolic heat is no problem. In the desert, however, ambient temperatures commonly exceed the usual levels of mammalian body temperature so the heat of metabolism cannot be lost except by the evaporation of water, and water is too scarce to squander.

A direct solution to this problem, the one shown by *C. leucurus*, is the acceptance of hyperthermia so that body temperature rises until it exceeds ambient temperature and metabolic heat can be lost passively without dependence on evaporative cooling. Captive *C. leucurus* show a conspicuously labile body temperature which increases regularly with ambient temperature (Fig. 2). Since it tolerates body temperatures as high as 43° C., *C. leucurus* can lose heat passively to the environment even when ambient temperatures are as high as 42° C. If it were to attempt to maintain body temperature constant at or near the resting level of 38° C. rather than to develop hyperthermia (assuming dry air and an ambient temperature of 42° C.), it would have to evaporate an amount of water equal to almost 13% of its body weight per hour, obviously an impossibility for an animal which may never have access to drinking water (Hudson, 1962).

Under natural conditions, antelope ground squirrels exploit their tolerance of hyperthermia in an intriguing way. Although they may sometimes be seen sitting quietly in the shade or perching in the open sparsely leafed creosote bushes (*Larrea tridentata*), they characteristically dash about furiously at high speed, often travelling a hundred yards or more from their home burrows. The furious activity, the burning soil and sand over which they travel, and the intense solar radiation soon cause them to heat up. Periodically throughout the day, therefore, the hyperthermic animal must unload its accumulated heat. This it does by going down a burrow, flattening itself against the relatively cool floor and losing heat to the substratum by conduction, and to the walls by radiation until its body temperature approaches its usual resting level. Thereupon the squirrel returns to the surface and forages until its hyperthermia again reaches the limits of tolerance, and heat is again unloaded underground (Fig. 1). The antelope ground squirrel has thus evolved a regulatory mechanism in which behaviour and physiology cause body temperature to oscillate over a range of 4 or 5° C. with a periodicity measured in parts of hours.

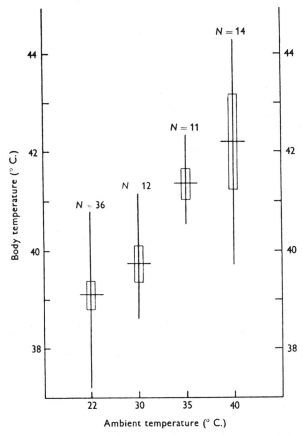

Fig. 2. The relation of body temperature to ambient temperature in the antelope ground squirrel. (Redrawn from Hudson, 1962.)

The critical element of this performance is the restriction of body temperature to a series of oscillations. The lower limit of these oscillations is set by the physiology of the mammalian temperature regulatory complex; the upper limit is not regulated by physiology, but by behaviour.

Reptiles

Among the small desert vertebrates which cope directly with high temperatures, lizards are the most abundant and conspicuous. Although lizards are of course poikilothermal, during activity each species maintains a characteristic level of body temperature (Cowles & Bogert, 1944). This pattern of thermal homeostasis is behaviourally controlled. By exploiting the variations in radiation and temperature in the local environment the lizard balances heat gain and heat loss with such precision that its body temperature is strikingly independent of ambient and black-bulb temperatures.

It is becoming apparent that the level of body temperature maintained by this behavioural thermoregulation has sufficient stability and sufficient antiquity to have associated with it a whole set of physiological adjustments to the temperature preferenda of the species in question (Dawson, 1960; Dawson & Templeton, 1963). This is shown dramatically by the desert iguana, *Dipsosaurus dorsalis*, a common lizard of the sand dunes in the deserts of south-western United States and northern Mexico, which is active during the intense heat of midday. This species is remarkable for its behavioural and physiological adjustments to high temperature. Much of the time its body temperature equals or exceeds that characteristic of birds and mammals. The mean body temperature of active animals is slightly above 42° C. (Norris, 1953). The body temperatures, sometimes in excess of 45° C., at which this species is normally active, are deleterious or lethal to many lizards. The tolerance to high temperatures of both the intact animal and its tissues *in vitro* is quite remarkable and is the most extreme reported for any terrestrial vertebrate (Dawson & Bartholomew, 1958). For example, (1) there is no apparent heat suppression of oxygen consumption at body temperatures as high as 45° C.; (2) body temperatures as high as 47° C. are tolerated; (3) spontaneous beating of excised auricles continues at temperatures between 46 and 47°; and (4) successive 5-min. exposures of strips of excised ventricle to temperatures between 42 and 44° C. do not cause irreversible effects on their capacity to respond to electrical field stimulation. Obviously these capacities fit the species nicely to its desert habitat, but such adaptations are not a necessary part of the adaptation of lizards to desert life. Other species of reptiles occupy the same sand dune habitat as *Dipsosaurus* even though they are much less heat-tolerant. They avoid high temperatures behaviourally by burrowing and by crepuscular and nocturnal activity. Thus, the spectacular physiological adjustments of *Dipsosaurus* to high temperatures are not a condition for existence in hot deserts, but an adjustment which allows it to exploit the desert environment during the daytime hours of intense heat when few other local reptiles are active.

Lizards show particularly clearly a condition which, from the ecological point of view, is as interesting as the 'steady-state' concept often equated with homeostasis and which is probably of basic evolutionary importance.

In many natural situations, animals do not maintain or achieve a steady physiological state, but rather their exercise of physiological control allows them only to modify the rates of change so as to extend survival or periods of activity. This is demonstrated by the body temperature of the large Australian agamid lizard, *Amphibolurus barbatus*. Although this animal, like other lizards, achieves its thermoregulation largely by behavioural means, it can control the rate of change in its body temperature by cardiovascular adjustments and to a lesser extent by metabolic rate (Bartholomew

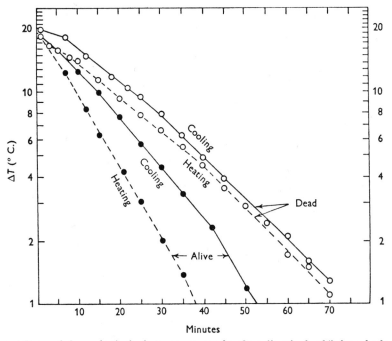

Fig. 3. Rates of change in the body temperature of a 480 g. lizard, *Amphibolurus barbatus*, while being heated from 20 to 40° C. and being cooled from 40 to 20° C. ΔT is the difference between body and ambient temperature. The living animal heats more rapidly than it cools, while the same animal, after being killed by a massive injection of nembutol, heats and cools at essentially the same rates. (Redrawn from Bartholomew & Tucker, 1963.)

& Tucker, 1963). The net effect of this control is to accelerate heating and retard cooling (Fig. 3). This type of control can have considerable utility for an animal such as *A. barbatus*, whose temperature preferendum is at 35° C., because it allows an increase in the amount of time which can be spent at temperatures near its preferendum. A similar capacity for the physiological modification of rates of heating and cooling has been found in the large monitor lizards (Bartholomew & Tucker, 1964) and such a capacity probably occurs widely in lizards. For example *Sauromalus varius*, a very large iguanid lizard which lives on the desert islands off the coast of

Mexico in the Gulf of California, has a similar capacity for physiological control of its rate of change in temperature (Bartholomew & Tucker, unpublished). It can heat slowly and cool fast as well as vice versa—a most convenient situation in a part of the world where, during much of the year, the problem is too much rather than too little heat. The evolutionary implications of this limited capacity for thermoregulation have been discussed elsewhere (Bartholomew & Tucker, 1963, 1964) and will not be considered here.

Birds

From the point of view of adaptations for desert life, birds are particularly interesting, not only because they are small and have high levels of endo-

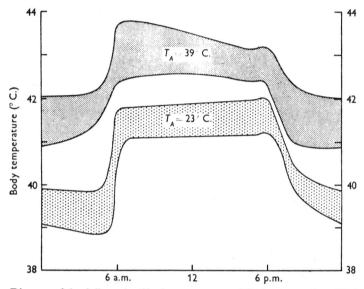

Fig. 4. Diagram of the daily cycle of body temperature of the brown towhee, *Pipilo fuscus*, at ambient temperatures of 23 and 39° C. showing the pronounced hyperthermia characteristic of all small birds when exposed to heat stress. The stippled areas include the intervals ±2 standard errors of the mean of the continuously recorded temperatures of six individuals. (Redrawn from Dawson, 1954.)

genous heat production, but because most of them are neither nocturnal nor fossorial and so must meet the desert challenge head-on.

Since the normal body temperatures of birds of most orders average 4 or 5° C. higher than mammals, they are pre-adapted to high environmental temperatures, although the intrinsic advantages of this situation are reduced by their high metabolic heat production. Under extreme conditions, however, desert birds, like animals of other groups, have no alternative but to resort to evaporative cooling. Like most other non-sweating animals they depend on evaporation of saliva by panting. However, the capacity to pant is a mixed blessing because panting movements involve heat production and the heat so produced adds to the total heat load which must be dissipated to the environment. In most adult birds for which data are available, panting can dissipate less than half the heat produced by metabolism; so, when environmental conditions make heat loss by radiation and conduction difficult, hyperthermia, even in the

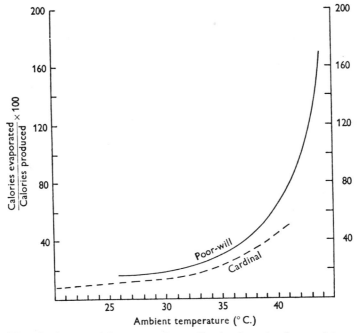

Fig. 5. The effectiveness of the evaporative cooling by the gular flutter of the poor-will as compared with the panting of the cardinal. (Data from Bartholomew, Hudson & Howell 1962; Dawson 1958.)

presence of intense panting, inevitably follows. However, in most circumstances this hyperthermia is self-limiting, because the elevation of body temperature makes possible passive heat loss to the environment and a steady state is re-established at a new and higher level. For example, Dawson (1958) has shown in the cardinal (*Richmondena cardinalis*) that at an ambient temperature of 40° C. the heat loss through evaporative cooling (8·7 cal./g./hr.) is less than basal heat production (12·7 cal./g./hr.). Initially, heat transfer to the environment cannot eliminate the remainder; the animal is forced to store heat and its temperature rises. By the time body temperature reaches 43° C. a balance between heat loss and heat production is attained and thermal homeostasis is achieved at the new level.

Almost invariably, birds of all orders appear to respond to heat stress by hyperthermia, and their tolerance of body temperatures as high as 45 or even 46° C. is a major factor in allowing them to exist in hot climates (Fig. 4). This mechanism for coping with environmental heat shows a pattern of homeostasis in which the actual level of body temperature is set by the inability of the animal to regulate at some lower level. In this situation the key is the capacity to tolerate high body temperature rather than to achieve a high level of heat dissipation at a fixed body temperature (Dawson, 1954; Bartholomew & Dawson, 1958).

Although there are probably a number of species of birds which can maintain body temperature below ambient temperature, to date this capacity has been demonstrated in only one form, the poor-will *Phalaenoptilus nuttallii*. This bird, like other members of the order *Caprimulgiformes*, has a relatively enormous mouth with a highly vascularized surface from which evaporation can readily occur. Little energy is required to flutter the extensive gular area vigorously, and as a result, in the presence of heat stress, the poor-will can increase its evaporation rate at a remarkably low metabolic cost. Not only can it dissipate by evaporation the heat produced by its own metabolism, but when ambient temperature exceeds body temperature it can unload much of the heat it gains from the environment (Fig. 5). Indeed, in an ambient temperature of 44–45° C. and a relative humidity of 10–20 %, it can maintain a stable body temperature of about 42° C. for many hours (Bartholomew, Hudson & Howell, 1962).

Desert islands

Problems of heat and aridity are not, of course, peculiar to extensive continental areas of desert. The physiological heat load due both to temperature and radiation can vary dramatically with local circumstances. Even in areas where the climate is equable, the extremes of the physical environment may be locally limiting. This situation is thrown into particularly sharp focus on low-latitude oceanic islands, partly because the vertebrates which visit them are primarily adapted to aquatic rather than desert conditions, and partly because they occupy these islands for reproduction, and the young of vertebrates characteristically have less homeostatic capacity than adults. In addition, since the thermoregulatory capacity of the young of aquatic birds varies markedly from species to species, they show almost diagrammatically the interplay between behaviour and physiology which allows the maintenance of homeostasis. For example, on Angel de la Guarda, a desert island in the Gulf of California, intense solar radiation and extreme aridity make overheating a primary physiological problem during the nesting period. Nevertheless, aquatic birds of three different orders, brown pelicans (*Pelecanus occidentalis*), great blue herons (*Ardea herodias*), and western gulls (*Larus occidentalis*) nest successfully in close proximity in completely exposed rookeries despite profound differences in the thermoregulatory capacity of their young. Nestling western gulls are, of course, precocial. They are covered with down at hatching and as soon as they dry out they can creep about and seek shelter from the sun. The parents are relatively inattentive. The newly hatched pelicans are naked, completely helpless, and have almost no capacity for thermoregulation. Their survival depends completely on the effectiveness with which they are shielded from the sun by their parents, who are in general extremely attentive. The young herons are altricial, although less so than pelicans, and in attentiveness the adults are intermediate between the gull and the pelican. So these three species are able to breed side by side in extremely unfavourable physical conditions in spite of the profound differences in the nestlings' capacity for temperature regulation, because the adults compensate behaviourally for the physiological limitations of their young (Bartholomew & Dawson, 1954).

Albatross are usually thought of as inhabitants of the cold and blustery high latitudes of the southern hemisphere, but two species, the Laysan albatross (*Diomedea immutabilis*) and the black-footed albatross (*Diomedea nigripes*) breed on atolls in the central Pacific, which, although characterized by a remarkably equable macroclimate, have some areas where the microclimate can be extremely demanding. For example, albatross nests which are located in the lee of sand dunes and are shielded from the trade winds frequently are exposed to air temperatures as high as 40° C. and black-bulb temperatures greater than 45° C. The downy young are brooded by the parents for 2–3 weeks and then are fed by the parents for approximately 5 months until they are able to fly. During the 5 months that they are unbrooded but dependent on the parents for food, many of the young remain near the nest where they were hatched, often on bare coralline sand, fully exposed to intense sun. They must depend entirely on their own capacity for heat loss to prevent overheating. Like all birds, they pant, but this presents problems because they have no water to drink and their food, mostly squid, is isotonic with the sea, and before the water in it can become physiologically available the excess salt must be excreted by the nasal gland (Frings & Frings, 1959). The birds, although covered with feathers, have a useful pair of feet, which are a strikingly effective adjunct to their capacity for heat dissipation. As described in detail elsewhere (Howell & Bartholomew, 1961), young albatross have large, webbed, heavily vascularized feet. Under conditions of heat stress they orient themselves with their backs to the sun, balance on their heels with their webbed toes spread and their feet held in the air in the shade of their bodies (Pl. 1). Their foot temperatures are higher than the air, but lower than the substratum. Consequently the birds can lose heat by radiation and conduction to the air, whereas heat would be gained if the feet rested on the ground. This method of heat dissipation is sufficiently effective, and week after week, for many hours during the heat of the day, the young birds rock gently backwards and forwards on their heels, with their feet in the air. In some cases, young albatross exposed to intense solar radiation remain crouched on the sand with the feet beneath the body and pressed against the sand. In this attitude, the sand beneath the birds is shielded from the sun and remains cooler than their feet. Consequently, the birds can lose the heat they gain from radiation and metabolism by conducting it from their feet to the substratum.

As indicated in the preceding discussion, mechanisms for heat dissipation that do not involve evaporative cooling should be advantageous, but most animals do not have morphological structures which can serve as effective heat exchangers. Among the birds best equipped morphologically for this purpose are members of the order *Pelecaniformes*, most of which have extensive naked gular areas. No quantitative partition between evaporation, radiation, and conduction is yet available but the heat loss via the gular pouch can clearly be of critical importance for the survival of pelicans and boobies. Downy, young white pelicans (*Pelecanus erythrorhynchos*) which nest on islands in the Salton Sea (a saline lake in the desert part of southern California) will die on a hot day if gular flutter is inhibited (Bartholomew, Dawson & O'Neill, 1953). Unbrooded downy young of the red-footed booby (*Sula sula*) in exposed nests on Midway Island in the central Pacific depend both on their feet and on gular flutter for heat dissipation, but if their beaks are taped shut so that gular flutter is prevented, they are unable to maintain a constant level of body temperature. In one case, when exposed to an air temperature of 29–31° C. and a black-bulb temperature of 44–46° C. with the bill taped shut to prevent gular flutter, body temperature rose from 39 to 44° C. in 1 hr. As soon as the bill was untaped, rapid gular flutter began and in a few minutes body temperature was back to normal (Howell & Bartholomew, 1962).

Basal metabolism

The conclusion of Scholander, Hock, Walters & Irving (1950) and Scholander (1955), that core body temperature and basal metabolic rate show negligible adaptive adjustments in homeothermic animals, has been generally substantiated for arctic forms by the additional data that have accumulated over the past decade. However, no such generalization can be made

PLATE I

Juvenile Laysan albatross resting on its heels with its feet in the air and serving as heat exchangers. For photographic clarity the bird was induced to turn so that its feet were in the sun; ordinarily they are kept in the shade of the body. From Howell & Bartholomew (1961).

about birds and mammals living under conditions of heat and aridity. *A priori*, a lower-than-normal metabolism should be advantageous to an animal living under conditions of high environmental temperatures, since such a reduced basal rate would minimize the production of endogenous heat and so reduce the burden of losing heat to a hot environment.

The data available at the present time, however, do not reveal any single adaptive pattern with regard to this problem. Desert passerine birds appear to have a metabolism appropriate to their size. The antelope ground squirrel (see p. 14) has a basal metabolism slightly higher than expected. On the other hand, rodents of the family *Heteromyidae* generally have a metabolic rate lower than that predicted on the basis of size, and McNab & Morrison (1963) have recently reported a similar situation in the desert species of the cricetid genus *Peromyscus*. Since all these mice are nocturnal, there are difficulties in interpreting their low basal metabolism as an adaptation to high environmental temperature. There is, however, at least one instance in which a conspicuously reduced metabolic rate appears to be clearly adaptive. The previously mentioned poor-will has a metabolic rate which in the thermal neutral zone is only about one-third of that of other birds of its size (40 g.). Not only is the poor-will one of the very few birds which can undergo long periods of dormancy, but over much of its summer range it is exposed to severe conditions of heat. It is a crepuscular feeder, but its habit of spending the daylight hours, even in the desert, sitting quietly in the open, completely exposed to the sun, imposes on it a much greater heat load than is faced by most other desert birds. The ability of this caprimulgid to tolerate exposure to such trying conditions depends on a complex of physiological, morphological and behavioural attributes (Bartholomew *et al.* 1962). These include a remarkably low basal metabolism, an energetically inexpensive gular flutter mechanism for evaporative cooling, a pattern of behaviour which leads it to remain motionless during the daylight hours, a zone of thermal neutrality which extends at least to 44° C., and a variable and labile body temperature (Fig. 6). (Alert poor-wills in good condition may have a body temperature anywhere between 35 and 43° C. Body temperature increases directly with ambient temperatures above 35° C., but is essentially independent of temperature between 0 and 35° C.) This complex of physiological and behavioural characteristics is reinforced by an almost unbelievably effective cryptic colouration which makes this bird extremely difficult to see even when it is crouching fully exposed on the bare desert floor.

The behavioural acceptance of hyperthermia makes an important though negative contribution to successful metabolic adjustment of a variety of desert vertebrates to high temperatures. By remaining quiet even in the face of greatly elevated body temperature an animal minimizes the metabolic contribution to its total heat load and this helps to extend the zone of thermal neutrality to impressively high temperatures. Indeed some species which accept hyperthermia calmly appear to have no clear-cut upper critical temperature. This behavioural adjustment to high temperatures has been observed in desert ground squirrels (Bartholomew & Hudson, 1961), desert white-footed mice, *Peromyscus eremicus* (Murie,

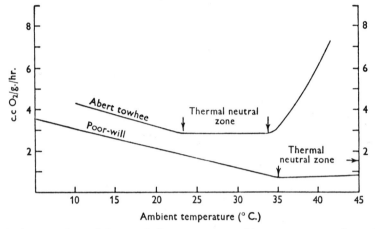

Fig. 6. A comparison of the metabolic response to ambient temperature of two birds, weighing between 40 and 50 g., which live in the deserts of the south-western United States. The poor-will has an unusually low metabolic rate and efficient evaporative cooling, and remains placid when exposed to high ambient temperatures. The towhee has a normal metabolic rate and depends on panting, which is relatively inefficient, for evaporative cooling. The arrows indicate upper and lower critical temperatures. (Based on Dawson 1954; Bartholomew *et al.* 1962.)

1961), an Australian flying fox, *Pteropus scapulatus*, which occupies desert areas (Bartholomew, Leitner & Nelson, 1964), and the dromedary camel (Schmidt-Nielsen & Schmidt-Nielsen, 1952), and it represents a widespread trait of adaptive importance.

Torpidity

Another modification of the energy metabolism of homeothermic animals which represents a major pattern of adjustment to seasonal heat and aridity is the phenomenon of aestivation. Current knowledge of this summer dormancy has recently been reviewed elsewhere (Hudson & Bartholomew, 1964) and need not be examined in detail here. However, at least one aspect of the phenomenon of mammalian torpidity at relatively high body temperatures is directly relevant to homeostatic control under desert conditions and merits brief consideration in this context.

A striking and rather unexpected feature of the physiology of heteromyid rodents, which are pre-eminently successful mammals of the deserts of North America, is their capacity to undergo periods of dormancy (usually lasting from a few hours to a day or two) during which body temperature falls to within a degree or less of that of their surroundings. These episodes of torpor can occur at temperatures anywhere between 5 and 30° C., and may take place at any time of year (Bartholomew & Cade, 1957; Bartholomew & MacMillen, 1961; Tucker, 1962). Similar but less well-documented instances of torpor appear to occur in some of the smaller cricetids which occur in arid regions, for example *Peromyscus* and *Reithrodentomys*.

Such brief, and often daily, periods of abandonment of homeothermy afford an obvious energy saving. Tucker (1962, and MS.) has shown that the California pocket mouse, *Perognathus californicus*, will maintain a regular daily cycle of torpor and arousal to normal activity indefinitely and at any time of year if its food ration is reduced slightly below its normal *ad libitum* consumption. Further, the duration of this daily torpor is inversely related to daily food ration. The mouse adjusts its periods of torpor to the food available: the smaller the food ration the longer the daily period of torpor. Thus, this mouse demonstrates a feedback mechanism coupling available food and daily metabolic cycle in such a way that it can strike a nice balance between food supply and energy expenditure. Carpenter (1962) has found a similar but less readily evoked response in a kangaroo rat, *Dipodomys merriami*. He has pointed out that although this response shows many physiological similarities to hibernation and aestivation, ecologically it represents an adjustment to chronic rather than seasonal food shortage. In deserts, rainfall is characteristically erratic and during some years may be completely lacking. Consequently the production of seeds, on which heteromyids are largely dependent, is also erratic. As a result, these rodents are less apt to face a complete seasonal absence of food than a chronic shortage of food which may last for many months or even years. Daily torpor, the occurrence and duration of which are determined by the availability of food, should allow these desert rodents to maintain themselves for prolonged periods on a reduced food supply without forcing any reduction in the intensity of their foraging activities.

It should be emphasized that the adjustment of heteromyids to food shortage has important non-physiological aspects. The desert genera, *Perognathus*, *Dipodomys*, and *Microdipodops*, all accumulate extensive stores of seeds, and all members of the family are equipped with external, hair-lined cheek pouches in which they carry seeds and other foodstuffs back from foraging areas to their food caches.

DISCUSSION AND CONCLUSIONS

If one adopts the point of view that organisms are interactions between self-sustaining physico-chemical systems and the environment, then at the level of the whole organism the maintenance of homeostasis involves all the behavioural, physiological, and morphological adjustments that control the direction and rate of interchange between these two inseparable components. Since each kind of organism is in equilibrium with only a limited part of the total physical environment, homeostasis from the point of view of physiological ecology can be most profitably examined in terms of that small fraction of the total possible environment with which the organism is in balance. For most vertebrates this effective environment is sufficiently limited so that, by their behaviour, they can assemble from the variety of physical conditions that exist in a given area an effective environment within which they can function adequately. This situation can be conveniently examined with regard to thermal exchanges in the desert environment where a prosperous fauna exists, although the macroclimate for much of the time exceeds the limits which vertebrates can tolerate. The patterns of adjustment which allow this situation to occur are varied and numerous. The ones for which documentation has been presented in the present paper are listed below:

(1) Complete avoidance of the problems of severe heat load by means of nocturnal and fossorial habits. This behavioural circumvention of the desert is the most common pattern found among small animals.

(2) Heat storage made possible through physiological tolerance and behavioural acceptance of hyperthermia. The periodicity of heat unloading may be daily, as in the dromedary camel, or it may be measured in fractions of hours as in the antelope ground squirrel.

(3) Behavioural control of heat exchange with the environment so that body temperature is maintained at or near some preferendum. This is a common pattern in reptiles, and the various taxa which have been studied show appropriate physiological adjustments to the behaviourally maintained levels of body temperature.

(4) Physiological control of rates of change without the attainment of a regulated steady state. At least some large lizards can regulate their rates of heating and cooling so as to extend the periods of time which can be spent at or near the temperature preferendum. This is a significant step in the direction of homeothermy.

(5) Sustained hyperthermia as a device for facilitating heat loss. Inadequacy of homeostatic control at one level of temperature results in hyperthermia which in turn facilitates regulation at a new and higher

body temperature. This pattern appears to be almost universal among birds.

(6) Reduced basal metabolic rate as a way of minimizing total heat load. The extent of this pattern is not yet known, but at least one bird, the poorwill, apparently employs it.

(7) Quietness in the face of hyperthermia. This is a very useful condition for extending the zone of thermal neutrality to high levels and occurs widely among desert birds and mammals.

(8) Behavioural compensation by adults for the inadequate regulatory ability of young animals. This is particularly well illustrated by aquatic birds which breed on desert islands. These thermoregulatory adjustments which involve the interaction of two or more individual organisms are precise enough to enable species whose young have widely divergent thermoregulatory capacities to breed side by side.

(9) Mechanisms for heat dissipation which involve a minimum of energy expenditure and/or a minimum of water loss. These are clearly advantageous to animals faced with a heavy heat load. For example, the webbed feet of young albatross serve as effective heat dissipators, as do the readily fluttered gular areas of birds of the orders *Caprimulgiformes* and *Pelecaniformes*.

Both seasonal and daily torpidity as adaptations to desert conditions have been demonstrated in mammals. In at least one species, the California pocket mouse, the duration of the period of daily torpor is inversely related to the daily food ration. This is interpreted as an adjustment to an erratic food supply in which food shortage may be of more than seasonal duration.

REFERENCES

BARTHOLOMEW, G. A. (1958). The role of physiology in the distribution of terrestrial vertebrates. In *Zoogeography* (ed. by Hubbs, C. L.), pp. 81–95. Publ. 51, AAAS. Washington, D.C.

BARTHOLOMEW, G. A. & CADE, T. J. (1957). Temperature regulation, hibernation, and aestivation in the little pocket mouse, *Perognathus longimembris*. *J. Mammal.* 38, 60–72.

BARTHOLOMEW, G. A. & CADE, T. J. (1963). The water economy of land birds. *Auk*, 80, 504–39.

BARTHOLOMEW, G. A. & DAWSON, W. R. (1954). Temperature regulation in young pelicans, herons, and gulls. *Ecology*, 35, 466–72.

BARTHOLOMEW, G. A. & DAWSON, W. R. (1958). Body temperatures in California and Gambel's quail. *Auk*, 75, 150–6.

BARTHOLOMEW, G. A., DAWSON, W. R. & O'NEILL, E. J. (1953). A field study of temperature regulation in young white pelicans, *Pelecanus erythrorhynchos*. *Ecology*, 34, 554–60.

BARTHOLOMEW, G. A. & HUDSON, J. W. (1959). Effects of sodium chloride on weight and drinking in the antelope ground squirrel. *J. Mammal.* 40, 354–60.

BARTHOLOMEW, G. A. & HUDSON, J. W. (1961). Desert ground squirrels. *Sci. Amer.* **205**, 107–16.

BARTHOLOMEW, G. A., HUDSON, J. W. & HOWELL, T. R. (1962). Body temperature, oxygen consumption, evaporative water loss, and heart rate in the poor-will. *Condor*, **64**, 117–25.

BARTHOLOMEW, G. A., LEITNER, P. & NELSON, J. E. (1964). Body temperature, oxygen consumption, and heart rate in three species of Australian flying foxes. *Physiol. Zoöl.* **37**, 179–98.

BARTHOLOMEW, G. A. & MACMILLEN, R. E. (1961). Oxygen consumption, estivation, and hibernation in the kangaroo mouse, *Microdipodops pallidus*. *Physiol. Zoöl.* **34**, 177–83.

BARTHOLOMEW, G. A. & TUCKER, V. A. (1963). Control of changes in body temperature, metabolism, and circulation by the agamid lizard, *Amphibolurus barbatus*. *Physiol. Zoöl.* **36**, 199–218.

BARTHOLOMEW, G. A. & TUCKER, V. A. (1964). Size, body temperature, thermal conductance, oxygen consumption, and heart rate in Australian varanid lizards. *Physiol. Zoöl.* **37** (in the Press).

CARPENTER, R. E. (1962). A comparison of thermoregulation and water metabolism in the kangaroo rats, *Dipodomys agilis* and *Dipodomys merriami*. Ph.D. Thesis. University of California, Los Angeles.

CHEW, R. M. (1961). Water metabolism of desert-inhabiting vertebrates. *Biol. Rev.* **36**, 1–31.

COWLES, R. B. & BOGERT, C. M. (1944). A preliminary study of the thermal requirements of desert reptiles. *Bull. Amer. Mus. Nat. Hist.* **83**, 261–96.

DAWSON, W. R. (1954). Temperature regulation and water requirements of the brown and Abert towhees, *Pipilo fuscus* and *Pipilo aberti*. *Univ. Calif. Publ. Zool.* **59**, 81–124.

DAWSON, W. R. (1955). The relation of oxygen consumption to temperature in desert rodents. *J. Mammal.* **36**, 543–53.

DAWSON, W. R. (1958). Relation of oxygen consumption and evaporative water loss to temperature in the cardinal. *Physiol. Zoöl.* **31**, 37–48.

DAWSON, W. R. (1960). Physiological responses to temperature in the lizard *Eumeces obsoletus*. *Physiol. Zoöl.* **33**, 87–103.

DAWSON, W. R. & BARTHOLOMEW, G. A. (1958). Metabolic and cardiac responses to temperature in the lizard *Dipsosaurus dorsalis*. *Physiol. Zoöl.* **31**, 100–11.

DAWSON, W. R. & TEMPLETON, J. R. (1963). Physiological responses to temperature in the lizard *Crotaphytus collaris*. *Physiol. Zoöl.* **36**, 219–36.

FRINGS, H. & FRINGS, M. (1959). Observations on salt balance and behaviour of Laysan and black-footed albatrosses in captivity. *Condor*, **61**, 305–14.

GORDON, M. S. (1962). Physiological ecology of osmotic regulation among littoral vertebrates. *Pubbl. Staz. Zool. Napoli*, **32**, suppl., 294–300.

HOWELL, T. R. & BARTHOLOMEW, G. A. (1961). Temperature regulation in Laysan and black-footed albatrosses. *Condor*, **63**, 185–97.

HOWELL, T. R. & BARTHOLOMEW, G. A. (1962). Temperature regulation in the red-tailed tropic bird and the red-footed booby. *Condor*, **64**, 6–18.

HUDSON, J. W. (1962). The role of water in the biology of the antelope ground squirrel *Citellus leucurus*. *Univ. Calif. Publ. Zool.* **64**, 1–56.

HUDSON, J. W. & BARTHOLOMEW, G. A. (1964). Terrestrial animals in dry heat: estivators. In *Handbook of Physiology*. Sect. 4, pp. 541–50. Amer. Physiol. Soc., Washington, D.C.

MCNAB, B. K. & MORRISON, P. (1963). Body temperature and metabolism in subspecies of *Peromyscus* from arid and mesic environments. *Ecol. Monogr.* **33**, 63–82.

Murie, M. (1961). Metabolic characteristics of mountain, desert and coastal populations of *Peromyscus*. *Ecology*, **42**, 723–40.

Norris, K. S. (1953). The ecology of the desert iguana *Dipsosaurus dorsalis*. *Ecology*, **34**, 265–87.

Schmidt-Neilsen, K. (1964). Desert Animals; physiological problems of heat and water.

Schmidt-Nielsen, K. & Schmidt-Nielsen, B. (1952). Water metabolism of desert mammals. *Physiol. Rev.* **32**, 135–66.

Schmidt-Nielsen, K., Schmidt-Nielsen, B., Jarnum, S. A. & Houpt, T. R. (1957). Body temperature of the camel and its relation to water economy. *Amer. J. Physiol.* **188**, 103–12.

Scholander, P. F. (1955). Evolution of climatic adaptation in homeotherms. *Evolution*, **9**, 15–26.

Scholander, P. F., Hock, R., Walters, V. & Irving, L. (1950). Adaptation to cold in arctic and tropical mammals and birds in relation to body temperature, insulation, and basal metabolic rate. *Biol. Bull., Wood's Hole*, **99**, 259–71.

Tucker, V. A. (1962). Diurnal torpidity in the California pocket mouse. *Science*, **136**, 380–1.

Part II
THE SEARCH FOR THE THERMOSTAT

Editor's Comments
on Papers 5 Through 8

5   OTT
Excerpt from *The Relation of the Nervous System to the Temperature of the Body*

6   BAZETT and PENFIELD
Excerpts from *A Study of the Sherrington Decerebrate Animal in the Chronic as Well as the Acute Condition*

7   RANSON
Excerpts from *Regulation of Body Temperature*

8   SATINOFF
Behavioral Thermoregulation in Response to Local Cooling of the Rat Brain

The papers in this section concern the question of the neural control of thermoregulation. They were all instrumental in focusing on the hypothalamus as *the thermostat* in the brain. This statement may be somewhat misleading because in the nineteenth and early twentieth centuries, no one considered the hypothalamus to be important for anything; it was known only as an anatomical area situated below the thalamus, and symptoms resulting from its damage such as obesity, polyuria and polydipsia, hypothermia, and genital atrophy were ascribed to damage to the nearby pituitary gland. In 1884 Isaac Ott in Philadelphia and Charles Richet in Paris almost simultaneously reported that temperature regulation could be deranged when the base of the brain was damaged, although neither used the term "hypothalamus" in their papers. As Ott (Paper 5) wrote: "All these experiments lead up to the conclusion that in the vicinity of the corpora striata are centers which have a relation to the temperature of the body." It soon became obvious that both rises and falls in body temperature could be obtained after various kinds of brain damage, and thus the idea of *two* centers was born. Much of the history of the work on temperature regulation can be seen as an attempt to localize and describe the properties of these centers. The German physiologist Hans Meyer (1913) was one of the first to propose a dual center hypothesis.

> The relationship of the chemical heat production to the physical heat loss determines the temperature of the body of an animal. Because of its constant temperature, the homeotherm must balance the positive and negative processes so that they can keep the absolute level of temperature steady. Thus, a central nervous regulation is necessary, with two antagonistic, correlated and interrelated centers, the one for warmth, and the other for cold. (Meyer 1913. Translated from the German.)

From this time, the search was on for the anatomical localization of these proposed centers. Sherrington (1924) showed that dogs whose spinal cord had been transected at a high cervical level were unable to shiver below the level of the transection, so the locus of control, at least for heat production, had to be in the brain. At about the same time Bazett and Penfield (Paper 6) reported that decerebrate cats, in whom all brain tissue above the pons had been removed, had no temperature control at all. (Even as late as 1922, when Bazett and Penfield published their studies, there was no mention of the hypothalamus in their paper.) Thus the critical areas for thermoregulation had to be above the level of the pons. Lilienthal and Otenasek (1937) and Clark, Magoun, and Ranson (1939) had demonstrated that decorticate cats, or cats with thalamic lesions, had only very minor disturbances in temperature control, and that left the base of the brain as the most likely place to contain the hypothetical centers.

The neurophysiologist Stephen W. Ranson (Paper 7) was one of the first to apply the technique of electrolytic lesions to the study of homeostatic mechanisms. He localized the two areas of Hans Meyer to the preoptic/anterior hypothalamus (the "heat-loss center") and the posterior hypothalamus (the "heat-production center"). He reported that lesions in the preoptic/anterior hypothalamus produced cats and monkeys that were unable to keep their body temperatures down in the heat, but were able to maintain near normal body temperatures in the cold. Thus this area of the brain was the "heat-loss center." If, however, lesions were made in the posterior hypothalamus, the animals were unable to regulate in the heat or in the cold. He theorized that the preoptic area sent efferents down to the posterior hypothalamus, and thus when that area was lesioned, all temperature control was lost. This separation, while aesthetically pleasing, is not borne out by Ranson's own data. Examination of Ranson's Table 11 shows that cats with symmetrical lesions in the rostral hypothalamus were *not* able to maintain their body temperatures adequately when placed in a cold box. Despite the contradictions in Ranson's own data, this paper was one of the most influential in the field of temperature regulation. Only much later did papers begin to appear that unequivocally demonstrated in several species that animals with preoptic lesions could not maintain their body temperatures in cold environments [cats (McCrum 1953, Squires and Jacobson 1968); goats (Andersson et al 1965); ground squirrels (Satinoff 1967); rats (Cytawa and Teitelbaum 1967, Satinoff and Rutstein 1970).] Other techniques also pointed to the hypothalamus as *the* thermostat of the brain. Magoun, Harrison, Brobeck, and Ranson (1938) showed that appropriate thermoregulatory responses could be activated when that area was heated (much of this work is discussed in Paper 7). They used radio-frequency current to heat discrete areas of the brain, and this was a tremendous advancement in technique. Previous workers had heated the whole head, or the carotid blood, and it was never possible to assure that the skin or cranial blood vessels were not being heated and carrying the heat to some unknown area of the brain. With radio-frequency current, the area heated had to be the neurons or blood vessels immediately between the two electrodes.

It was assumed that if the preoptic area was the thermostat, it should contain thermosensitive cells, that is, cells that alter their firing rates with changes in their temperature. In 1963 such units were identified by Nakayama, Hammel, Hardy, and Eisenman who recorded single-unit activity in this area in anesthetized cats during local heating and cooling. They found that about 20% of the 1000 neurons tested increased their firing rate when local brain temperature was increased over a range of 32-41°C. These units had Q_{10}'s of 5-15. (This means that the units fired from 5 to 15 times their initial discharge rate as a result of a 10°C increase in their temperature. Thermally insensitive neurons have Q_{10}'s of 1.) Soon after, Hardy, Hellon, and Sutherland (1964) reported the existence of cool-sensitive cells in the preoptic area, that is, cells that increase their firing rates with decreases in their temperature.

In all the studies described, only physiological means of heat production and heat loss had been measured. None of the animals had been given an opportunity to regulate their body temperatures behaviorally. Then in 1964 Evelyn Satinoff reported that local cooling of the preoptic region in rats elicited not only reflexive thermoregulatory responses but also the behavioral urge to keep warm, demonstrated by the rats' willingness to press a bar to turn a heat lamp on much more when their brains were being cooled than when they were not. Thus motivated, nonreflexive behavior designed to conserve heat was also influenced by the temperature of the preoptic region.

In summary, almost everything up to the middle 1960s pointed to the preoptic area as *the thermostat.* Lesions disrupted normal thermoregulation. Both behavioral and reflexive thermoregulatory responses could be elicited by local cooling or warming. And a neuronal basis for these phenomena had been established with the discovery of an abundance of temperature-sensitive cells in the area.

REFERENCES

Andersson, B.; Gale, C.; Hokfelt, B.; and Larsson, B., 1965. Acute and chronic effects of preoptic lesions. *Acta Physiol. Scand.* **65**:45-60.

Clark, G.; Magoun, H. W.; and Ranson, S. W., 1939. Temperature regulation in cats with thalamic lesions. *J. Neurophysiol.* **2**:202-207.

Cytawa, J., and Teitelbaum, P., 1967. Spreading depression and recovery of subcortical functions. *Acta Biol. Exper.* (Warsaw) **27**:345-353.

Hardy, J. D.; Hellon, R. F., and Sutherland, K., 1964. Temperature-sensitive neurons in the dog's hypothalamus. *J. Physiol.* (London) **175**:242-253.

Lilienthal, J. L., and Otenasek, F. J., 1937. Decorticate polypneic panting in the cat. *Bull. Johns Hopkins Hosp.* **61**:101-124.

Magoun, H. W.; Harrison F.; Brobeck, J. R.; and Ranson, S. W., 1938. Activation of heat loss mechanisms by local heating of the brain. *J. Neurophysiol.* **1**:101-114.

McCrum, W. R., 1953. A study of diencephalic mechanisms in temperature regulation. *J. Comp. Neurol.* **98**:233-281.

Meyer, H. H., 1913. Theorie des Fiebers und seiner Behandlung. *Verhandl. Dtsch. Bes. Inn. Med.* **30**:15.

Nakayama, T.; Hammel, H. T.; Hardy, J. D.; and Eisenman, J. S., 1963. Thermal stimulation of electrical activity of single units of the preoptic region. *Am. J. Physiol.* **204**:1122-1126.

Richet, C., 1884. La fièvre traumatique nerveuse et l'influence des lésions du cerveau sur la température générale. *C. R. Soc. Biol.*, Series 8 **1**:189-195.

Satinoff, E., 1967. Aberrations of regulation in ground squirrels following hypothalamic lesions. *Am. J. Physiol.* **212**:1215-1220.

Satinoff, E., and Rutstein, J., 1970. Behavioral thermoregulation in rats with anterior hypothalamic lesions. *J. Comp. Physiol. Psychol.* **71**:77-82.

Sherrington, C. S., 1924. Notes on temperature after spinal transection, with some observations on shivering. *J. Physiol.*, (London) **58**:405-424.

Squires, R. D., and Jacobson, R. H., 1968. Chronic deficits of temperature regulation produced in cats by preoptic lesions. *Am. J. Physiol.* **214**:549-560.

THE RELATION OF THE NERVOUS SYSTEM TO THE TEMPERATURE OF THE BODY.

By ISAAC OTT, M.D.

THAT the nervous system has an influence upon heat-production has been known since the time of Brodie, who called the attention of the profession to the rise of temperature seen after a lesion of the spinal cord. He believed the central nervous system to have an immediate influence on heat-production. The proofs that the nervous system has a direct action on the heat of the body may be classed, for purposes of study, into thermometric, chemical, and calorimetric. Thermometric facts: Tscheschichin was the first to announce the existence of an inhibitory heat-centre in the nervous system. In rabbits, after he divided the spinal cord, there was a fall of temperature, but after he divided the medulla from the pons there was a rise. He believed the rise of temperature to be due to the removal of a nerve-centre in the higher parts of the brain inhibiting the thermogenic centres. Nauyn and Quincke found in dogs that after section of the spinal cord there was at first a fall of temperature, which was followed by a rise. They attributed this to an actual increase of heat-production, which, in time, overcame the dissipation engendered by vaso-motor paralysis. Parinaud found, after section of the spinal cord, in rabbits, a continuous fall in temperature, especially of the deeper parts of the body. Schreibler discovered, after injury of the pons in all its parts, the cerebral peduncles, the cerebrum, and cerebellum, an increase of the temperature of the body, when heat-dissipation was prevented by artificial means. The rise of temperature was constant after an injury between the medulla and the pons; he used rabbits. Eulenberg and Landois found an increase of temperature after removal of the cortex cerebri about the sulcus cruciatus. Prof. Hitzit has arrived at similar conclusions. Chemical facts: In dogs, Leyden and Fränkel found the carbonic acid increased in fever. Colasanti and Pflüger found cold to increase the amount of carbonic acid and the absorption of oxygen, which has been confirmed by Sanderson, Röhrig, and Zuntz. It was discovered by the two latter observers, that in poisoning with urari there was a marked diminution in the consumption of oxygen and the excretion of carbonic acid; the bodily temperature also fell, even when prevented by external preventive measures. In the uninjured animal the external cold excites some sensory nerves which, conveying impressions to the central nervous apparatus, cause it to send out afferent impulses to the muscles, increasing tissue-metamorphoses, and thus furnishing more heat. But when the nerves going to the muscles are paralyzed by urari, then these efferent impulses can not reach the muscles, and no increment of tissue-change takes place upon the application of cold. The experiments of Pflüger and his pupils tend to show that there is a nervous apparatus by which external cold increases the tissue-metamorphoses, and thus generates more heat. Hence chemical data support the thermometric, and the view of Tscheschichin, that there is a nervous apparatus to inhibit thermogenic centres. From this stand-point fever is supposed to cause a paresis of the inhibitory mechanism allowing the thermogenic centres to come into play of exaggerated activity. Colasanti found in the guinea-pig that, in pyrexia, the usual reaction did not ensue upon the application of external cold. Finkler has made a very exhaustive study of the phenomena of fever, and deduced the law that the consumption of oxygen is increased during fever, and that there is an increase of carbonic acid due to increased heat-production. He also arrives at the conclusion that fever is a neurosis, mainly a disease of a nervous system regulating the temperature. Calorimetric facts: Profs. Senator, Leyden, and Wood have studied the matter by means of the calorimeter. Senator found in fever an increased heat-production, which has been confirmed by Prof. Wood's numerous experiments on the subject. Wood is the only one who has calorimetrically studied the effect of sections in a transverse direction of the nervous system from the spinal cord upward at different levels. He confirms Tscheschichin's theory, that the inhibitory heat-centre is probably in the pons varolii. Hence thermometric, chemical, and calorimetric proofs go to show that the phenomena of heat are dominated by the nervous system.

My plan has been to make transverse sections of the brain, not from below but from before backward. The experiments were made on rabbits and cats. The apparatus used was d'Arsonval's calorimeter and Voit's respiration-apparatus. The calorimeter is composed of two concentric conical cylinders forming two cavities: the central one is the inclosure for the animal; the other annular cavity is filled with distilled water, which uniformly distributes the heat around the inclosure containing the animal, and prevents sudden variations of temperature. The external copper wall of the calorimeter has a tube on one side which communicates with the water cavity, and is closed externally by a vertical rubber membrane, which is the only portion of the calorimeter that can be pushed outwardly by the variations in the volume of the water, the apparatus being closed. To the side-tube of the calorimeter, is attached the gas-mechanism. The gas is carried by a tube which normally opens at the centre of the rubber membrane. At a small distance from the gas-tube there is an exit-tube, which permits the gas to escape from the brass box to the burners, which heat the calorimeter. The gas-tube and rubber membrane form a sort of stop-cock of a very sensitive nature, and whose amplitude of openness is dependent upon the variations in the volume of water, which only allows the gas-burners to consume a quantity necessary to compensate the dissipation of heat. In this combination the combustible warms directly the water, which, being the regulator, in its turn, reacts directly upon the combustible gas. Between the two concentric cylinders, passing through the water, is a hollow tube conveying air, one end of which opens externally, and the other in the inclosure containing the animal. In my method of experimentation, part of the air supplied the animal was conveyed to it through this tube.

The management of the apparatus was as follows:

The annular space between the two copper cylinders was filled with water by the external orifice. In this orifice a thermometer was placed, it being arranged so as to allow the escape of the heated water, not closing the aperture; the tubes of rubber going to the burners are adjusted, the gas-tube is screwed against the rubber membrane; when the thermometer attains the desired temperature, it is removed and the space occupied by enough water to fill the cavity made vacant by the thermometer. Then the orifice is closed either by a rubber stopper containing in it a thermometer-bulb, or a rubber stopper perforated with a hollow, glass tube permitting the water to rise in it. The apparatus is now regulated for this temperature by the following mechanism: one tube, which conveys the gas, carries a small movable disk, which applies itself against the rubber membrane, and tends without cessation to remove it from the orifice of the gas-tube, this action being due to a small spring, upon which the disk rests. The water, enlarging in volume, mounts up the glass tube, and this column of water exerts upon the rubber membrane a pressure more or less strong, which gradually antagonizes the elasticity of the spring, and brings the membrane more and more against the orifice from which the gas escapes. There is a constant play between the variations in the volume of the water and the amount of gas given to the gas-burners: the more the water expands the less gas is consumed, and *vice versâ*. To obtain the full sensitiveness of the apparatus, a solid rubber cork is used, instead of one with a glass tube, for the external orifice. This apparatus, upon testing, will remain nearly fixed in temperature for a long time. Geissler's thermometers were used, after having been carefully tested at the Yale Observatory and the necessary corrections made. The temperature of the animal was taken by a rectal thermometer, which was inserted fifty millimetres at each trial.

To close the space occupied by the animal a brass door inclosing a glass and lined at its edge by rubber was used. The closure was made air-tight by screws and bolts.

The calorimeter had its central exit-tube attached to the main air-tube of Voit's respiration-apparatus, through which the air of the calorimeter was aspirated by means of the water-wheel driving the great meter. To describe Voit's apparatus would require too much space. My instrument was made under Prof. Voit's direction, and its management as well as the estimates of carbonic acid and water were carried out as laid down by him in the *Zeitschrift für Biologie*, Band xi., Heft, 4. The method pursued was as follows: The animal was placed in the calorimeter and the change in the rectal thermometer noted as well as the amount of air aspirated from the calorimeter. After an hour or two the animal was removed, etherized, and a transverse section in the brain was made. When it had recovered from the ether it was again placed in the calorimeter and the same changes as heretofore noted. All experimentalists hitherto have made transverse sections of the central nervous system from the spinal cord upward. The plan I pursued was to commence at the anterior part of the brain and to go backward. The skull was broken up by a bone forceps and the "seeker" introduced, the section being made by its blunt-edge. The small "seeker" penetrated the cortex with but little injury to it, and when it reached the base of the brain was drawn transversely so as to divide the parts without injuring the cortex of either side. At the side of entrance the blunt instrument made an injury of but a few millimetres in extent.

In *Experiment* 1 the temperature at the end of the experiment was the same as at the end of the normal observation—that is, 104°. Here the olfactory bulbs were divided. When a transverse section was made behind the corpora striata the temperature was as follows: In exp. 3 the rise was $4\frac{7}{8}°$, in exp. 4 the rise was $4\frac{2}{8}°$, in exp. 7 the rise was 1°, in exp. 8 the rise was $1\frac{7}{8}°$, in exp. 10 the rise was $2\frac{2}{8}°$, in exp. 11 the rise was $2\frac{1}{16}°$, whilst in exp. 5 the fall of temperature was $\frac{2}{8}°$, probably due to shock. When a transverse section was made through the middle of the corpora striata, in exp. 2 there was a rise of 7°, in exp. 6 a rise of $1\frac{5}{8}°$, whilst in exp. 12 there was a fall of $6\frac{3}{8}°$. If an observation was made on the day following a section behind the corpora striata, the rise in exp. 11 was $1\frac{14}{16}°$, showing that the rise of temperature is not wholly temporary. That this rise is not due to injury to the convolutions of the brain is shown by the following experiments: In exp. 9 the convolutions of one hemisphere were destroyed and the fall of temperature was $3\frac{2}{8}°$, whilst where the sulci cruciati were destroyed, as in exp. 14, there was no change in the temperature, but when a section was made through the middle of the corpora striata the rise of temperature was $\frac{4}{8}°$, notwithstanding the removal of the sulci cruciati.

In exp. 13 the convolutions of both hemispheres were destroyed anterior to the sulci cruciati, and the temperature fell $\frac{7}{8}°$

All these experiments lead up to the conclusion, that in the vicinity of the corpora striata are centres which have a relation to the temperature of the body.

Whether the rise of temperature here is due to heat-dissipation being lessened or to increased heat-production I am not able to say. By a different arrangement of the calorimeter I propose to determine it. The analyses of carbonic acid were too few to draw conclusions from.

It is known in man that a very extensive lesion of the internal capsule is often followed by a very marked rise in the temperature of the body.[1]

[*Editor's Note:* Material has been omitted at this point.]

---

[1] Ranney: The internal capsule reprint. 1883.

6

Copyright © 1922 by Macmillan Journals Ltd.

Reprinted from pages 185–194, 205–214, 237–239, 251–252, and 257–262 of *Brain* 45:145–265 (1922)

A STUDY OF THE SHERRINGTON DECEREBRATE ANIMAL IN THE CHRONIC AS WELL AS THE ACUTE CONDITION

BY H. C. BAZETT, M.D.OXON., F.R.C.S.ENG.
Professor of Physiology, University of Pennsylvania,

AND

W. G. PENFIELD, M.D., B.Sc.OXON.
Beit Memorial Research Fellow; Associate in Surgery, Columbia University, N.Y.[1]

(*From the Physiological Laboratory, Oxford, and the National Hospital, Queen Square, London.*)

	PAGE
INTRODUCTION	186
CHAPTER I.—HISTORICAL NOTE ON DECEREBRATION	187
CHAPTER II.—TECHNIQUE	188
(a) Operative	189
(b) Post-operative	192
CHAPTER III.—DESCRIPTION OF CHRONIC DECEREBRATE PREPARATIONS	194
(a) Detailed description of two animals kept in baths	194
(b) General description of posture	197
(c) Magnus position-reflexes	202
(d) Opisthotonos	203
(e) Observations on temperature control and respiration with description of an incubator experiment	205
(f) Pseudaffective reflexes, vocalization and mastication	214
(g) Spontaneous movements	222
(h) Acoustic reflexes	224
(i) Nystagmus	225
(j) Glycosuria	226
CHAPTER IV.—UNILATERAL DECEREBRATION	227
(a) Operative technique	227
(b) Forced movements and attitude of head	229
(c) Side of permanent extensor rigidity	230
(d) Variations after posterior semi-decerebration	232
(e) Flexor rigidity	232
(f) Eye signs	233
(g) Temperature control	234

[1] In general the experimental work was done by H. C. B., and Chapters II, III and VI (b) were written by him. The semi-decerebrations and all the anatomical examinations were made by W. G. P., and Chapters IV, V and VI (a) were written by him.

		PAGE
CHAPTER V.—ANATOMICAL INVESTIGATIONS		237
(a) Tracts and centres of the mid-brain		237
(1) Development		237
(2) Anatomy		238
(b) Results of anatomical examinations		239
CHAPTER VI.—DISCUSSION OF RESULTS		252
(a) Decerebrate rigidity		252
(b) Temperature control		257
CONCLUSIONS		259

INTRODUCTION.

THE researches on which this paper is based were commenced originally in a search for a preparation which would allow experiments on the origin of surgical shock, and also on the effect of temperature on the resistance to infection in animals not under an anæsthetic and yet without there being any possibility of pain. The Sherrington preparation seemed ideal for this purpose, since the question of any conscious sensation could be completely excluded, and in addition the preparation would remain quiet during experimentation. The main object has, therefore, been to develop a technique which would allow the preservation of the decerebrate animal for many days, in spite of the removal not only of the cerebrum, but also of the basal ganglia and most of the mid-brain. This has necessitated the use of 124 animals, and in the course of the work some points of interest have arisen in connection with the reflexes and the histological anatomy. In many experiments our data are not as complete as we would have wished, but it was found that the process of examining the animals diminished their chance of survival and consequently the observations had to be curtailed, if the main object of the research was to be attained.

In spite of these difficulties a certain number of important facts were demonstrated. One of the most noticeable was the very remarkable similarity between the decerebrate preparations which had been kept "alive" for two or three weeks, and the acute preparations described by Sherrington on examination an hour or so after the operation. The decerebrate rigidity often continues (with only brief intermissions) for weeks, so that there can be no possibility of the condition being due to stimulation of the cut nerve fibres at the site of operation, and since in this time the damaged nerves are undergoing degeneration it is possible to exclude definitely some structures from any participation in this reflex. On the other hand, the chronic preparations do show certain definite differences from the acute; for instance, after the first two or three days a small percentage of the animals will occasionally make definite sounds, and these sounds appear to be reflex since they occur usually as the result of some definite stimulus. The most common of these sounds was purring occurring immediately after the stomach had been filled with milk by a stomach tube, while mewing has only been observed in two animals, and in these it occurred only once or twice and could not be made to recur by applying stimuli which would be painful to a conscious animal.

Extensive histological studies have only been possible on a few of the preparations, but in these the results are of interest since we believe that never before have preparations been maintained in a sort of life so as to allow degeneration to occur after such complete removal of all the higher nervous centres. In many other cases the nervous tissue was useless for detailed examination as the result of incubation for some hours after death and consequent rapid post-mortem changes.

Chapter I — Historical Note on Decerebration.

The method of removing parts of the brain in order to study function in the central nervous system has been employed since Rolando's experiments on birds in 1809. Flourens, indeed, succeeded in keeping pigeons alive for months after the removal of the fore-brain. But Longet [32], in 1842, first discriminated carefully as to what brain structures he actually removed. He noticed that many "reactions" remained if he removed all of the brain down to and including the colliculi—"chose remarquable!"; but that after removal of the pons all reactions disappeared, excepting only respiration and circulation.

Goltz kept three decerebrate dogs alive for months. The brain-stem of one animal most carefully studied was examined by Gordon Holmes [21], who found that a considerable amount of thalamus and all the cranial nerves were intact, although there seemed to be some softening of the corpora quadrigemina. There was no loss of temperature control and the animal shivered in a cold room. It could walk and would eat when brought close to food. Rothmann [51] [52] likewise produced a chronic preparation; a dog from whom all or most of the brain above the optic thalamus was removed. This animal ate and drank normally, walked and ran and showed a limited ability to learn. He likewise had normal temperature control and apparently normal muscle tone. Karplus and Kreidl [25] extirpated the hemispheres of monkeys and kept one alive as long as twenty-six days. More recently, Dusser de Barenne [2] has kept one cat alive six months, and another animal eleven months, after removing the fore-brain. A careful examination by Brouwer [7] of the post-mortem material showed the thalamus, mid-brain, and rhinencephalon to be intact in one, and most of the thalamus and mid-brain in the other. These preparations did not show decerebrate rigidity and were able to maintain a normal temperature. Only the animal with a portion of the fore-brain intact would eat spontaneously.

As indicated briefly above, in all previous *chronic* decerebrate preparations much of the thalamus was intact as well as the mid-brain. The animals possessed normal temperature control and have been able to execute complicated co-ordinated movements.

Decerebration at the level of the mid-brain was shown by Sherrington in 1896 to produce "decerebrate rigidity." This condition in *acute* preparations has been the subject of much study since that time by Horsley, Thiele, Weed, Magnus, Cobb, *et al.* Their conclusions will be discussed in another part of this communication. Recently, signs and symptoms in a group of patients considered to be analogous to decerebrate animals have been described by Wilson [63]. Riddoch and Buzzard [49] have also described such symptoms in man.

Boyce [6], in 1895, removed the brain on one side down to the middle of the mesencephalon and kept the animals (cats) alive as chronic preparations. His principal interest was in the descending degenerations thus produced. Probst [46], in 1904, made sections through one lateral half of the mid-brain without removing any tissue and studied the general behaviour of the animals as chronic preparations.

This brief sketch makes no pretence to enumeration of all of the valuable contributions to neurology made by this experimental method. In general, preparations produced by decerebration at higher levels and extending down as far as the medulla oblongata have been studied as *acute* experiments. In the case of all previous *chronic* experiments, the level of removal has been at least well above the mid-brain and most of the thalamus.

Chapter II.—Technique.

The animals used have been cats in all cases. The general procedure has been to decerebrate the animals aseptically under an anæsthetic and then to keep them under conditions which will maintain by artificial means their temperature constant. In the main the technique has followed the lines described by one of us (H. C. B.) [4] in a preliminary communication,[1] but as it has been modified in details it will be advisable to give here a full description. The whole process is founded on Sherrington's original (and older) method developed in 1896 [53] [54].

(a) Operative.

Pure chloroform has been used as the anæsthetic since we have had no success with ether. The preparations have considerable difficulty in getting rid of mucus and this probably accounts for the difficulties we have had. Subcutaneous injection of 0·5 mgm. of atropine has been given immediately after induction of anæsthesia in some animals, perhaps with beneficial effect. During the actual decerebration a total cessation of respiratory movements for a short while is not uncommon, and causes the loss of some animals. To overcome this intratracheal insufflation has been tried both with chloroform and ether. With chloroform insufflation we have had some success, but the greater irritation to the trachea and the consequent risk of infection seem to more than counterbalance the diminished risk of the actual operation. Cessation of respiration is less liable to occur if the animal is deeply anæsthetized at the time the section is made, so that with a good anæsthetist trouble of this kind is rare, even with an ordinary inhalation method, and it is this that we have mostly used.

The skin on the back of the head is shaved, cleaned and painted with an alcoholic solution of iodine, and in the later technique the animal is then turned over and the front of its neck is also shaved. A small incision is made (1 to 2 cm. in length) in the mid-line at the level of the thyroid cartilage, and from this both carotid arteries can be reached; a thick ligature is passed around each of them, and is then tied to form a loop so that traction can be made on the vessels to obstruct the blood-flow as desired. These two loops are brought out through the middle line and then the wound is closed with two or three waxed silk sutures. Hæmostat forceps hanging on these loops during the actual process of decerebration will prevent most of the bleeding and allow much more accurate work. It does not seem necessary to take any aseptic precautions after this neck wound has once been stitched up, and we have had no serious infection of these wounds.

The animal is then again turned over and an incision made on the

[1] In the protocols "earlier method" indicates that the operation was performed as described in this communication.

back of the head. A skin flap is turned back on one side, the temporal muscle is separated from the bone, the skull is trephined and the hole enlarged to a good size. While the actual decerebration is performed the weights on the carotid are allowed to hang free and constrict the vessels and the anæsthetist compresses the vertebral arteries just below the transverse processes of the atlas. The dura is opened and with a blunt spatula the occipital lobe on the near side is removed so as to bring into view the corpora quadrigemina. The division across the brain-stem can then be made at any desired level; this is done with the same blunt spatula, carrying the cut right through the brain to the base of the skull. It has proved much more satisfactory to make absolutely certain of preserving the pituitary gland intact, as will be explained later, and this has been attained most easily by the careful choice of the plane of the section and by carrying a horizontal cut forwards with a specially bent and rather blunt scalpel, starting from the transverse incision and passing forwards through the base of the brain a millimetre or two above the level of the pituitary. The whole of the brain matter above these two cuts is then scooped out and the cavity is lightly plugged with cotton wool. After this the vertebral arteries may be released and a little later the carotid blood-flow may be allowed to return. The whole procedure of removing the brain should only take about a minute or two.

The packing in the cavity is left for five or ten minutes and is then carefully removed. If there is any bleeding a small pad of muscle is put over it and packing is repeated and maintained for another five minutes. When the cavity is quite dry the closing of the wound is commenced. It is important to obliterate the large dead space of the cranial cavity in order to minimize the chance of infection. This has been done usually by filling it with a mixture of equal parts of vaseline and paraffin wax (melting 55° C.), the mixture setting at about 45° C. It is sterilized by heating to 140° C. for three-quarters of an hour on three successive days. If the hæmorrhage has been difficult to stop it is usually safer first to spread over the base of the skull a thin layer of gauze soaked in saline, the end of the gauze protruding through the wound, the temporal muscle being split vertically so as to make this easier, and then pour in the wax at a temperature of about 47° C., allowing it to set in the cranium. The wax does not stick to the wet gauze, so that when the wax is set the gauze may be gently withdrawn, or if hæmorrhage is feared it may be left in place until the following day. In the latter case a corner of the gauze may either be allowed to protrude through a counter-incision in the skin, or it may lie just beneath the skin which has been closed by a suture above it, so that on the next day the stitch may be removed and the gauze withdrawn without any trouble. It is probable that it is best to leave always a little gauze in place while the wax is setting, since on removal it leaves a small cavity and so allows room for some swelling of the remaining parts. In the earlier experiments where the whole brain matter to the level of the tentorium was scraped away trouble from this factor was not conspicuous, but with the modification of the operation as here described a post-operative rise in intracranial pressure may play a more important rôle.

In the few animals in which only one hemisphere has been removed the use of wax has proved unsatisfactory, probably owing to the much greater chance of swelling of the remaining parts in these animals. In the unilateral decerebrations in consequence the cavity has been merely filled with saline and it is conceivable that this might also be satisfactory even in the complete decerebrations. The other variations in the technique in the unilateral operation are described later (Chapter IV), since all these animals have been operated on by W. G. P. An attempt has also been made to get rid of the dead space in the animals by removing the whole cranial vault and drawing together the two temporal muscles over the base of the skull. Unfortunately it is not possible to make these muscles cover the space completely, nor is it possible to leave a smooth surface at the base of the skull without opening the nasal sinuses, and in our experience this had led to an infection of the wound with a regularity not seen in the paraffin method. For experiments lasting only three or four days, however, it is satisfactory.

The main difference between the later technique and the earlier one previously described is the horizontal section just above the level of the pituitary gland, allowing the blood-supply of this gland to remain intact. The contrast between the two methods may be seen by reference to figs. 9 and 10, which show photographs of a brain-stem operated on by the later method and comparing with them fig. 22, which illustrates the brain-stem left by the older technique. As in the earlier experiments the stalk of the gland was cut across, the blood-vessels supplying the anterior lobe must have been divided, and the gland itself may easily have been damaged and perhaps subjected to some pressure. In spite of this several animals survived for periods of two or three weeks, but as a rule they would not live more than three or four days. In these death was preceded usually by convulsive movements without signs of infection, or else was due to broncho-pneumonia. .The convulsions occurring within three or four days were apparently similar to those described by Cushing [12] [13] as the result of removal of the pituitary, and consequently the modified operative technique was developed. This necessitates a more ragged incision with greater hæmorrhage, so that temporary control of the carotid arteries had to be also adopted. Following this change in the technique death on the third or fourth day preceded by convulsions has no longer been seen, and the animal's resistance to broncho-pneumonia has also seemed greater, but on the other hand, many more animals have been lost during the first day from reactionary hæmorrhage, and infection of the wound itself has also been more common.

The later technique has also given rise to some trouble, since the changed slope of the section has often been overdone, resulting in damage to the fifth nerve. Such damage must be avoided since it leads to the development of ophthalmia and pressure sores around the mouth, complicating the nursing and greatly increasing the risk of broncho-pneumonia.

(b) Post-operative.

The wound requires no covering, especially if it be stitched with waxed silk to prevent infection tracking inwards along the sutures, which may otherwise occur if the animal be kept in a bath. The wound may be left dry or covered with mercury ointment.

The main difficulty to be overcome is the maintenance of a constant temperature. For the most part the animals have been kept in a water bath, through which water was kept circulating and in which they were supported on a sling three-quarters immersed in the water. Stirrers maintained a constant mixing of the water, and a toluol regulator maintained the bath with a variation of only 0.2 to 0.4 C. in twenty-four hours at its best. Unfortunately, it often fell below this standard, as owing to post-war conditions the gas pressure sometimes fell so low as to be incapable of heating the large volume of water circulating through the bath.

In a bath the preparation readily gains heat from or loses heat to the circulating water, so that its rectal temperature rarely differs from that of the bath by more than $1°$ C. and practically never by more than $1.5°$ C. The bath has been usually kept at a temperature between $38°$ and $39.5°$ C. This method is possibly the simplest that will allow the maintenance of a constant body temperature in a preparation where the metabolism shows rapid variations, as it must in any which happen to show strong running reflexes. No ill effects from the prolonged immersion in water have been seen, even after two or three weeks, except a gradual shedding of the hair.

The disadvantages of the bath method are obvious, and it is this method that is mainly responsible for the paucity of our observations on the reflexes. Rough examination of the animal is possible as it is turned each day in the bath to avoid pressure sores, but in order to make any careful examination, it has to be taken out, washed in hot water, washed again with alcohol, roughly dried with a towel and finally dried in a hot incubator. Even with this technique, which requires four or five hours, the body temperature is apt to drop to 35° C. or lower, and though no ill results are noticed at the time, pneumonia often supervenes. Consequently any such examination will greatly decrease the preparation's chance of survival.

Owing to this many attempts have been made to keep them in an incubator. An ordinary incubator is useless, since the heat loss varies with the moisture in the air. Haldane [18] has shown the importance of the wet bulb temperature, but even an incubator controlled on a wet bulb basis is insufficient. The reason for this is that a very exact balance between heat loss and heat production has to be attained in these animals, and once this balance is upset the discrepancy tends to grow, since it is probable that in these preparations the metabolism falls with fall of body temperature instead of showing a compensatory increase. An incubator method has, however, proved possible by using an incubator heated by live steam escaping within it, so that heating occurs through condensation on its walls. In this way the wet and dry bulb temperatures are identical, the air being completely saturated with moisture. Even in such an incubator the control has to be altered during the first few days to match the gradual loss of the insulating action of the fur as it becomes penetrated with moisture, or this variation must be overcome by destroying these heat insulating properties of the fur by soaking the animal immediately after operation in petroleum oil. Under these conditions the animal loses heat mainly by radiation and conduction and usually keeps a rectal temperature about 4° C. above that of the incubator, so that this should be kept at about 34° or 35° C. The proper mixing of the air is ensured by an electrically driven fan. Further details of incubator experiments are given in Chapter III.

Food is given twice a day by stomach tube, a full-size cat requiring about 100 c.c. of milk at each feed. Great care must be taken to keep the mouth, eyes and nose clean, and the animal must be placed in a position which would be comfortable if it were conscious.

The chief causes of failure are infection, usually respiratory, less commonly infection of the wound, post-operative hæmorrhage, and in the earlier technique probably also pituitary insufficiency. No testicular atrophy nor adiposity has, however, been observed. One animal (previously reported) lost 10 per cent. of its weight in a week, but another (cat II) after twelve days "life" showed no change whatever in its body weight. Further experiments on the functioning of the pituitary in these animals are planned.

The prevalence of pneumonia in these preparations is of some interest in that it was commoner if the animals had been previously exposed to conditions giving them a low body temperature, even if this occurred a considerable time after their anæsthetization. An instance of this is seen in the case of cat II, which is described later.

[*Editor's Note:* Material has been omitted at this point.]

(e) *Temperature Regulation.*

There seems to be no doubt that after operation these preparations have no power of temperature control whatever, and that they never regain it however long they may survive the operation. If they are kept in an incubator the setting of the incubator temperature has to be exactly right, so that the heat loss balances the heat production, in order to maintain a constant temperature, and the setting of the incubator required for different animals is not the same. Even for the same animal the balance has to be hit exactly, taking into account the degree of the dryness or dampness of the fur, and to strike such a balance becomes a difficult problem even if the air be maintained saturated with moisture.

Not only are they unable to adapt themselves to even small changes in temperature but they have never shown normal reactions to these temperature changes. If their body temperature is dropped suddenly no shivering has ever been observed, nor does their respiratory rate increase to a normal extent if their body temperature is quickly raised. It is true that few people have seen a normal cat shiver in cold weather or pant under the influence of heat, but anyone who has tried washing a cat will have observed the shiver (see also animal V), and if a wet cat be dried under sufficiently warm conditions the respiratory rate may easily rise to 70 or 80 a minute. The ordinary cat does not usually show either symptom, since it is careful to keep itself under comfortable conditions and has a thorough insulation in its fur. The metabolism of these preparations has not so far been investigated by us, but we have no reason to doubt the low metabolism figures obtained on the acute preparation by Roaf [47], since the difficulty of keeping the preparations warm has always been remarkable. The stretching out of the limbs in extension no doubt increases the heat loss, but even so it is surprising that keeping the animals on a warmed operation table, surrounding them with hot-water bottles and covering with a blanket, is usually only just able to keep the animal's temperature normal if the room is at all cold, i.e., 15° C. There is also some evidence that the metabolism is greater the higher the temperature and that it decreases with a fall in temperature, so that there is no effort at a control of body temperature through a control of metabolism. Thus in an incubator animal with its fur greased and in an atmosphere saturated with moisture (when the loss of heat is accomplished by radiation and conduction), the difference in temperature between the animal and the incubator is usually greater the higher the temperatures happen to be. Thus with a high rectal temperature a difference of 5° C. above the temperature of the incubator may be found, and with a low rectal temperature the difference between the two may be only 3° C. Thus there is a tendency for the rectal temperature to swing more widely than that of the incubator. The resulting difficulty in finding a correct setting for the incubator is readily seen by reference to a few examples.

The general correspondence between the animal's temperature and that of the incubator is well shown in fig. 1, which represents the temperatures observed in cat VII, of which some description has already been given. This animal was kept immediately after operation in an incubator in which the air was constantly saturated with moisture by having the wall covered with damped cloth. Considerable difficulty was found in getting a correct setting of the incubator and it is seen that between eight hours and twelve and a half hours on the first day the rectal temperature of the preparation dropped 2·1° C. (from 37·2° to 35·1° C.), corresponding with a fall of 1·5° C. in that of the incubator. It is also noticeable that after three and a half hours on the first afternoon the rectal temperature was 39° C. and was rising rapidly as the result of an incubator temperature of 29·7° C. dry and 29° C. wet bulb, while twenty-seven hours after operation the rectal temperature was fairly steady at only 36·2° C., though the incubator setting had now been adjusted to give a temperature of 30·3° C. dry and 29·8° C. wet bulb. The difference in the two results is probably due to the gradual loss of the insulating action of the fur as the effect of the penetration of the moisture through it. This was demonstrated accidentally on the second afternoon. At 3.10 p.m. the animal was taken out of the incubator and was examined for Magnus reflexes in front of a gas stove until 4.25 p.m. During this examination its rectal temperature rose to 39·7° C., its fur became dry and in good condition, and it was then returned to the incubator in the belief that its temperature would gradually drop again to 36·2° C., since the incubator setting had not been altered. The

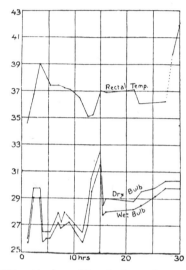

Fig. 1.—Experiment VII. Abscissa : Time in hours from end of operation. Ordinate : Temperature degrees centigrade. Lowest curve : Wet bulb temperature of incubator. Middle curve : Dry bulb temperature of incubator. Upper curve : Rectal temperature of animal. Dotted part of curve : Change when animal was out of incubator being examined in front of a gas stove.

observer was interrupted by teaching work and when the animal was again examined, one hour and thirty-five minutes later, it was found to be dead, with a post-mortem rectal temperature of 42·1° C. This case, therefore, illustrates very well the difficulties of incubator preservation of these preparations, and the numerous factors that have to be taken into consideration, as well as demonstrating the absence of any control.

The same points, and also the general course of such an incubator experiment, may be seen in the case of cat VIII (also previously described in relation to the Magnus reflexes). The general history of its temperature changes is illustrated in fig. 2 ; but a more or less detailed account may also be given to serve as an example of an incubator experiment.

Animal VIII.—Female. (For post-mortem see p. 252, and previous description p. 203). Decerebrated by later method so as to preserve stalk of pituitary. Rather ragged section and considerable hæmorrhage afterwards. Large piece of gauze left in, but in spite of this a considerable hæmatoma formed shortly after the end of the operation. Placed in steam-heated incubator at 30·8° dry and 30° C. wet, i.e., about temperature at which cat VII had a rapid rise of temperature on its first day. At three and a quarter hours the rectal temperature was still below 35° C.; at four and a quarter hours it was still only 34° C., though the incubator temperature had been still further raised. The incubator was then set to give 33·5° dry and 32·8° wet bulb, and the rectal temperature gradually rose to reach 38° C. after about nine hours. The

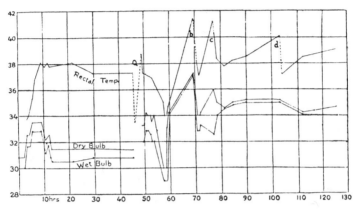

FIG. 2.—Experiment VIII. Abscissa: Time in hours from end of operation. Ordinate: Temperature. Lowest curve: Wet bulb temperature of incubator. Middle curve: Dry bulb temperature of incubator. Top curve: Rectal temperature of animal. Dotted parts of curve: Animal out of incubator—(a) taken out, wound dressed; as bodily temperature fell fur treated with paraffin and animal then warmed in front of gas stove; (b) and (c) cooled by being kept in a room for a short while; (d) examined in front of stove for Magnus reflexes.

incubator temperature was then gradually reduced and regulated to give a temperature of about 31·5° dry and 30·5° wet bulb, when the rectal temperature reached a steady level at about 37·3° C. During this time the animal had had good rigidity of all limbs; twelve hours after operation it showed brisk corneal and pinna reflexes, and some mastication reflex, and when the ear was twisted it moved its head and fore limbs slightly and showed a definite increase in decerebrate rigidity.

Thirty hours after operation reflexes as before but mastication no longer seen; respiration rate 21, temperature 37·3° C. Lying on right side it now had good rigidity of all four limbs, but on turning on to left side showed some flexion of back, good rigidity of hind limbs, little rigidity in the left fore, and it held the right fore rigid, but flexed and placed forward across the left, with the ankle joint acutely flexed and the claws extended. It kept its head turned to the right if put into a sitting posture. The tail was intensely rigid, and was wagged whenever the animal was touched.

Forty-six hours after operation taken out of incubator, and gauze plug removed from wound; hæmatoma found infected. Animal kept on warmed table but room temperature only 9° C. (owing to post-war conditions), and the rectal temperature soon dropped to below 35° C. Animal then warmed in front of gas stove and lubricating oil rubbed into fur to diminish its heat insulation value.

Returned to incubator at 2 p.m. (i.e., fifty hours after operation) with temperature of 38.7° C. The respiration rate was then 20, having dropped to 15 per minute when the body temperature was at its lowest. Fig. 2 shows the changes in temperature occurring after this and the close correspondence between the rectal and incubator temperatures. When the rectal temperature had risen to 41.6° C. the respiratory rate was 40, dropping again to a rate of 31 when the rectal temperature had reached 38.1° C. on taking the animal out of the incubator and keeping in a cold room. This drop of temperature occurred in about thirty-eight minutes, and yet there was never any sign of shivering. The posture of the animal when lying on the left and right sides was as before. The wound was opened up and some pus evacuated.

Sixty-six hours after operation the animal was again examined. It now had good extensor rigidity of all four limbs when lying on the *left* side, and on being turned to the *right* side there was good rigidity of the hind limbs and extrusion of claws, and on stimulating by twisting the ear the fore limbs assumed a rigid flexed position with extrusion of the claws, and maintained it.

One hundred and three hours after operation the rectal temperature was 38.5° C., and the respiratory rate 32. The upper fore limb was kept flexed on whichever side it was lying. On touching the animal there was reflex extrusion of claws and either retraction or prodding forward of the limbs.

One hundred and eighteen hours after operation the animal was much as above, but when traction was made on its flexed upper fore limb, it withdrew it with a clawing movement. It also made a cry—an inco-ordinated mew—when being turned.

One hundred and twenty-seven hours after operation was as above but on various stimuli, such as mere touching, would often extend a limb and spread out claws, in the position a cat holds its limbs when stretching after waking up.

At the end of the sixth day the position of the animal on whichever side it was lying was very definite and has already been described in connection with the Magnus reflexes. It had no auditory reflexes, but by this time some mastication had again returned. The respiration rate was 33 with a rectal temperature of 39.1° C. There was a discharge of pus and wax with disintegrating hæmatoma from the wound, and in the middle of the seventh day the animal died as the result of a secondary hæmorrhage within the cranium. Urine passed during the third day contained no trace of sugar.

The effect of body temperature on the respiratory rate is also well shown in the case of cat XII kept in an incubator with a wet bulb control, but in which the air was not saturated with moisture and where the insulating action of the fur was not interfered with. The changes in body temperature and of the incubator for thirty hours, starting seven hours after the end of the operation, are shown in fig. 3; the wet and dry bulb readings were taken at frequent intervals and the rectal temperatures were taken with an electrical resistance thermometer and were recorded every minute on a Gamgee [16] thread recorder as

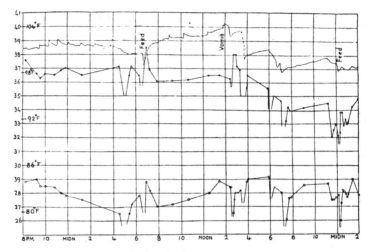

Fig. 3.—Experiment XII. Abscissa: Time: operation finished seven hours previous to commencement of record. Ordinate: Temperature. Lower curve: Wet bulb temperature of incubator. Middle curve: Dry bulb temperature of incubator. Upper curve: Rectal temperature of animal (record with resistance thermometer). Gaps in records, opening of incubator door.

described by Woodhead and Varrier Jones [65]. In all, more than forty-five observations of the respiratory rate at varying body temperatures were made on this animal, though the results were somewhat modified by the gradual development of mucus in the trachea giving some irregularity in the respiration with a later development of lung congestion. If all the figures for the respiratory rate during the first few days are averaged the following results are obtained:—

Temperature	Number of observations	Average rate per minute
36°,1 to 37°	4	23
37° to 38°	18	28
38° to 39°	16	33
39° to 40·1°	7	36

If, however, only the earlier figures are taken, before much tracheitis had occurred, the respiratory rates are rather slower but the relative differences are much the same.

Average temperature	Number of observations	Average rate per minute
36·1°	1	15
37·2°	4	19
37·8°	5	25
38·0°	2	29
38·3°	4	31
38·7°	4	32
39·5°	3	33

The brain section in this animal was from the anterior end of the inferior colliculi to 1 mm. in front of the pons, but a detailed histological examination was not made. It will be noticed that in this

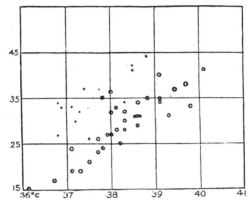

FIG. 4.—Experiment XII. Abscissa: Rectal temperature. Ordinate: Respiratory rate per minute. Variations in respiratory rate with changes in temperature. o = Observations during first twenty-nine hours. . = Observations made later than this.

animal as well as in cat VII there was a very definite dependence of the respiratory rate on the body temperature. This is still more clearly seen in fig. 4 in which the respiratory rate observations are plotted against the body temperatures. The figures are obviously affected by other factors, such as the mucus already referred to, and quite possibly the respiratory rate is different for any given temperature, if the temperature is then falling, compared to what it would be if the temperature were rising. In spite of these variable factors it is clear from a consideration of these two examples that there is a relationship, and that the temperature coefficient is rather high, probably about 4·0, possibly as high as 5. But in any case the effect is much less than in the normal cat where a slight rise in body temperature may easily give a respiratory rate of 80 [20] [30] [48]. That the temperature coefficient is so high may be partly due to the fact that not only might the change in temperature affect the respiratory centre directly, but it might also do so indirectly through altering the metabolism.

In only one instance was a really fast respiration seen with a raised body temperature. In this animal the temperature rose to 42·8° C. in an incubator through an accident. At the time of observation it was moribund and its respiratory rate was about 60, but this animal had had a considerable reactionary hæmorrhage, with marked opisthotonos and convulsive spasms, and its fast respiratory rate may have been due to other factors. In any case the rate observed is not very different from what might be expected at this body temperature, judging by the other animals. The section in this animal was also rather anterior and very ragged.

With regard to the respiration it should be noted that as a rule with a normal body temperature the respiratory rate was about 24 per minute, which is about the normal rate for a cat. Irregular and spasmodic breathing described by Macleod [34] [35] as occurring spontaneously in some animals after decerebration was not seen except where there was fairly definite evidence of hæmorrhage, as in cat X, nor can we agree with the provisional statement made by Macleod that "when the posterior corpora quadrigemina are even slightly wounded spontaneous respiration is seldom, if ever, observed." We have repeatedly made sections through this level and observed normal respirations, and we have even made sections through the very posterior edge of the inferior colliculi without having trouble with the respiration (e.g., cat XXI described later, survived over twenty-two days with a normal respiratory rate, and cats XX and XXIII furnished other examples of this). The contrast between our experience and that of Dr. Macleod may depend on the fact that our animals immediately after the operation were left quite quiet without further experimentation, while such a course is impossible in dealing with the acute preparation. Any manipulation or stimulus must increase the risk of reactionary hæmorrhage either below the tentorium or into the substance of the medulla, and that this was a common occurrence even in our animals may be seen by reference to the anatomical descriptions. Clearly the macroscopical inspection of the brain-stem does not allow the physiological level of the decerebration to be determined, and consequently discrepancies between different observers are to be expected according to the relative probability of post-operative hæmorrhage. A similar explanation may well account for the discrepancies between Trevan [58] and ourselves as to the effect on the respiratory rate of section at various levels.

Fig. 3 demonstrates well how sensitive the body temperature was to the changes in the incubator produced by opening the door for a few minutes, as for instance for feeding.

That the wet bulb temperature is very important is shown by the fact that animal VII required a wet bulb temperature of about 29° C. with a dry bulb reading of about 30° to maintain a normal temperature, while animal XII required a wet bulb temperature of between 28° and 29° with a dry bulb of about 34° for the same result. And yet it was not the only factor; the rectal temperature showed a rise in cat XII between 8 a.m. and 2 p.m. with a rise in the wet bulb temperature, but it also showed a fall between 4 p.m. and 10 p.m. although the incubator wet bulb temperature had altered very little, while the dry bulb fell considerably. Also the difference seen between animals VII and VIII, one requiring a wet bulb temperature of 29° and the other of 31° (before the fur was greased) with almost complete saturation with moisture, to maintain normal conditions, requires explanation. Cat VII had an excellent coat and cat VIII was rather mangy, and this may well have been the cause.

The incubator temperatures in the case of cat XII were rather variable; the reason for this was that the outer door of the incubator was left open throughout the experiment to allow the repeated examinations of the respiratory rate on this animal, and the glass door did not give adequate insulation. The sudden peaks occurring in the rectal temperature record in this figure are also of interest. They were always seen when an animal attempted to defæcate against the resistance of the thermometer; since they show fairly steep falls directly the efforts ceased they may represent a change in temperature which was chiefly local.

The absence of shivering on cooling was specially tested in one animal (cat XIII) four hours after operation. This animal's reflexes were brisk and the section was an anterior one passing through the superior colliculi to 2 mm. in front of the pons; it was operated on by the older method. Though its temperature was reduced from 39·5° to 37·8° by wetting with cold water (the normal temperature for a cat is about 38·5°) within five minutes, yet no sign of shivering was seen. In another animal which has already been referred to (cat V) the preparation was kept in an incubator for the first nine hours; it was constantly making running movements, so that it was not found easy to balance the heat exchange. At the end of this nine-hour period its rectal temperature was 40·9° C.; then it was cooled down by holding under a tap delivering water at 16·5° C., the water being rubbed into the flanks, so that the temperature fell to 37·8° within sixteen minutes, and yet no sign of shivering could be observed. That shivering does not occur either even in animals that have survived much longer is clear in that no such movement was ever seen in animals, such as cat II which was taken from a bath and had its temperature dropped rapidly from a normal level to one much below normal during the process of drying.

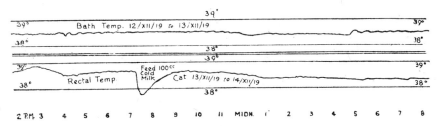

Fig. 5.—Experiment XIV. Abscissa: Time. Ordinate: Temperature. Lower curve: Rectal temperature of animal in bath, December 13 to 14. Upper curve: Temperature record of bath December 12 to 13.

Fig. 5 gives examples of records with a resistance thermometer of the rectal temperature of animal XIV (described a little later) kept in a water bath, as well as a record of the bath itself, the two records being taken on successive days, and the part reproduced includes the early morning hours when the greatest fluctuations of the bath temperature usually took place. The relative steadiness of the rectal temperature by the use of this method is very noticeable, and also the rapid drop of nearly 1° C. in the rectal temperature on feeding this animal (of 2·1 kilo weight) with about 90 c.c. of cold milk by stomach tube. The effect of feeding was always marked, if the food were cold, so that in incubator experiments it was necessary first to warm the milk to the body temperature, since under these conditions recovery of a normal temperature took place much more slowly.

[*Editor's Note:* Material has been omitted at this point.]

CHAPTER V —MICROSCOPICAL ANATOMY OF SPECIMENS.

(a) *Development of the Mid-Brain.*

In lower vertebrates (e.g., fishes) the mid-brain is a centre of great importance, controlling the individual response to environment. With the development of the optic thalamus and the geniculate bodies (reptiles) the importance of the mid-brain for co-ordinating visual stimuli is decreased, and in man with the elaboration of the visual cortex the anterior corpora quadrigemina of the mid-brain become rudiments. There is likewise a forward displacement of the sensory centres to the ventral part of the thalamus, which continues to develop along with the sensory cortex. The auditory paths move out from the midbrain to the internal geniculate body and auditory cortex. The midbrain has therefore a progressively decreasing functional importance proportionally during phylogenetic development.

Microscopically the red nucleus of a lower vertebrate is composed of large cells of motor type from which arise the rubrospinal tracts. Higher in the phylogenetic scale there appear smaller cells in the red nucleus and, as von Monakow [43] has pointed out, it is possible to distinguish an orally placed group of smaller cells, the nucleus parvicellularis of Hatschek [19] or chief nucleus, from the more caudal giant cells, or nucleus magnocellularis. The small-celled nucleus develops *pari passu* with the thalamus and cerebral cortex and is connected with them. In anthropoids and man this nucleus of small cells makes up the major part of the red nucleus, while the giant-celled part has become very small and, as a result, the rubrospinal tract has greatly diminished in volume. Nevertheless, because of the great increase in size of the small cell part, the red nucleus of man is almost twice as large as that of the cat in proportion to the vertical diameter of the mid-brain of each species [43].

(b) Anatomy of the Mid-Brain.

The anatomy of the mid-brain of a cat may be quickly reviewed by reference to fig. 8, which is a drawing from a lateral longitudinal section through a cat's brain. It will be seen that a section through the mid-brain may leave or remove according to its level a number of

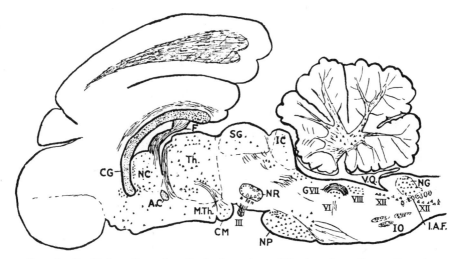

Fig. 8.—Sagittal section of cat's brain a few millimetres from the median plane, drawn from two adjacent sections, one stained by Loyez's hæmatoxylin method and the other by Nissl's method. This as well as all the other microscopical drawings was made with the help of an Edinger Drawing Apparatus kindly loaned by Dr. Gordon Holmes. Diagrams in succeeding figures are taken from this drawing reduced to natural size. C.C., corpus callosum; N.C., nucleus caudatus; A.C., anterior commissure; F., fornix; Th., thalamus; M.Th., mamillo-thalamic tract or bundle of Vicq d'Azyr; S.C., superior colliculus; I.C., inferior colliculus; N.R., nucleus ruber; N.P., nuclei pontis; III, root of III nerve; G VII, genu of VII nerve; VI, nucleus and root of VI nerve; VIII, dorsal nucleus of VIII nerve; XII, nucleus of XII nerve; V.Q., fourth ventricle; I.A.F., internal arcuate fibres splitting XII nucleus.

structures, the anterior and posterior corpora quadrigemina, the third and fourth cranial nerve nuclei, the whole of the tegmentum, the red nucleus, &c. The red nucleus is circumscribed by a felt-work of fibres. The caudal two-thirds of the nucleus is made up of giant cells as indicated. The oral one-third only is made up of small cells.

According to Hatschek, this phylogenetically younger nucleus parvicellularis is in relation with the nucleus dentatus of the cerebellum, while the older nucleus magnocellularis is interposed between the cerebellar nucleus globosus and emboliformis above and the descending rubrospinal tract. He concludes that this close anatomical relationship of the nucleus magnocellularis with the cerebellum and Monakow's bundle constitutes the primitive path of nerve impulses for automatic movements. Other tracts converge on the red nucleus. Von Monakow describes a fronto-rubro-tegmental pathway. The fibres thus indicated arise in the frontal cortex and pass down via the red nucleus to the tegmentum of the mid-brain and formatio reticularis of the pons, from where they are supposed to send a relay to the spinal cord. Wilson [62] has shown that there is a considerable tract of fibres from the globus pallidus which passes across Forel's field to the anterior capsule of the red nucleus. The principal red nucleus efferents are the rubrospinal tracts of Monakow which cross the mid-line in the fountain decussation of Forel. This tract arises wholly in the giant cells, according to Cajal [10]. Von Monakow describes some fibres as coming from the cells of medium size. Certainly the tract arises chiefly from the nucleus magnocellularis, and comparative anatomical studies show that this part of the red nucleus varies directly in size with that of the tract in most animals, and both are small in man. Efferent fibres of importance also pass from the red nucleus to the tegmentum and formatio reticularis, as well as to the thalamus and frontal cortex.

Because of its venerable phylogenetic history and its wide anatomical connections, the red nucleus has been rather heavily burdened with theoretical functions. It is therefore of importance to note carefully the state of preservation of this nucleus in the experimental animals.

[*Editor's Note:* Material has been omitted at this point.]

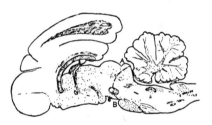

Fig. 29.—Experiment VII.

Animal VII (p. 202).—Death twenty-eight hours after decerebration, due to hyperpyrexia after warming before stove. No evidence of infection. A small amount of isolated brain tissue remained over the infundibulum. Top of brain-stem was smooth, symmetrical, and convex in outline, beginning in the middle of the inferior colliculi dorsally and the upper border of the pons ventrally. (Fig. 29, line A—B).

To microscopical examination, part of the red nucleus was present on either side and many of the cells were normal. Several small areas of hæmorrhage extended downward a few millimetres. The pons was otherwise normal.

Animal VIII (p. 203).—Death on the seventh day after decerebration due apparently to secondary hæmorrhage from the brain-stem. There was pus and blood-clot between the wax and the brain-stem. Progressive microscopical sections showed that there was no encephalitis below the superficial layers. To microscopic examination the level of decerebration passed between the superior and inferior colliculi on the right and through the middle of the inferior colliculus on the left, to end at the upper border of the pons ventrally. See figs. 30 and 31. A small amount of brain matter remained above the infundibulum. The tissue corresponding to the dotted areas was softened and filled with compound granular corpuscles.

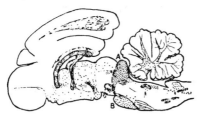 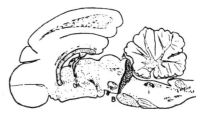

Fig. 30.—Experiment VIII. Right side.   Fig. 31.—Experiment VIII. Left side.

There was no remnant of red nucleus on the left side and only a few cells belonging to this nucleus remained on the other side. The interpeduncular ganglion and most of the substantia nigra cells were normal as well as the pontile nuclei. Most of the formatio reticularis was normal. With the exception of a small proportion of the cells of the right red nucleus, the level of decerebration was below the mesencephalon.

[*Editor's Note:* Material has been omitted at this point.]

(b) *Temperature Control.*

The whole literature on the subject of temperature control and the nervous mechanisms which are concerned has been so carefully reviewed recently by Barbour [1] that no attempt need be made here to give any detailed account, and reference may only be made briefly to such papers as bear most closely on the results here reported.

It has been shown by Rogers [50] that in pigeons section of the brain-stem below the level of the thalamus completely removes all temperature control, while section above the thalamus does not, so that even in birds, in which the development of the corpus striatum is very different from that of mammals, section of the brain-stem in the region we have studied destroys all temperature control. The similarity with the results obtained in cats is the more marked when the experiments of de Barenne [2] [7] are also taken into consideration. He removed practically all the brain above the thalamus in the second of the two cats described in the above paper, and yet found it to have a normal temperature control. Therefore, in cats also the control of temperature may be definitely located between the level of the middle of the superior colliculi and 2 mm. in front of the pons below and the upper limit of the thalamus above. Such a situation for the centres controlling temperature regulation is in entire accord with the work of Barbour on the effect of localized heating of the brain.

Isenschmidt and Krehl [22] in 1912 showed that rabbits in which the brain-stem was divided at the level of the junction of the diencephalon and mesencephalon lost all power of temperature control, while if the hemispheres alone were removed control was not lost. But their animals without temperature control only survived a few days, and considering how long reactions may be depressed following a lesion, the possibility of a later return of temperature control was not excluded.

Later Isenschmidt and Schnitzler [23] made various sections in the mid-brain at different levels, these sections being often one-sided and partial, and they came to the conclusion that the heat regulation centre was in the region of the tuber cinereum. This would be consistent with our results, but their conclusions cannot be regarded as firmly established since again they were unable to keep alive for more than a few days any animals which lost their power of temperature control. Consequently they could not elaborate their results by examination for nerve degeneration. The "Zwischenhirnstich" of Leschke [31] gave similar results.

The animals that we report offer therefore strong confirmatory evidence as to the position of the centres controlling temperature, and it should be much easier to determine whether these are in the thalamus or tuber cinereum by the employment of the methods we have developed for nursing animals which have lost their power of temperature regulation.

It will be evident from the experiments that have been described that no sign of temperature control remains and that no such control returns at any rate within two weeks. The complete absence of shivering in the bilaterally operated animals and its ready occurrence after the unilateral operation is very striking, and it is of considerable interest that in the latter case shivering occurs equally on the two sides of the body. With the unilateral decerebrations it was noticed that if the remaining brain matter was exposed to some pressure (as occurred when wax was used in these cases) the raised intracranial pressure readily destroyed the power of temperature control, and it was also noticed that in the other animals which showed deficiency of temperature control the lesion had slightly damaged the left side of the brain-stem as well as the right.

It seems probable, therefore, that tracts which carry either sensory or motor fibres intimately concerned in temperature control run in the mid-brain close to the mid-line, possibly for instance in the posterior longitudinal bundle, though we have no direct evidence of this.

The animals with bilateral decerebration showed a considerable change in the respiratory rate with change in body temperature in spite of the apparent absence of temperature control. This however appears to be adequately explained by assuming that a rise in body temperature alters the respiratory rate by affecting the respiratory centre directly, just as it alters the heart rate by direct action on the heart, and that the respiratory rate may be still further quickened by the indirect action of a metabolism which has also been increased as the result of the raised body temperature. In any case the quickening of the respiratory rate with raised body temperature is probably considerably less than that occurring in the intact animal [20] [30] [48].

While up to the present these preparations have given no great additions to our knowledge of the mechanisms concerned in temperature control, the possibility of the preservation over long periods of animals, in which this control has been lost, would seem to offer many new opportunities for the investigation of this problem.

Conclusions.

(1) A technique is described which allows cats in which all the brain above the level of the pons has been completely removed to be preserved for days and occasionally as long as three weeks. The essential factor is shown to be the maintenance artificially of a normal body temperature.

(2) Chronic decerebrate preparations obtained in this way are found to maintain throughout this period the general characteristics described by Sherrington. The extensor rigidity is usually maintained with brief intermissions, and further proof is thus offered that it is a "release phenomenon." Rigidity continues in many cases after full degeneration of the divided tracts has occurred.

(3) Decerebrate rigidity is described lasting for more than two weeks in animals in which all the red nuclei can be shown histologically to be no longer functioning. Decerebrate rigidity can therefore occur in the absence of the red nucleus, and although many reactions and position-reflexes disappear in the more posterior decerebrations, the character of the extensor rigidity seems to be the same. It was

sometimes found that sensory stimuli would increase the rigidity already present, or cause it to reappear if it were absent.

(4) Decerebrate rigidity is probably dependent on some centre or reflex arc at about the level of Deiters' nucleus. None of our histological examinations showed any lesion to this nucleus nor to the ventral spino-cerebellar tract of Gowers, and the rigidity did not disappear after mid-line sagittal section of the pons and medulla.

(5) Head retraction accompanies extreme rigidity, but is only seen with other evidences of an irritative lesion such as may be produced by progressive hæmorrhage. It then forms a part of what may be called a tonic extensor fit which on the day of operation occasionally developed into a well marked opisthotonos. Spontaneous nystagmus sometimes appears also as the result of this irritation.

(6) The level of the section was generally rather oblique and usually more was removed on the left side; many animals showed an asymmetric head position, usually with a tendency to bend it to the left side. In these animals the section level suggested a possibility of some of the right red nucleus cells still functioning.

(7) Certain differences between the acute and chronic states were noticed. In the later stages there is a much greater tendency for flexed positions to be assumed, and these are often asymmetrical and sometimes flaccid and sometimes rigid. Of these rigid flexion of the terminal joints of the fore-limbs with extrusion of the claws is very common.

(8) The preparations show no sign of temperature control and never shiver. Bilateral shivering is however seen after unilateral decerebration where temperature regulation is preserved. Variations in the respiratory rate are seen after bilateral decerebration as the result of differences in body temperature, but these variations are probably not so great as in the intact animal.

(9) The respiratory rate is usually that normal for a cat, i.e., about 24. Normal respirations have been observed for days even when the section was at the lower margin of the inferior colliculi.

(10) Pseudaffective reflexes were usually much more developed after the first few days than immediately after operation, and new responses were seen. Besides those described in the succeeding section, movements of the head away from certain nocuous stimuli were noticed and occasionally a movement of the fore-limbs as though directed to the protection of the head.

(11) Vocalization was noted in ten of the 124 complete decerebra-

tions. This was rarely seen except after three or four days, and consisted most commonly in purring after the stomach had been filled with milk by a stomach tube. When purring was observed the animal was a kitten or a young female. Other cats growled, such sounds being never repeated, and in this case the sounds were only observed in old males. A true mew has only been observed in one animal which made such a sound on two occasions, but would not do so in response to nocuous stimuli. An ill co-ordinated cry, resembling the sound a cat may make while anæsthetized, has been observed in five animals, but it too is apparently sporadic and is rarely repeated.

(12) Many preparations show a well developed mastication reflex with chewing of the food (or tube), swallowing of the bolus and with a final nose and mouth-licking reflex. This may be initiated by mere opening of the mouth sometimes or by insertion of food into the back of the mouth. A stomach tube is merely swallowed during the first day, is often also chewed later, while after four or five days food may be well swallowed while a stomach tube is received with apparent objection, the head being moved about and attempts being made to spit it out.

(13) Movements are seen in animals with a section between the superior and inferior colliculi and these increase after the first few days. They consist of movements as described under (10) and (14), and also of spontaneous progression movements and of kicking when touched. They were distinct from the violent movement found with more anterior sections (already described by others). It has not so far proved possible to preserve animals for any time when the section has been at the level of the anterior border of superior colliculi, owing to the complications introduced by these violent reactions.

(14) Acoustic reflexes become more brisk after the first few days. Lifting of the head to one side (usually the left) has been often seen, and would sometimes occur in reaction to such slight sounds as were made for instance by entering the room quietly, and this in spite of the fact that the bath stirrer was maintaining a constant noise. Though the constant noise gave no reaction, it caused a reaction when the motor was started after a quiet interval.

(15) Glycosuria persisting for some time after decerebration (as described by J. Mellanby) is confirmed, but this glycosuria disappears again on the second or third day and does not reappear. The cessation of glycosuria is not caused by disappearance of glycogen.

(16) Unilateral decerebration was carried out on a series of seventeen

cats, the removal extending back to the level of the pons in some, to the anterior part of the mid-brain in others.

(17) In these semi-decerebrate animals forced movements were always directed away from the side of removal. The head was rotated with occiput away from the side of removal and the neck flexed and retracted so as to carry the head over to the sound side. In some of the more anterior removals a transient stage of forced movements to the ipsilateral side followed operation, to be succeeded invariably by the opposite.

(18) In experiments in which the semi-decerebration passed through the anterior part of the mesencephalon, extensor rigidity was found in the contralateral limbs. This was preceded sometimes by ipsilateral or bilateral extensor rigidity. When the semi-decerebration was more posterior and passed between mid-brain and pons, there was considerable variation in the rigidities and a tendency for them to be asymmetrical. The anatomical deduction is that the pathway whose interruption is responsible for extensor rigidity decussates in the midbrain.

(19) A condition of increased resistance of the flexor muscles to passive extension was described, both in complete decerebrations and semi-decerebrations. In the latter, when the removal was anterior, this flexor rigidity appeared usually in the ipsilateral limbs, though it might be seen in the contralateral limbs for a short time after operation. Thus it alternated with extensor rigidity.

(20) Flexor rigidity as an entity may be described as tonic innervation of those muscles which are subjected to tonic inhibition during extensor rigidity. It is possible to speak of two types of decerebrate rigidity.

In conclusion we would express our thanks to Professor Sherrington for his continued help and advice throughout the course of the experiments, and to Dr. J. G. Greenfield for the hospitality extended to us in his pathological laboratory at Queen Square.

The greater part of the expenses of this research was met by a grant from the Medical Research Council, to whom we are much indebted for kind assistance.

REFERENCES

1. Barbour, H. G., 1921. The heat-regulating mechanisms of the body. *Physiol. Rev.* 1:295.
2. Barenne, J. G. Dusser de, 1919. Recherches expérimentales sur les fonctions du système nerveux central, faites en particulier sur deux chats dont le neopallium avait été enlevé. *Arch. Neerl. de Physiol. de l'homme et des Anim.* 4:31
4. Bazett, H. C., 1919–1920. The use of decerebrate preparation for experiments extending over several days. *J. Physiol.* 53, *Proc. Physiol. Soc.*, xiii.

6. Boyce, R., 1895. A contribution to the study of descending degenerations in the brain and spinal cord. *Phil. Trans.* **186**:B, 321.
7. Brouwer, B., 1919-1920. Examen anatomique du système nerveux central des deux chats décrits par J. G. Dusser de Barenne. *Arch. Neerl. de Physiol. de l'homme et des Anim.* **4**:124.
10. Cajal, S. Ramon, 1909. *Système Nerveux de L'homme et des Vertébrés.* Maloine, Paris.
12. Crowe, S. J.; Cushing, H.; and Homans, J., 1910. Experimental hypophysectomy. *Bull. of Johns Hopkins Hosp.* **21**:127.
13. Cushing, H., 1912. *The Pituitary Body and Its Disorders.* Philadelphia.
14. Forbes, A., and Sherrington, C. S., 1914. Acoustic reflexes in the decerebrate cat. *Am. J. Physiol.* **35**:367.
16. Gamgee, A., 1909. On methods for the continuous and quasi-continuous registration of the diurnal curve of the temperature of the animal body. *Phil. Trans.* **200**:B, 219.
18. Haldane, J. S., 1905. The influence of high air temperatures. *J. of Hyg.* **5**:494.
19. Hatschek, 1907. Zur vergleichenden anatomie des nucleus ruber. *Neurol. Centrlbl.* **26**:870.
20. Heymans, J. F. 1919. Iso-, hyper-, et hypothermisation des mammifères par calorification et frigorification du sang de la circulation carotidojuglaire anastomosée. *Arch. Int. de Pharmacodyn. et de Ther.* **25**:1.
21. Holmes, Gordon, 1901-1902. The nervous system of a dog without a forebrain. *J. Physiol.* **27**:1.
22. Isenschmidt, V. R., and Krehl, L., 1912. Ueber den einfluss des gehirns auf die wärmeregulation. *Arch. f. Exper. Pathol. u. Pharm.* **70**:109.
23. Isenschmidt, V. R., and Schnitzler, W., 1914. Beitrage zur lokalisation des der wärmergulation vorstehenden zentralapparates in zwischenhirn. *Arch. f. Exper. Path. u. Pharm.* **76**:202.
25. Karplus, J., and Kreidl, A., 1914. Ueber totalexstirpationen einer und beider grosshirn hemisphären an affen. *Arch. f. Anat. u. Physiol. Physiol. Abth.*, p. 155.
30. Lefevre, J., 1911. *Chaleur Animale et Bioénérgetique.* Paris, p. 483.
31. Leschke, E., 1913. Ueber den einfluss des zwischenhirns auf die wärmeregulation. *Z. f. exper. Path. u. Ther.* **14**:167.
32. Longet, F. A., 1842. *Anatomie et Physiologie due Système Nerveux.* Paris.
34. Macleod, J. R., 1919. Respiration after decerebration in cats. *Trans. R. Soc. of Can.* **5**:85.
35. Idem., 1921. Periodic breathing in decerebrate cats. *Am. J. Physiol.* **55**:175.
43. v. Monakow, C., 1910. *Rote Kern, die Haube und die Regio Hypothalamica bei einigen Säugentieren und beim Menschen.* Bergmann, Wiesbaden.
46. Probst, 1904. *Jahrb. f. Psych. u. Neurol.*, 24.
47. Roaf, H. E., 1912. The influence of muscular ridigity on the carbon dioxide output of decerebrate cats. *Q. J. Exp. Physiol.* **5**:31.
48. Richet, C., 1898. Chaleur. *Dict. de Physiol.*, Paris **3**:178.
49. Riddoch, G., and Buzzard, E. F., 1921. Reflex movements and postural reaction in quadriplegia and hemiplegia with special reference to those of the upper limb. *Brain* **44**:397.
50. Rogers, F. T., 1919. Regulation of body temperature in the pigeon and its relation to certain cerebral lesions. *Am. J. Physiol.* **49**:271.
51. Rothmann, M., 1909. Der hund ohne grosshirn, *Gesellschaft deutscher Nervenärzte.* Auto-referat *Neurol. Centralbl.* **28**:1045.

52. Idem., 1912. Demonstration des sektionsbefundes des grosshirnlosen hundes. *Berliner Gesellschaft f. Psychiatrie und Neurologie. Neurol. Centralbl.* **31**:867.
53. Sherrington, C. S., 1897. On reciprocal innervation of antagonistic muscles. *Proc. R. Soc.* **60**:414.
54. Idem., 1897-1898. Decerebrate rigidity and reflex co-ordination of movements. *J. Physiol.* **22**:319.
58. Trevan, J. W., 1915-1916. The control of respiration by the mid-brain. *J. Physiol.* **50**, *Proc. Physiol. Soc.*, xliii.
62. Wilson, S. A. K., 1913-1914. Anatomy and physiology of the corpus striatum. *Brain* **36**:427-492.
63. Idem., 1920. On decerebrate rigidity in man and the occurrence of tonic fits. *Brain* **43**:220-269.
65. Woodhead, G. S., and Warrier-Jines, P. C., 1916. Clinical thermometry. *Lancet*, i, pp. 173-180, 281-288, 338-340, 450-454, 495-502.
66. Woodworth, R., and Sherrington, C., 1904. A pseudoaffective reflex and its spinal path. *J. Physiol.* **31**:234-243.

REGULATION OF BODY TEMPERATURE[1]

S. W. RANSON, M.D.

INTRODUCTION

THE remarkably stable body temperature of warm blooded animals is evidence of an efficient thermostatic control. Heat is continuously formed in the body, even during rest, and the amount thus formed is very greatly increased by muscular activity. To maintain an even temperature the rate of heat loss must be adjusted to the rate of heat formation and this adjustment is complicated by changing environmental temperature. A high external temperature reduces heat elimination, and vigorous muscular exercise increases heat production, causing the body temperature to rise and thus starting heat loss activities such as dilatation of the cutaneous blood vessels and sweating. In carnivores panting is the chief means of heat elimination. Exposure to cold causes shivering, which increases heat production, and constriction of the cutaneous vessels which reduces heat loss by diminishing the flow of warm blood to the skin.

A distinction is sometimes made between physical means of regulating body temperature (vasoconstriction, vasodilation, sweating and shivering) and chemical means. Very little is known about chemical regulation. DuBois (1936) could find no evidence of any increase in metabolism as a result of exposure to cold until shivering began. On the other hand Cannon (1939) has shown that heat loss evokes activity of the adrenal medulla and the extra output of adrenalin accelerates metabolism. Under otherwise similar conditions cats shiver more when the adrenals are out of function. In man icewater taken into the stomach may cause a marked and prolonged increase in basal metabolic rate without shivering. The thyroid is stimulated by cold and cooperates with the adrenals but in a less quickly responsive manner. It hypertrophies in ani-

[1] Institute of Neurology, Northwestern University Medical School, Chicago, Ill. Aided by a grant from the Rockefeller Foundation.

mals exposed to cold and apparently plays a part in slow adaptations to climate (Burton, 1939). Uotila (1939) has suggested that this compensatory activity of the thyroid is brought about through the hypophysial stalk and the thyrotropic hormone of the hypophysis.

In addition to these automatic adjustments other more or less conscious reactions take place in the interest of temperature regulation, such as seeking a cool place and remaining quiet and relaxed when external temperatures are high, or seeking a warm spot and reducing the amount of exposed surface or engaging in vigorous activity when the temperature is low.

We shall consider the automatic regulation of body temperature in so far as this is controlled by the hypothalamus.

[*Editor's Note:* Material has been omitted at this point.]

LOCALIZED HEATING OF THE BRAIN

Vasodilation, vasoconstriction, sweating and shivering can be produced reflexly from the skin. It is also known that the brain contains a mechanism that can be activated directly by the temperature of the blood. Sherrington (1924) has shown that, if the hind quarters of a dog with spinal cord transected in the lower cervical region are immersed in ice water under conditions that exclude the possibility of the cold stimulating any sensory nerve endings connected with the cerebrospinal axis in front of the lesion, shivering only occurs in the anteriorly situated non-paraplegic muscles.

Warming the carotid blood entering the head has been shown (Kahn, 1904; Moorhouse, 1911; Hammouda, 1933) to cause sweating, peripheral vasodilatation and hyperventilation in experimental animals, and the authors have inferred that these effects were produced by direct activation of regulating centers in the brain by the rising temperature of the blood. Support for this view has been provided by experiments in which parts of the brain have been heated directly, either by warm water passing through a closed tube inserted in or applied to the desired region, or by open irrigation with warm saline. Barbour (1912), Hashimoto (1915) and Prince and Hahn (1918) found that a fall in body temperature resulted from heating the corpus striatum in the rabbit and cat, and the first two workers described a peripheral vasodilatation during this procedure. Moore (1918) confirmed the antipyretic action of warming this region but showed that the corpus striatum was not specifically related to the effect, and Sachs and Green (1917) could not observe any constant result of warming or irritating the corpus striatum in various other ways. It has been pointed out by Bazett (1927) that the effects obtained by Barbour and others were probably produced at some distance from the site of heating in the corpus striatum because of the high temperatures which had to be used to obtain reactions. Hasama (1929) found a fall in body temperature to result from warming the base of the hypothalamus and preoptic region in the cat and observed a profuse sweating on the footpads during this procedure. Irrigation of the third ventricle with warm saline was shown to produce polypnea and panting in the dog by Hammouda (1933). These observations are not altogether consistent but appear to indicate that a rising intracranial temperature is able to activate the heat loss mechanism and to this extent fall in line with the experiments described below in which a high frequency alternating current was used as the source of heat.

It has been shown that heat loss mechanisms, including panting and sweating, can be activated by local heating of the preoptic and supraoptic regions in cats under urethane anesthesia (Magoun, Harrison, Brobeck and Ranson, 1938). The heat was provided by a high frequency alternating current of low voltage passing between two electrodes, consisting of straight lengths of 22 gauge nichrome wire, insulated to within 2 mm. of their tips and inserted into the brain, 4 mm. apart and parallel to each other, with the aid of a multiple needle carrier and the Horsley-Clarke instrument. The electrodes were inserted vertically from above downward and stopped every 2 mm. on the way down to allow for the passage of the heating current at each of these levels and for the observation of its effect. By repeated punctures at 2 mm. intervals from before back, some medially and some laterally placed, it was possible to explore the brain and to locate any regions which were specifically responsive to heat. In experiments, in which reactive regions were being examined, not more than 2 or 3 pairs of punctures were made in any one animal. The heating current oscillated at 1,000,000 cycles and the voltage across the electrodes varied between 11 and 14.3 V. Thermocouple readings taken between the electrodes within the brain and 1 mm. from the nearest electrode showed a rise in temperature of 6.6° to 11.3° according to the voltage. After the current was turned on it took one minute or more for the tissue to reach its highest temperature. Since the rectal temperatures of the cats averaged 97.9° an increase of 6.6° would bring the region 1 mm. from the electrode to a temperature of about 104.5°. An increase of 11.3° would bring the temperature to about 109°, which would account for the localized damage done in some of these experiments.

In a series of experiments extensive exploration of the forebrain and midbrain of the cat revealed only a limited region from which responses to heating could be obtained. Local heating of this reactive area caused a marked acceleration of respiration, the excursion of which became very shallow. Soon the mouth was opened and rhythmic movements of the nostrils, angles of the mouth and tongue appeared with each respiratory excursion. When at their peak these respiratory alterations constituted a characteristic polypneic panting and were frequently, but not invariably, accompanied by the appearance of sweat on the pads of the feet. The responses were in all points similar to those obtained by heating the entire animal; but the rectal temperature of the animals was always low, usually between 95.9 and 99.5°, and never rose during the heating of the brain.

The region from which these reactions were elicited was sharply localized and is indicated schematically in Fig. 207 on a paramedian sagittal section through the cat's brain. The rostral portion of the reactive area (cross-lined in the figure) is located in the preoptic and supraoptic regions between the optic chiasma and the anterior commissure and extends forward beyond them. Continued backward from this region through the dorsal part of the hypothalamus and ventral part of the thalamus is another, outlined by broken lines, which gives reactions of the same kind but which is much less responsive. In several respects the responses obtained by heating the region inclosed by dotted lines were weaker than those obtained from the preoptic and supraoptic regions. They were less abrupt in onset, required a longer time to reach their peak and were quantitatively smaller. the average maximum respiratory rate reached being 155 per min. Panting sometimes failed to occur and usually required the facilitation provided by pulling down the lower jaw. In contrast, heating the preoptic and supraoptic region induced prompt acceleration of respiration to an average rate of 255 per min. Panting appeared spontaneously and continued as long as the heat was applied and for a brief period after it was discontinued. Usually the acceleration of respiration began between 0.5 and 1.5 min. after the onset of heating. Once begun the increase in respiratory rate was fairly abrupt and the peak rate was usually reached between 2 and 3 minutes after the onset of heating. Panting usually began 1 to 2.5 min. after the onset of heating

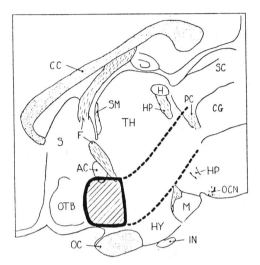

Fig. 207. Schematic outline of the region reactive to heating, projected on a paramedian sagittal section through the brain of the cat. (Reproduced from the *J. Neurophysiol.*) Abbreviations for Fig 207 and 208 are as follows:

AC, anterior commissure
C, caudate nucleus
CC, corpus callosum
CG, central gray matter
E, entopeduncular nucleus
F, fornix
GP, globus pallidus
H, habenula
HP, habenulo-peduncular tract
HY, hypothalamus
IC, internal capsule
IN, infundibulum
LV, lateral ventricle
M, mammillary body

MB, midbrain
MFB, medial forebrain bundle
MT, mammillo-thalamic tract
OC, optic chiasma
OCN, oculomotor nerve
OT, optic tract
OTB, olfactory tubercle
PC, posterior commissure
S, septum
SC, superior colliculus
SM, stria medullaris
TH, thalamus
3V, third ventricle

and stopped 0.5 to 2 min. after the cessation of heating. After the respiration had returned to normal, acceleration and panting could again be obtained by repeating the heating. In many instances the respiratory alterations were accompanied by the appearance of sweat on the pads of the feet.

The results obtained by heating this anterior region are

illustrated in Fig. 208. The electrodes were inserted on either side of and 2 mm. from the midline. In Fig. 208A, the small letters *a*, *b*, *c* and *d* represent the position of the bare portions of the electrodes for each of 4 levels at which heat was applied. In *B* the same letters indicate the respiratory responses obtained from these levels. The duration of current flow at *a* was 5 minutes, other times were in proportion to the length of the horizontal line immediately above the letter. At *a* there

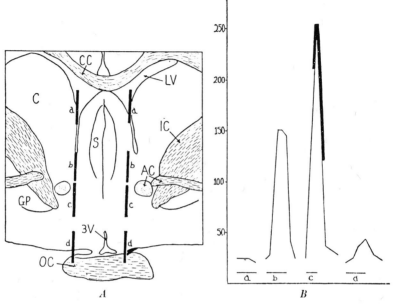

Fig. 208A. Transverse section through the brain of Cat 1. The positions of the two electrodes during heating are shown at a, b, c and d.

Fig. 208B. Chart showing the respective respiratory responses obtained from heating at positions a, b, c and d, Fig. 208A. The rate of respiration is shown on the ordinate; panting is shown by a heavy line. Period a represents 5 min., other times are in proportion. (Reproduced from the *J. Neurophysiol.*)

was no response, at *b* the respiratory rate was accelerated to 150 without panting, at *c* with a shorter flow of current the rate increased from 25 to 252 per min. and panting occurred as indicated by the heavy portion of the line. At *d* there was only a slight increase in respiratory rate. The maximum reaction was obtained at level *c* in the preoptic region. The acceleration of respiration obtained at *b* was due in part at least to extension of the heating into the region *c*.

The reactions to heating were always obtained from within

4 mm. of the midline in the preoptic region and diencephalon, never from the lateral parts of the hemisphere.

Histological examination of the brains showed that in the majority of instances the heat had not damaged the brain. In some cases when 14.3 V. were used there was some damage to the brain around and between the electrodes. Even in the cases where this damage occurred the usual responses were obtained and presumably were elicited from the adjacent reactive region outside the area damaged.

The reactions obtained were due specifically to local heating. The high frequency current used could not stimulate. Moreover faradic stimulation of the preoptic region usually causes slowing of respiration; and while faradic stimulation of the supraoptic region may cause acceleration of respiration, it rarely if ever causes panting, and the acceleration is accompanied by dilatation of the pupils, erection of hair and other signs of generalized excitation. On the other hand heating of this region causes only acceleration of respiration, panting and sweating. The responses were in all respects similar to those obtained by heating the entire animal except that much less time was required to induce the response.

The obvious interpretation of these results is that in these experiments there have been set into play by artificial heating the same mechanisms in the same regions of the brain which are activated in the normal animal when the temperature of the blood rises too high.

The results of these experiments should not be taken to indicate that the region responsive to heating contains the final efferent collection of supranuclear neurons for coordinated polypneic panting, for decorticate panting still occurs after the area here delimited has been largely destroyed (Lilienthal and Otenasek, 1937). This motor mechanism occupies a more caudal position perhaps in the mesencephalon (Keller, 1933). The responsive area found in these experiments contains elements sensitive to heat and capable of exciting efferent groups of neurons situated at more caudal levels.

"It is said that the activation of heat loss activities by the hypothalamus serves as evidence that parasympathetic centers are located in this part of the brain (Fulton, 1938). But panting is not an autonomic reaction and, as might be expected, is not affected by large doses of atropine. Sweating is mediated through the sympathetic system. The nervous mechanism involved in vasodilatation is not well understood; but, if an active vasodilatation as distinct from a central inhibition of vasoconstriction forms part of the heat-loss reaction, it must be mediated either by the dorsal roots or by the sympathetic system and can be classified as parasympathetic only by analogy. In view of the fact that some sympathetic fibers are cholinergic it is scarcely justifiable to regard sweating and cutaneous vasodilation as parasympathetic on the basis of the action of drugs." (Ranson and Magoun, 1939, p. 147.)

When a dog is exposed to the hot rays of the sun it begins panting before there is any increase in rectal temperature (Bazett, 1927). This panting exhibits the characteristics of a conditioned reflex; and conditioned panting can be established experimentally (Hammouda, 1933). This reflex panting is brought about in a different manner than the panting caused by heating the brain, and we have not investigated the nervous pathways by which it is mediated.

When taken by themselves the panting and sweating caused by heating the preoptic and supraoptic regions serve only as presumptive evidence that the same region is activated normally by heat when the temperature of the blood rises too high. Later we shall reinforce this evidence by showing that this region, which is specifically sensitive to heat, normally plays an essential part in temperature regulation and that cats in which this region has been destroyed do not pant when overheated.

IMPAIRMENT OF TEMPERATURE REGULATION BY LESIONS IN THE HYPOTHALAMUS

Transection of the brain stem behind the hypothalamus or extensive injury to the hypothalamus in experimental animals causes a loss of ability to regulate body temperature. Isenschmid and Schnitzler (1914) showed that small bilateral lesions in the region of the mammillary bodies and the posterior part of the tuber cause loss of the capacity to guard the body against abnormal drops in body temperature. More recent work has established beyond dispute the importance of the hypothalamus for regulation of temperature, but there is a great deal of difference of opinion as to the exact location of the center or centers involved (Alpers, 1936; Barbour, 1939; Bazett, Alpers and Erb, 1933; Bazett and Penfield, 1922; Davison and Selby, 1935; Dworkin, 1930; Frazier, Alpers and Lewy, 1936; Glaubach and Pick, 1933; Keller, 1933). The work of Keller (1933, 1938) and of Thauer and Peters (1937) which lead to contrary results must be discounted because these investigators did not remove the brain in front of the transection and there is no adequate guarantee that the transections were complete.

In acute experiments such as those of Isenschmid and Schnitzler it is difficult to be sure that the shock of the operation has not temporarily abolished functions which would return later if the animals were kept alive. For this reason effort has been made in recent years to keep the animals alive as long as possible. Bazett, Alpers and Erb (1933) removed the cerebral hemispheres and part or all of the diencephalon in cats and kept the animals alive for periods ranging from 3 to 7 days. One decerebrate cat lived 19 days. They found that while transections behind the hypothalamus abolished temperature regulation, cats in which the caudoventral parts of the diencephalon were left intact were able to keep the body temperature up to normal in a comfortably warm room. In these cats the transections passed through the hypothalamus a short distance behind the optic chiasma and the supraoptic and preoptic regions were cut away. It is important to note that while the ability to resist chilling was retained, ability to react to overheating by panting was abolished.

Observations made on animals decerebrated at various levels are not altogether satisfactory because so much of the brain is removed, because the preparations thus obtained differ greatly from normal animals and because parts of the brain stem remaining in place may have been rendered functionless by anemia, hemorrhage and inflammatory reactions. Observations made on animals with lesions placed with the Horsley-Clarke apparatus furnish more reliable information because the lesions are sharply defined, while all the rest of the brain is left undamaged, and because such animals can be kept and studied for as long as desired and are normal except for the disturbances in temperature control, changes in emotional reactions and a tendency to drowsiness, all of which symptoms result directly from the hypothalamic lesions. When this method is used the lesions must not be too large or else the animals will not live very long and the observations are likely to be misleading (Frazier, Alpers and Lewy, 1936).

In a large series of cats lesions have been placed in the hypothalamus with the Horsley-Clarke apparatus and the resulting disturbances have been studied in some detail (Clark, Magoun and Ranson, 1939a). The lesions were of varying size and location; but in every case the attempt was made to produce lesions which were as nearly as possible bilaterally symmetrical.

Preceding the operation, daily observations were made of the cat's rectal temperature for a week or more and tests were made of the ability of the animal to regulate its body temperature in the cold box and in the hot box. After the operation the cats were kept for one or more days in an incubator set to run at about 86° and thereafter at room temperature which averaged around 75°. Daily observations were made of the rectal and environmental temperature for 2 weeks or more. In the following paragraphs we have used the terms hypothermia and hyperthermia to designate conditions of abnormally low or abnormally high body temperature when the environmental temperature is within the usual comfortable range. For experiments on cats and monkeys we may arbitrarily set this range between 70 and 90°.

Tests were made in a hot box and cold box about a week after the operation and repeated on the second and fourth weeks and in some cases after much longer intervals. The cold box was the same as that used by Teague and Ranson (1936) and its temperature ranged from 34° to 46° but was usually around 44°F. The hot box was altered by inserting a fan, which allowed more even regulation of the temperature within it, but because of the increased movement of the air, favored the evaporation of perspiration. The temperature in the hot box was 103 or 104°F. The cats were kept in the box until they panted or until the rectal temperature reached 106°.

Large lesions. Very large lesions in the hypothalamus resulted in the death of most of the animals. When these lesions destroyed the posterior part of the hypothalamus the cats' temperature was low on the first postoperative morning (92° to 95°) and the cats died early, usually within 5 days. When the lesions destroyed the middle or anterior part of the hypothalamus the temperature did not as a rule fall so low and might even be normal or above normal. These cats lived somewhat longer and usually died between the fifth and tenth day. When removed from the incubator on the first morning after the operation the two cats with lesions anterior to the anterior commissure had rectal temperatures above 104.5°; 4 of the 7 with lesions at the level of the anterior commissure had temperatures of 104.5° or higher and the lowest temperature in this group was 99°; 3 of the 11 with lesions in the anterior hypothalamus had temperatures above 104.5° and 4 had temperatures below 98°. One of these continued to run a subnormal temperature and was dead on the fifth day. The temperature of the other 3 was normal or above normal on the second and third days in the incubator. While a majority of the cats of this anterior group ran temperatures which were normal or slightly above on the second and third days some of these developed subnormal temperatures before the tenth day.

Because of the early death or poor condition of a majority of these cats only 8 were tested in the hot box. These tests showed that in this group of animals with large anteriorly placed lesions there was a marked loss in ability to regulate against heat (Table 6). In all but one cat the rectal tempera-

TABLE 6

Hot box tests on cats with large anteriorly placed lesions. Rectal temperature in F.°

(Reproduced from the J. Neurophysiol.)

Cat no.	Days after operation	Level of lesion	Temp. at end of test	Rise in temp.	Final respiratory rate	Panting	Sweating
53	35	A.A.C.	104.5	3.6	140	yes	yes
51	38	A.C.	106.4	4.3	72	no	not tested
94	30	A.C.	106.2	13.8	16	no	no
96	81	A.C.	106.4	8.3	22	no	no
97	64	A.C.	106.2	7.8	24	no	not noticed
12	104	A.H.	106.3	4.3	36	no	no
81	71	A.H.	106.0	3.2	24	no	no
83	28	A.H.	106.2	5.6	32	no	no

ture rose above 106° without causing panting or much increase in respiratory rate. In 6 of the cats the rate did not exceed 36 per min. One cat (53) panted at a rectal temperature of 104.5° and it is interesting to note that this was the only one in which the lesions were situated so far forward as to spare most of the anterior commissure and much of the preoptic region (Fig. 209). In Cat 51 the lesions involved the anterior commissure and the brain ventral to it thus largely destroying the preoptic region (Fig. 210). In the hot box test made on this cat 38 days after the operation the rectal temperature reached 106.4° without causing panting or an increase in respiratory rate above 72 (Table 6). The difference between the reactions of these 2 cats in the hot box, the one with the preoptic region largely intact showing a normal reaction and the one with this region largely destroyed showing a failure to pant, fits very well with the results obtained by local heating of the brain. The anterior limits of the area concerned with

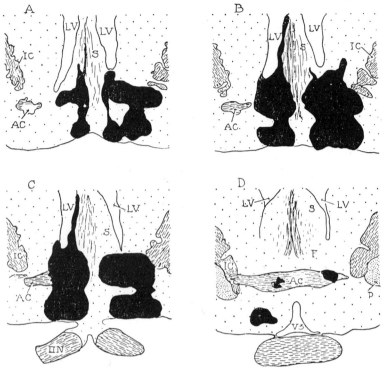

Fig. 209. The lesions in Cat 53 indicated in solid black on four drawings from transverse sections through the brain at the level of and in front of the anterior commissure, lettered in order from before backward. (Reproduced from the *J. Neurophysiol.*) Abbreviations for Fig. 209 to 215 are the same as in Fig. 207 and 208 with the following additions.

A, B, C, D, order of sections caudalward
CP, cerebral peduncle
H, field H of Forel
HG, habenular ganglion
Hy, hypophysis
ME, median eminence
MLF, medial longitudinal fasciculus
MP, mammillary peduncle
MT, mammillothalamic tract

NII, optic nerve
NIII, third nerve
P, globus pallidus
PC, posterior commissure
S, septum
SN, substantia nigra
Sth, subthalamic nucleus
V3, third ventricle

heat loss activity as determined by heating the brain coincide within half a millimeter with the limits of this area as determined by lesions.

TEMPERATURE REGULATION

Moderate sized lesions offer much more information about functional localization than do the massive lesions considered in the preceding section. These smaller lesions have been so placed as to destroy selectively rostromedial, rostrolateral, caudomedial or caudolateral portions of the hypothalamus.

Moderate sized lesions in medial part of anterior hypothalamus in 17 cats produced no easily recognizable symptoms. On the

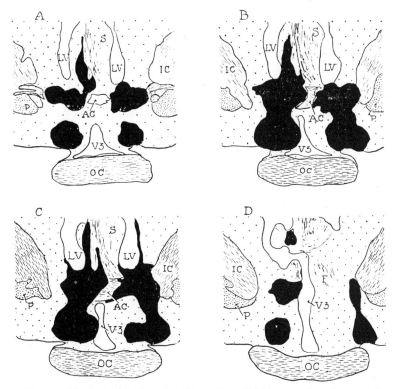

Fig. 210. The lesions in Cat 51 indicated in solid black on four drawings from transverse sections through the brain at the level of and behind the anterior commissure. (Reproduced from the *J. Neurophysiol.*)

morning after the operation these animals were alert and active. They usually ate spontaneously on the first or second postoperative day. Their daily temperatures subsequent to the operation were normal except that a considerable number of them had temperatures in the range of 104 or 105° on the day following the operation. Eight of these cats on which hot and cold box tests were made are listed in Tables 7, 8 and 9.

Their reactions in the cold box were entirely normal. In the hot box tests made one month after the operation their records

TABLE 7

Postoperative rectal temperatures in F.° of cats with moderate sized hypothalamic lesions. The figures in parentheses indicate the temperature of the room or incubator. The cats are listed in four groups: those with anteromedial lesions are designated by the letters AM; those with anterolateral lesions, AL; those with posteromedial lesions, PM; and those with posterolateral lesions, PL.

(Reproduced from the J. Neurophysiol.)

Cat	1st day	3rd day	5th day	7th day	10th day
A. M.					
6	103.2 (84)	101.7 (75)	102.0 (75)	101.2 (77)	101.0 (77)
11	105.0 (84)	103.0 (75)	103.3 (76)	101.9 (76)	103.0 (75)
13	102.6 (84)	102.5 (76)	103.0 (76)	102.3 (75)	103.2 (75)
15	103.0 (87)	103.2 (76)	104.0 (75)	103.8 (75)	
16	105.4 (87)	104.9 (76)	102.9 (75)	103.3 (75)	
17	103.5 (87)	103.6 (75)	104.1 (75)	D.C.	
18	104.6 (87)	102.0 (75)	101.8 (75)	D.C.	
22	104.4 (82)	102.2 (74)	101.3 (75)	101.9 (74)	101.6 (73)
A. L.					
8	107.9 (84)	103.7 (76)	103.5 (77)	102.1 (74)	101.8 (76)
20	94.0 (78)	102.1 (85)	101.0 (76)	99.3 (76)	100.7 (75)
40	104.6 (82)	106.0 (74)	98.2 (74)	104.9 (72)	100.1 (74)
23	104.6 (85)	104.7 (76)	103.6 (76)	103.2 (75)	102.5 (75)
39	102.1 (82)	101.9 (74)	102.7 (74)	102.0 (72)	100.7 (74)
42	105.1 (85)	103.8 (76)	102.7 (76)	101.3 (75)	102.5 (75)
43	105.2 (82)	99.6 (74)	98.7 (74)	97.3 (72)	101.3 (74)
55	106.0 (82)	100.1 (76)	101.7 (76)	102.4 (76)	
P. M.					
103	102.1 (86)	103.6 (76)	102.1 (76)	102.2 (76)	103.3 (74)
25	101.6 (81)	104.0 (77)	103.2 (75)	101.2 (75)	101.4 (75)
38	101.8 (81)	103.5 (77)	103.5 (75)	103.8 (75)	103.5 (75)
105	97.1 (84)	103.1 (77)	103.8 (77)	102.5 (77)	103.2 (77)
P. L.					
46	90.6 (83)	102.9 (82)	97.6 (74)	99.6 (74)	99.3 (77)
47	87.0 (83)	105.4 (82)	97.6 (74)	102.6 (74)	103.7 (77)
100	91.0 (82)	100.2 (?)	98.8 (77)	105.7 (79)	102.6 (77)
107	94.7 (82)	97.1 (79)	103.6 (77)	102.9 (75)	102.5 (75)
109	93.4 (88)	99.1 (90)	101.8 (91)	96.4 (73)	98.2 (74)
110	95.6 (84)	99.4 (86)	99.5 (84)	99.5 (76)	101.1 (74)
114	96.0 (85)	98.6 (84)	99.9 (73)	102.3 (75)	101.4 (77)

were only slightly abnormal as shown by comparison with the similar preoperative tests recorded in the same table. There was some elevation of the panting temperature level and decrease in final respiratory rate as judged by the preoperative controls; but they all panted at rectal temperatures below 106°.

The lesions lay close to the midline and in Cat 13 extended from the optic chiasma to the fornix and caudalward over the median eminence (the expanded end of the infundibulum where this is attached to the tuber) as shown in Fig. 211. In this cat the daily temperature remained within normal limits (Table 7). In some of the others temperatures of 104 and 105° were encountered. The important thing is that at no time did any of these cats show a loss in capacity to keep the body temperature up to the normal level. Five days after the operation Cat 13 was in the cold box for 3 hours at a temperature

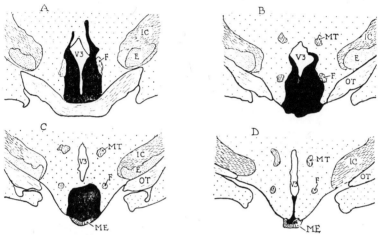

Fig. 211. The lesions in Cat 13 indicated in solid black on four drawings from transverse sections through the brain at the level of and behind the optic chiasma. (Reproduced from the *J. Neurophysiol.*)

of 42° without any significant drop in its rectal temperature (Table 8). Other cats of this group also reacted normally in the cold box during the first postoperative week.

When tested in the hot box one month after the operation Cat 13 panted at a rate of 132 per min. at a rectal temperature of 104.1 as compared with a respiratory rate of 230 per min. and a temperature of 101.7° in the preoperative test (Table 9). In hot box tests made one week after the operation 2 of these cats with medial lesions failed to pant at temperatures below 106°. Of these 2, one had a respiratory rate of 78 at 106.1° and the other, a rate of 212 at 106.2°. At the end of a month, however, both of these cats (11 and 22) had merely an increase in panting level of about 1.5° above the preoperative level (Table 9). The cats in this group showed somewhat less disturbances in regulation against heat than did those reported by Teague and Ranson (1936) and this is to be correlated with the fact that the lesions did not extend quite as far lateralward (Fig. 211).

TABLE 8

Cold box tests on cats with medial lesions in rostral part of hypothalamus. Rectal temperature in F.°

(Reproduced from the *J. Neurophysiol.*)

Cat no.	Preoperative tests					Tests made during the first postoperative week				
	Temp. at start	Temp. at end	Change	Shivering	Av. temp. of box	Temp. at start	Temp. at end	Change	Shivering	Av. temp. of box
6	100.8	100.8	0	yes	41	102.5	102.1	−0.4	yes	44
11	101.8	101.9	+0.1	yes	46	103.3	103.8	+0.5	yes	44
13	101.7	101.2	−0.5	yes	45	101.6	101.0	−0.6	yes?	42
15	101.3	101.0	−0.3	yes	45	102.4	102.1	−0.3	yes	42
16	101.3	99.6	−1.7	yes	43	104.5	103.6	−0.9	yes	42
17	102.5	101.1	−1.4	yes	46	102.9	102.1	−0.8	yes	43
18	101.7	101.7	0	yes	43	101.5	101.2	−0.3	yes	43
22	101.0	99.5	−1.5	yes	44	101.8	100.0	−1.8	yes	45

TABLE 9

Hot box tests on cats with medial lesions in rostral part of hypothalamus. Rectal temperature in F.°

(Reproduced from the *J. Neurophysiol.*)

Cat no.	Preoperative tests					Tests one month postoperative				
	Temp. at end of test	Rise in temp.	Final respiratory rate	Panting	Sweating	Temp. at end of test	Rise in temp.	Final respiratory rate	Panting	Sweating
6	102.7	0.3	184	yes	no	103.0	2.2	146	yes	no
11	104.1	2.4	112	yes		105.7	4.2	186	yes	yes
13	101.7	0.9	230	yes	no	104.1	2.9	132	yes	no
15	103.0	0.5	160	yes	yes	105.9	3.5	126	yes	no
16	105.2	2.5	224	yes	no	104.1	2.0	196	yes	yes
17	102.1	0.8	216	yes	yes	105.3	4.0	180	yes	no
18	103.9	0.9	184	yes	yes	105.0	2.2	180	yes	no?
22	102.8	0.4	260	yes	no	104.3	3.8	216	yes	yes

Moderate sized lesions in lateral part of anterior hypothalamus caused much greater disturbance in temperature regulation than did those more medially placed. There were 8 cats in this series and of these, 3 (8, 20, 40) had symmetrically placed lesions in the extreme lateral part of the hypothalamus, reaching and often damaging the medial edge of the basis pedunculi and internal capsule (Fig. 212). The region, between the fornix and internal capsule above the optic chiasma, which contains the medial forebrain bundle was destroyed on both sides of the brain. The remaining 5 cats had lesions asymmetrically placed so that a considerable part of the medial forebrain bundle escaped damage on one or the other sides.

On the morning after the operation all 8 of these animals were alert and active. In all but one case the rectal temperature was normal or above normal on the morning following

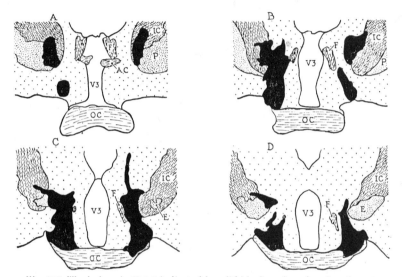

Fig. 212. The lesions in Cat 8 indicated in solid black on four drawings from transverse sections through the brain at the level of the optic chiasma. (Reproduced from the *J. Neurophysiol.*)

the operation. The exception, Cat 20, had a subnormal temperature on the first postoperative morning but on the second and third days it was normal (Table 7). Six cats on the morning following the operation had temperatures above 104.5 and in one it reached 107.9°. The second day 3 had temperatures above 104.5. Soon thereafter all temperatures became normal and remained so except for Cat 20 which for 23 days (from the 22nd to the 45th postoperative day) ran a subnormal temperature as low as 97°F. In another series of experiments lesions were placed bilaterally in the anterior part of the lateral hypothalamus in cats under ether anesthesia and a continuous record taken of their rectal temperature for 20 hours. Seven out of 11 showed postoperative hyperthermia, the temperature reaching 104.8 to 106.8 within 6 hours after the operation (Ranson and Clark, 1938).

The 3 cats with symmetrical lesions (8, 20, 40) suffered a marked loss in ability to prevent overheating as measured by the respiratory responses to hot box tests one month after the operation (Table 10). In each of these the rectal temperature was raised to 106° without causing panting and without causing much increase in the respiratory rate. Later tests (Cat 40, 6 weeks; Cat 20, 10 weeks and Cat 8, 16 weeks) re-

TABLE 10

Hot box tests on cats with lateral lesions in rostral part of hypothalamus. Cats 8, 20 and 40 had symmetrical lesions, the other four had asymmetrical lesions. Rectal temperature in F.°

(Reproduced from the *J. Neurophysiol.*)

Cat no.	Preoperative tests					Tests one month postoperative				
	Temp. at end of test	Rise in temp.	Final respiratory rate	Panting	Sweating	Temp. at end of test	Rise in temp.	Final respiratory rate	Panting	Sweating
Sym.										
8	104.5	3.2	176	yes	no	106.2	3.8	70	no	yes
20	104.4	2.3	230	yes	yes	106.0	7.7	26	no	no
40	103.3	1.3	246	yes	yes	106.0	2.8	36	no	no
Asym.										
23	103.1	−0.2	206	yes	yes	103.9	2.6	220	yes	yes
39	102.7	0.4	252	yes	yes	105.9	3.7	140	yes	no
42	104.0	1.4	176	yes	yes	104.9	3.8	120	yes	yes
43	102.7	0.6	240	yes	yes	105.9	2.6	222	yes	no

vealed little improvement in ability to resist overheating except that Cat 8 showed some increase in respiratory rate. However, 27 weeks after the operation Cat 8 was able to pant at 105.3°, thus showing a delayed and partial recovery of the capacity to regulate against heat

The cats with asymmetrical lateral lesions showed an early loss in the ability to regulate against heat. In tests made one week after the operation Cats 39 and 43 reached temperatures above 106° without panting and with almost the same respiratory rates as at the beginning of the tests. The other 2 (23 and 42) did not pant at this temperature although the respiratory rate was high. However, one month after the operation

Cats 23 and 42 had so far recovered that the panting levels were less than 1° above the preoperative levels and Cats 39 and 43 panted at 105.9°. Cold box tests on the cats with symmetrical lesions showed greater than normal drops in temperature, though in the tests made one month after the operation the final temperatures reached were not abnormally low. The cats with asymmetrical lesions gave normal reactions in the cold box (Table 11).

It is evident from the data presented that symmetrical lesions in the lateral part of the rostral hypothalamus are much more effective than asymmetrical lesions in causing disturbances in temperature regulation but even symmetrical

TABLE 11

Cold box tests on cats with lateral lesions in rostral part of hypothalamus. Cats 8, 20 and 40 had symmetrical lesions. The other four had asymmetrical lesions. Rectal temperature in F.°

(Reproduced from the *J. Neurophysiol.*)

	Tests one week postoperative					Tests one month postoperative				
Cat no.	Temp. at start	Temp. at end	Change	Shivering	Av. temp. of box	Temp. at start	Temp. at end	Change	Shivering	Av. temp. of box
Sym.										
8	102.6	97.4	−5.2	yes	40	103.0	100.6	−2.4	yes	39
20	99.5	97.1	−2.4	yes?	44	102.1	99.6	−2.5	yes?	42
40						102.7	99.7	−3.0	no	43
Asym.										
23	101.7	100.4	−1.3	yes?	45	102.4	101.8	−0.6	yes?	39
39						102.1	101.8	−0.3	yes?	43
42	103.2	102.4	−0.8	yes?	44	101.8	102.4	+0.6	yes?	40
43						101.9	102.0	+0.1	yes	44

lesions do not cause great loss in the capacity to regulate against cold. But such symmetrical lesions cause a profound and prolonged loss in the ability to regulate against heat.

Laterally placed moderate sized lesions in posterior part of hypothalamus cause very great impairment in the ability to regulate against both heat and cold. The 7 cats in this series had lesions at the level of the mammillary bodies extensively damaging the lateral hypothalamic area. In all but one the lesions were fairly symmetrical bilaterally and in most instances they extended far enough dorsally to involve the fields of Forel and to interrupt fibers which may run through the supramammillary decussation. In Cat 47 (Fig. 213) the lesions were situated somewhat farther rostrally than in the others and extended from the level of the infundibulum to the level of the middle of the mammillary bodies. In Cat 110 the lesions were placed farther caudally (Fig. 214). In one cat (114) the lesions were quite asymmetrical one being in the lateral hypothalamic area on the left side, the other being in the midline leaving the right lateral hypothalamic area largely intact. These lesions would have interrupted in addition to the fibers descending on the left side into the mesencephalic tegmentum any fibers which may enter the central gray matter through the supramammillary decussation. This cat showed

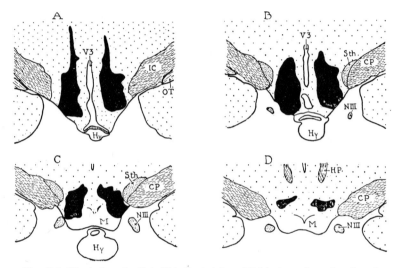

Fig. 213. The lesions in Cat 47 indicated in solid black on four drawings from transverse sections through the brain at the level of and in front of the mammillary body. (Reproduced from the *J. Neurophysiol.*)

marked disturbances in temperature regulation, a month after the operation but tests were not made after 2 or 3 months to determine to what extent recovery may have taken place.

On the morning after the operation these animals were lethargic. There was considerable extensor tonus especially in the hind legs and the cats could be molded into various bizarre postures. Usually within 4 days after the operation the catalepsy disappeared and the cats became alert and active and would eat spontaneously. On the morning after the operation they had, when removed from the incubator, rectal temperatures of 96° or less (Table 7). On the seventh day they all were able to maintain normal temperatures in the warm animal room except Cat 109.

In cold box tests all of these cats showed a marked decrease in ability to prevent a loss of body heat (Table 12). Five of the 7 showed in one or more of the tests drops of more than 7° as a result of 3 hours exposure to temperatures ranging from 34 to 44°. In 2 (Cats 100, 107) there was evidence of partial recovery of ability to resist chilling with the lapse of time. But in the others practically no recovery occurred within the

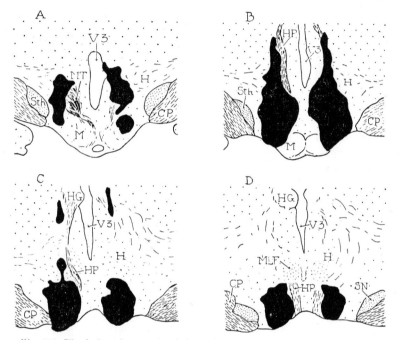

Fig. 214. The lesions in Cat 110 indicated in solid black on four drawings from transverse sections through the brain at the level of and behind the mammillary bodies. (Reproduced from the *J. Neurophysiol.*)

limits of time intervening between the first and last tests. Cat 47 showed a drop of 9.7°, 123 days after the operation; Cat 109 showed a drop of 7.8°, 68 days after the operation; and in Cat 110 the temperature dropped to 96.6° in the test made 62 days after the operation. It will be obvious, therefore, that while most of the cats run an approximately normal temperature when kept in a warm room a week after operation, this cannot be regarded as evidence for a return of normal temperature regulation.

The postoperative hot box tests showed a marked loss in ability to regulate against overheating (Table 13). With rectal temperatures of 106° none of the cats panted and the respiratory rate remained slow. The information at hand does not necessarily show that the ability to pant was abolished by these lesions. Certainly the threshold for panting was raised above 106°, but in those cats in which the rectal temperature was forced sufficiently high, panting occurred. In a test made

TABLE 12

Cold box tests in cats with lateral lesions in caudal part of hypothalamus. Rectal temperature in F.°

(Reproduced from the *J. Neurophysiol.*)

	Preoperative tests					Postoperative tests					
Cat no.	Temp. at start	Temp. at end	Change	Shivering	Av. temp. of box	Days after operation	Temp. at start	Temp. at end	Change	Shivering	Av. temp. of box
46	102.3	99.7	−2.6	yes	44	6	100.3	96.9	−3.4	yes	42
						41	100.4	97.0	−3.4	yes	40
47	101.3	100.0	−1.3	yes	44	18	102.6	93.6	−9.0	no?	44
						123	102.6	92.9	−9.7	no?	39
100	101.5	101.8	+0.3	yes	46	20	102.3	94.6	−7.7	yes	41
						55	102.6	98.4	−4.2	yes?	40
107	101.8	102.1	+0.3	yes	41	26	103.8	96.2	−7.6	no	37
						84	102.8	100.0	−2.8	no	35
109	101.9	101.7	−0.2	no	42	33	102.3	98.0	−4.3	no	37
						68	102.5	94.7	−7.8	no	36
110	101.1	100.3	−0.8	yes	36	32	101.0	96.7	−4.3	no	41
						62	100.1	96.6	−3.5	no	36
114	101.7	101.0	−0.7	no?	34	33	102.7	94.6	−8.1	no	34

on Cat 46, 49 days after the operation, the box temperature was raised to 114° and panting began when the rectal temperature reached 109.7°. This test was made long after the operation and it is possible or even probable that had a similar effort been made to force panting within a week or so after the operation the animal's temperature would have reached a fatal level before panting occurred.

Medially placed moderate sized lesions in caudal part of hypothalamus caused little disturbance in temperature regulation. There were 3 cats with unilateral lesions destroying one mammillary body and one cat (103) with bilateral lesions destroying both mammillary bodies. The temperatures of all these cats during the postoperative days were normal except that one cat (105) with a unilateral lesion had a subnormal temperature on the first postoperative morning. Hot and cold box tests made from 2 to 4 weeks after the operation were normal. Earlier tests were not made in any of these cats except in one (Cat 25) with unilateral destruction of the mam-

TABLE 13
Hot box tests on cats with lateral lesions in caudal part of hypothalamus. Rectal temperature in F.°
(Reproduced from the J. Neurophysiol.)

Cat no.	Preoperative tests					Days after operation	Postoperative tests				
	Temp. at end of test	Rise in temp.	Final respiratory rate	Panting	Sweating		Temp. at end of test	Rise in temp.	Final respiratory rate	Panting	Sweating
46	105.1	2.1	208	yes	no	7	106.2	8.4	42	no	no
						20	106.0	7.5	30	no	no
47	103.4	1.6	300	yes	no	7	106.0	3.2	26	no	no
						42	106.0	3.0	28	no	no
100	102.0	−0.1	240	yes	yes	28	106.1	3.5	66	no	no
107	103.3	1.1	230	yes	no	26	106.3	6.2	30	no	no
						42	106.3	3.4	40	no	no
109	103.4	1.1	180	yes	no	33	106.0	5.4	20	no	no
						66	106.0	4.1	36	no	no
110	101.8	0.9	210	yes	yes	32	106.3	3.9	30	no	no
						56	106.0	4.4	30	no	no
114	102.5	0.9	240	yes	yes	35	106.0	1.9	140	no	no

millary body. On the second postoperative day this cat failed to pant in the hot box although its respiration was 156 when its temperature reached 106°. In a later test this cat reacted normally in the hot box and the poor performance on the second day is to be explained by a transient impairment of function in regions outside the anatomical lesion.

The chief interest lies in Cat 103 with bilateral destruction of the medial part of the caudal hypothalamus (Fig. 215). The lesions destroyed all of both mammillary bodies except the rostral tip of the one on the right and every thing dorsal to the mammillary bodies as far as the floor of the third ventricle. In this cat the lesion was in a position to interrupt any fibers which descend from the hypothalamus through the central gray matter of the aqueduct; but the main pathway which runs lateral and dorsolateral to the mammillary body was intact on both sides. This cat showed no disturbance in temperature regulation so far as could be determined by the daily

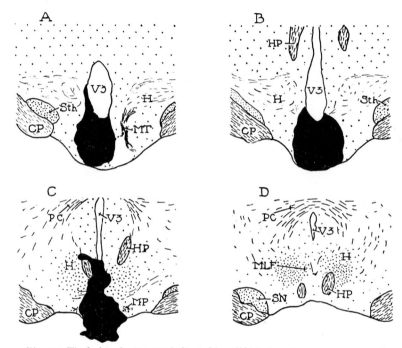

Fig. 215. The lesions in Cat 103 indicated in solid black on four transverse sections through the brain at the level of and behind the mammillary bodies. (Reproduced from the *J. Neurophysiol.*)

temperature records and by the hot and cold box tests made 30 days after the operation. It is quite possible that had these tests been made within a day or two after the operation transient abnormalities would have been detected.

Discussion. It is difficult to make accurate observations on sweating and shivering. Sweating, which in the cat occurs only on the pads of the feet and is never profuse, is often obscured by evaporation. In the preoperative tests shown in Table 13 only 3 of the cats were observed to sweat and 4 were not.

Hence no great importance can be attached to the fact that in the postoperative tests shown in the same table none of the cats were observed to sweat. It is not always easy to tell whether or not a cat is shivering. In the postoperative tests shown in Table 12, Cat 47 showed some twitching of the muscles which did not feel to the observer's hand like shivering and the observation was entered as a questionable negative. Shivering was not always detected in the preoperative tests. Two of the cats with lateral lesions in the caudal part of the hypothalamus did shiver during the postoperative tests. Hence it is not possible to say that these lesions abolished shivering although shivering occurred less often in the operated than in the unoperated cats.

Subnormal temperatures were observed on the first postoperative morning in none of the cats with medially placed lesions and in only one of the cats with laterally placed lesions in the anterior hypothalamus (Cat 20, Table 7). The cats with moderate sized laterally placed lesions in the posterior part of the hypothalamus showed a marked hypothermia the first morning after the operation, but there was a rather rapid recovery so that on the tenth day only 2 of them had temperatures below 101° (Table 7). It would appear that cats recover from the hypothermia caused by hypothalamic lesions more rapidly than do monkeys; but as will be shown later monkeys regain the ability to maintain normal body temperatures under ordinary room conditions in a few weeks.

Rectal temperatures of animals kept under ordinary room conditions do not furnish a satisfactory measure for their ability to regulate against cold. For this purpose cold box tests are required. Such tests made during the first postoperative week gave normal results in cats with medially placed anterior lesions (Table 8) and abnormally large drops in cats with symmetrical lateral lesions in the anterior part of the hypothalamus. Even after one month these cats with symmetrical anterolateral lesions still showed falls in temperature which were somewhat greater than normal (Table 11). By far the greatest chilling, however, was seen in the cold box tests on cats with laterally placed lesions in the posterior part of the hypothalamus (Table 12).

It is particularly significant that these cats with posterolateral lesions showed little tendency to recover the ability to prevent chilling in the cold box. In 2 of the cats listed in

Table 12 there was evidence of some recovery. But in one of these (Cat 100) the lesions were somewhat asymmetrical and in the other cat (107) the lesions were much smaller than in the other cats of this group. The recovery of ability to resist chilling in the cold box shown by these cats was evidently due in large part to the incompleteness of the lesions.

Hyperthermia was present in many of the cats on the morning after the operation. In the group with large anterior lesions 9 out of 20 cats had temperatures above 104.2°. Four out of 8 cats with medially placed moderate sized lesions in the anterior hypothalamus and 6 out of 8 cats with laterally placed moderate sized lesions in the anterior hypothalamus also had temperatures above 104.2 on this first morning. These high temperatures can scarcely be attributed to nonspecific results of the operation, because in a group of cats in which similar lesions were made in the thalamus by the same method the temperatures on the first postoperative morning were 102.1, 102.3, 103.6, 103.7 and 104.1° respectively. In these as in all the cats with hypothalamic lesions the rectal temperatures were taken on the first morning at the time the cats were removed from the incubator in which they had been kept over night so that the factor of environmental temperature was essentially the same in all cases. Under these conditions normal cats regulated their temperatures perfectly and many of the cats with hypothalamic lesions showed subnormal temperatures. The hyperthermia seen in some of the cats cannot, therefore, be attributed entirely to the high temperature of the incubator.

The location of the lesions in the animals with high temperatures on the first postoperative morning have varied a good deal. Large lesions in front of the anterior commissure (Cats 53 and 54), at the level of the anterior commissure (Cats 52 and 92) or in the suprachiasmatic hypothalamus (Cats 27 and 81) and smaller lesions either medially placed at the level of the infundibulum (Cats 11 and 18) or in the anterior part of the lateral hypothalamus (Cats 8 and 42) have frequently caused hyperthermia. It would seem most reasonable to attribute the rise in temperature to an irritation of the mechanism for heat conservation and heat production since in none of these animals was most of the hypothalamus destroyed and the most distant lesions were within 2 mm. of it. Many of these cats showed impaired capacity to regulate against heat as measured by the hot box tests, and this probably was a factor in permitting the high temperatures to develop. But this factor is not in itself sufficient to cause hyperthermia. In Cats 8, 20 and 40 the impairment of capacity to prevent overheating persisted for at least a month but hyperthermia lasted for only a few days in 8 and 40 and did not appear in 20.

The capacity to regulate against overheating as measured by the hot box test was very seriously disturbed in all but one of the cats with large anteriorly placed lesions which were subjected to this test. It is significant that in all but this one the region dorsal to the optic chiasma was extensively damaged. A month or more after the operation these cats failed to pant or to show much of an increase in respiratory rate when their temperatures were raised to 106° or higher (Table 6). The exception was Cat 53 in which the lesions were situated in front of the anterior commissure and a large part of the preoptic region remained intact. Cats from which the frontal lobes had been removed for another purpose (Magoun and Ranson, 1938) also showed normal reactions in the hot box.

Cats with smaller medially placed lesions in the anterior part of the hypothalamus showed after one month only a moderate increase in the panting level and in respiratory rate (Table 9). But cats with moderate sized symmetrical lesions in the lateral part of the anterior hypothalamus failed to pant and showed little increase in respiratory rate when their rectal temperatures had been raised to 106° one month after the operation (Table 10, Fig. 212). All of the cats with laterally placed lesions in the posterior part of the hypothalamus also showed greatly impaired capacity to regulate against heat (Table 13). This is in part a confirmation and in part an extension of the observations of Teague and Ranson (1936).

It has been shown that a region specifically responsive to heat exists above the chiasma and in the neighborhood of and below the anterior commissure. The results of the investigation on the effects of lesions agree very well with those obtained by heating the brain if it is assumed that the "center" in the suprachiasmatic and preoptic region, which is sensitive to heat, sends fibers backward by way of the medial forebrain bundle in the lateral hypothalamus, which are interrupted by such lesions as those illustrated in Fig. 212, 213 and 214.

EFFECTS OF THALAMIC LESIONS ON TEMPERATURE REGULATION

Since in some of the experiments discussed in the preceding section the lesions extended dorsally beyond the confines of the hypothalamus into the thalamus it became necessary to check again the work of earlier workers showing that the thalamus is not essential for temperature regulation.

With the aid of the Horsley-Clarke instrument, moderate sized lesions were placed bilaterally in various parts of the thalamus in 9 cats. In 6 other cats large lesions were made and in 3 of these in order to secure the maximum destruction without killing the animals, the operation was performed in two stages, first on one side and then the other side of the thalamus, with a period of 8 to 16 days intervening. A needle-like bipolar electrode with the bare tips of the constituent wires separated by 2 mm. along the long axis of the needle was used and through it a direct current of 3 ma. was passed for 1 minute to produce a lesion. In the 3 cats in which the greatest injury was inflicted 9 punctures were made on each side and 3 lesions were placed along each of the 3 lateral punctures and 2 along each of the other 6. The 21 lesions all fused together to form one very large lesion which after the second operation was united with a similar one on the opposite side (Clark, Magoun and Ranson, 1939b).

Daily records were made of the rectal and environmental temperature and at varying times after the operation tests were made to determine the ability of the animals to withstand heat and cold. Before they were sacrificed each of the 6 cats with large lesions were decorticated under ether and as they recovered from the anesthetic they were watched for decorticate panting. They were then killed and the brains prepared for microscopical study.

The 9 cats with moderate sized lesions may be considered together as group A and the 6 with large lesions as group B. In both groups the rectal temperatures were above the normal average on the first postoperative morning. In group A the temperatures were around or slightly above the upper limits of normal. In group B the 3 cats with the largest lesions had temperatures ranging from 104.4 to 105.1°F. on the first morning following each of the 2 operations. In no case was a subnormal temperature encountered. By the third day the cats began to eat voluntarily. In hot box tests all the cats of both groups panted in response to heat and although the panting level was in most instances higher than the average for normal cats it was in each case within the range of normal variation. The tests in the hot box, therefore, revealed no significant disturbance in temperature regulation. All of the cats reacted normally in the cold box and after 3 hours exposure to a temperature around 40°F. only one cat had a temperature as low as 99.8°F.

As a final test the 6 cats of group B were decorticated under ether anesthesia 28 to 53 days after the last operation. During recovery from the anesthesia decorticate polypneic panting occurred in all but one. Another panted only when the mouth was forced open. The failure of one cat to show decorticate panting, although it did pant in response to heat, cannot be attributed to the lesion, for the area destroyed in this cat was also destroyed in 3 other cats which did pant following decortication.

In the cats with the most extensive damage all of the medial portion of the thalamus was destroyed from the level of the anterodorsal nucleus backward to the posterior commissure and in one instance to the border of the superior colliculus (Fig. 216C, D). The medial nuclei of the thalamus, the habenular nuclei and posterior commissure were destroyed. The medial nuclei are the ones most intimately associated with the hypothalamus and it would be to them that one would look for any control which the thalamus might exert over body temperature. The lateral nuclei which serve to relay impulses to the cerebral cortex would not be likely to be concerned in temperature regulation. Moreover decorticate cats in which the lateral thalamic nuclei had been removed or had undergone degeneration regulated body temperature in an apparently normal manner although they were more inclined to shiver than normal cats in a cool environment (Pinkston, Bard and Rioch, 1934; Bard and Rioch, 1937). These decorticate cats panted when they became overheated.

There would seem, therefore, to be no reason to attribute any essential part of temperature regulation to the thalamus if it were not for the observations of Lilienthal and Otenasek (1937) who found that the polypneic panting, which occurs in acutely decorticate cats, was not abolished by removal of the hypothalamus to the level of the caudal border of the mammillary bodies so long as the caudodorsal part of the thalamus

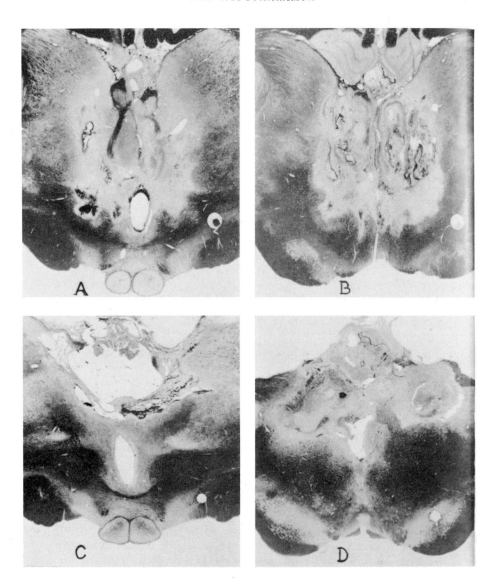

Fig. 216. A and B represent photomicrographs from transverse sections at the level of the lower border of the mammillary bodies (A) and at a level slightly caudal to the mammillary body (B) of the brain of a cat with thalamic lesions. C and D represent photomicrographs from transverse sections at the level of the mammillary body (C) and at the level of the third nerve (D) of the brain of another cat with thalamic lesions. (Reproduced from the *J. Neurophysiol.*)

remained intact. Removal of this part of the thalamus abolished panting. On the basis of these observations they postulated the existence of a center for polypneic panting in "the caudodorsal portion of the thalamus, an area which lies below the habenular complex and surrounds the anterior part of the iter." But, since our cats in which this part of the thalamus was destroyed, panted in a normal manner in hot box tests and showed typical polypneic panting when decorticated, it cannot be said to contain a center which is essential for panting.

Special interest attaches to the cat with the lesions shown in Fig. 216A, B. These lesions left only the most ventral part of the zone of transition between the hypothalamus and mesencephalic tegmentum intact. Farther forward the fields of Forel were destroyed but the region dorsolateral to the mammillary bodies was bilaterally intact. The fact that this cat showed no disturbance in temperature regulation is to be attributed to the integrity of these regions. Although the lesions in the hypothalamus which most seriously disturbed temperature regulation (Fig. 213 and 214) extended dorsally into the fields of Forel and even into the thalamus proper, the normal temperature regulation in the cat with the lesions illustrated in Fig. 216A, B show that these more dorsal parts may be destroyed without impairing thermostasis. The pathways essential for temperature regulation occupy a ventral position at the level of transition from the hypothalamus into the mesencephalon. Temperature regulation persists after destruction of the central gray matter at this level.

OBSERVATIONS ON MONKEYS

Since cats depend on panting instead of sweating as a means of eliminating body heat and in this respect differ fundamentally from primates, one cannot draw conclusions in regard to temperature regulation in man from observations on cats alone. For this reason information about temperature regulation in monkeys is of special importance. Unfortunately our observations on the monkey (Ranson, Fisher and Ingram, 1937) antedated those on the cat and are less complete than they would have been if we had had more experience at that time in this line of investigation.

[*Editor's Note:* In the material omitted here, Ranson details the results of his experiments on monkeys.]

From all these observations it becomes clear that postoperative hyperthermia develops in the monkey when bilateral lesions are made in the lateral part of the rostral portion of the hypothalamus. Hypothermia develops when the bilateral lesions are situated dorsolateral to the mammillary bodies. The more caudally placed lesions cause impairment of ability to regulate against both heat and cold.

DISCUSSION AND SUMMARY

The observations which have been presented in the preceding pages indicates that temperature regulating functions are more or less segregated anteroposteriorly so that anteriorly placed lesions are likely to impair regulation against heat while regulation against cold remains intact. The disturbance after caudally placed lesions involves regulation in both directions. Evidence in favor of a somewhat similar anteroposterior segregation of function has been presented by Barbour (1939) and by Erickson (1939).

The neural mechanism which is concerned in preventing overheating of the body includes the region above and in front of the optic chiasma and below the anterior commissure shown in Fig. 207. Since local heating of this region causes panting and sweating it is reasonable to assume that similar heat-loss activities are initiated when it is warmed by overheated blood. Damage to this region impairs the cat's ability to protect itself from overheating and its temperature may reach 106° or higher without panting. Since medially placed lesions in the hypothalamus behind the chiasma do not prevent panting, the descending pathway from this heat-sensitive region, or at least the major part of this pathway, does not run backward through the medial part of the hypothalamus. On the contrary it must enter the lateral part of the hypothalamus far forward and run backward through it, since it is interrupted by lesions which destroy the lateral hypothalamus either at the level of the optic chiasma (Fig. 212) or at the level of the mammillary body (Fig. 214). Lesions in the caudal part of the lateral hypothalamus are most effective in eliminating regulation against heat and this may indicate that the path from the part of the responsive area enclosed in the dotted lines (Fig. 207) also descends through the lateral hypothalamic area. The motor mechanism responsible for coordinated panting movements is certainly not destroyed by any of these lesions. It lies caudal to the hypothalamus probably in the mesencephalon (Clark, Magoun and Ranson, 1939b). This motor center is still capable of functioning after the heat-sensitive center has been cut away. It is well known that, immediately after decerebration through the caudal part of the hypothalamus, this motor panting center is freed from cortical inhibition, and decerebrate panting may occur although the body temperature is subnormal.

The neural mechanism which protects the body against chilling is not seriously impaired by moderate sized lesions in the hypothalamus unless these are bilateral and are located in the caudal part of the lateral hypothalamus as illustrated by cats 47 and 110. Very large lesions anywhere in the hypothalamus cause a loss of the capacity to keep the body temperature up to normal, and this loss is more pronounced when the lesions are caudally placed. We interpret these observations as meaning that the neurons concerned are distributed through a large part of the hypothalamus and are perhaps in part at least identical with those which are responsible for vasoconstriction and piloerection and some other sympathetic functions. The fact that the chief efferent sympathetic path from the hypothalamus lies lateral and dorsolateral to the mammillary bodies in the region, the destruction of which with moderate sized lesions causes marked impairment in temperature regulation, fits in perfectly with this conception. It cannot, however, be definitely asserted that medially placed lesions do not interrupt any descending fibers concerned with regulation against cold. We can only say that the interruption of any such medially placed fibers as may exist, does not cause an obvious impairment of regulation against cold under the conditions of our experiments.

From the information at hand the most probable explanation of the transient hyperthermias seen in cats, monkeys and man after lesions in the anterior hypothalamus and the suprachiasmatic region is that these lesions cause an irritation of the centers for heat formation and heat conservation in the hypothalamus and that the resultant rise in temperature fails to activate the heat loss mechanism since the corresponding center has been destroyed or at least cut off from lower lying centers (Ranson and Magoun, 1939). Bilateral destruction of the caudal part of the lateral hypothalamus by large lesions like those shown in Fig. 213 largely destroys the centers for heat formation and conservation and interrupts the descending paths from these centers as well as those from the center for regulation against heat in the suprachiasmatic and preoptic region. Such lesions cause a loss in the capacity to regulate against both heat and cold.

Cats and monkeys with lesions in the lateral part of the

posterior hypothalamus showed after days or weeks more or less recovery of the ability to maintain normal body temperatures when the environmental temperature was not too low. This was probably in part due to the incompleteness of the lesions and in part also to decrease in peripheral vasodilation and to compensatory endocrine activity. A study of temperature regulation in spinal cats has shown that a very slow adjustment of temperature can occur. Unlike the nervous regulation this requires many hours or even days and it is possible that stimulation of the thyroid by subnormal body temperature is a factor.

The question may be raised how far are the conclusions reached as a result of investigations on cats applicable to man. In the first place it must be recognized that, while the processes involved in heat production and heat conservation are essentially alike in the cat and man, the processes involved in heat loss are different. Man with his hairless skin and abundant sweat glands is able to lose heat rapidly through the skin; but the cat depends on panting and the evaporation of water from the tongue, mouth and respiratory passages. It sweats only on the pads of the feet. Because of this difference in the peripheral mechanism of heat loss it is not safe to assume that the central control is the same in the two forms. More work needs to be done on the monkey which sweats but does not pant.

Nevertheless, some indication that the center controlling heat loss may have a location in the monkey similar to that in the cat, is that in the monkey just as in the cat lesions in the anterior part of the lateral hypothalamus cause hyperthermia.

There are better reasons for the belief that the center which protects against chilling has the same location in primates as in carnivores. In the first place the peripheral mechanism is alike. In the second place it is known that in monkeys as in the cat the lesions which cause the greatest loss in ability to prevent chilling are located in the caudal part of the lateral hypothalamus (Ranson, Fisher and Ingram, 1937). A mechanism which is alike in the cat and monkey is not likely to be greatly different in man.

Clinicians know that extensive damage to the hypothalamus, such as is sometimes produced by tumors, causes subnormal body temperature. But autopsy findings in such cases have revealed diffuse lesions, which are not well placed for an analysis of the problem of temperature regulation. An instructive case was reported by Davison and Selby (1935). During the patient's stay in the hospital he had a consistently low temperature. During the greater part of one month the temperature remained at about 92.4°. At autopsy an angioma was found which largely destroyed the hypothalamus. In the figure representing a section through the anterior part of the mammillary bodies, these bodies can scarcely be recognized; and it is stated that the lateral hypothalamic area was destroyed. In a section through the caudal part of the mammillary bodies "the same changes were noted as in the preceding sections, except that both mammillary bodies were easily identified at this level." This extensive damage to the hypothalamus, extending far enough back to destroy the lateral hypothalamic area at the level of the mammillary bodies was, we believe, the cause of the hypothermia.

Two very instructive cases of a different kind have been reported by Alpers (1936). In both cases operations on suprasellar tumors were followed by rapidly developing hyperthermia and death. In both cases there was extensive destruction of the gray matter surrounding the third ventricle behind the optic chiasma. These cases show that this part of the hypothalamus cannot be responsible for heat production and conservation or else its destruction would have caused a fall instead of a rise in temperature. On the basis of the evidence furnished by the experiments on cats and monkeys, by the hypothermic patient reported by Davison and Selby and by Alpers' two cases of hyperthermia we may safely conclude that the same part of the hypothalamus protects the body against chilling in carnivores and primates.

CONCLUSIONS

The autonomic regulation of body temperature is for the most part under the control of the CNS; and the hypothalamus is the chief center for the integration of this control. All of the brain in front of the hypothalamus can be removed in animals without seriously affecting this nervous regulation but if the hypothalamus is removed this regulation is very greatly impaired or entirely abolished. It is also very greatly impaired by transection of the cervical spinal cord since this interrupts the descending paths from the hypothalamus to the sympathetic system which controls piloerection, vasoconstriction and sweating.

The sympathetic centers in the hypothalamus which regulate the caliber of the cutaneous blood vessels and the erection of the hair are in control of the conservation of the body heat. Bilateral lesions in the hypothalamus which extensively damage these sympathetic centers impair the ability of the body to prevent chilling on exposure to cold. This impairment is greatest when these bilateral lesions are placed in the caudal part of the lateral hypothalamus so as to interrupt the descending sympathetic pathways from the hypothalamus.

The rate of heat loss can be increased in carnivores by panting and in primates by sweating. In cats panting can be induced by local heating of the region between the optic chiasma and anterior commissure and destruction of this region greatly impairs the cat's ability to pant and thus to prevent overheating of the body. Bilateral lesions in the lateral hypothalamus anywhere from the level of the optic chiasma to the level of the caudal border of the mammillary bodies has the same effect, probably because of the interruption of fibers which descend from the region between the optic chiasma and the anterior commissure to a panting center which lies somewhere behind the hypothalamus. In cats bilateral lesions dorsolateral to the mammillary bodies greatly impair the animal's ability to protect itself against both heat and cold, while bilateral lesions in the lateral hypothalamus above the optic chiasma interfere with regulation against heat but leave regulation against cold relatively intact.

It seems not unlikely that the region between the optic chiasma and anterior commissure may preside over heat loss by sweating in the primates but no satisfactory evidence for this has been presented. Hyperthermia and hypothermia may be used to designate body temperatures which run respectively above and below normal when the body is at rest in a comfortably warm room. In the cat, monkey and man lesions in the neighborhood of the anterior commissure are likely to produce a transient hyperthermia which is best explained as due to an irritation of the sympathetic hypothalamic centers associated with damage to the region which is responsible for controlling heat loss or of the descending paths from this region in the anterior part of their course. In man as in the animals lesions in the caudal part of the lateral hypothalamus cause hypothermia due to the destruction of the descending sympathetic paths from the hypothalamus. In such cases the associated damage to the heat loss mechanisms does not come into evidence unless the body is exposed to excessive external heat or is engaged in vigorous muscular exercise.

Nervous regulation of body temperatures as effected through changes in the caliber of the cutaneous blood vessels, shivering, sweating and panting bring about rapid adjustments which almost immediately compensate for rapid changes in environmental temperature as well as for the increased heat production involved in increased muscular activity. In addition to this rapid regulation there is a supplementary very slow adjustment apparently hormonal in nature which enables cervical spinal animals to maintain normal temperatures when variations in environmental temperature occur very gradually and within the narrow range ordinarily regarded as comfortable.

[*Editor's Note:* The discussion has been omitted.]

REFERENCES

Alpers, B. J. *Arch. Neurol. Psychiat.*, Chicago, 1936, *35*, 30–42.
Barbour, H. G. *Arch. exp. Path. Pharmak.*, 1912, *70*, 1–26.
Barbour, H. G. *Amer. J. Physiol.*, 1939, *126*, 425P.
Bard, P., and Rioch, D. McK. *Johns Hopk. Hosp. Bull.*, 1937, *60*, 73–147.
Bazett, H. C. *Physiol. Rev.*, 1927, *7*, 531–599.
Bazett, H. C., Alpers, B. J., and Erb, W. H. *Arch. Neurol. Psychiat.*, Chicago, 1933, *30*, 728–748.
Bazett, H. C., and Penfield, W. G. *Brain*, 1922, *45*, 185–265.
Burton, A. C. *Ann. Rev. Physiol.*, 1939, *1*, 109–130.
Cannon, W. B. *The wisdom of the body*, New York, Norton and Co., 1939.
Clark, G., Magoun, H. W., and Ranson, S. W. *J. Neurophysiol.*, 1939a, *2*, 61–80.
Clark, G., Magoun, H. W., and Ranson, S. W. *J. Neurophysiol.*, 1939b, *2*, 202–207.
Davison, C., and Selby, N. E. *Arch. Neurol. Psychiat.*, Chicago, 1935, *33*, 570–591.

DuBois, E. F. *Basal metabolism in health and disease*, Philadelphia, Lea and Febiger, 1936.
Dworkin, S. *Amer. J. Physiol.*, 1930, *93*, 227-244.
Erickson, T. C. *Brain*, 1939, *62*, 172-190.
Frazier, C. H., Alpers, B. J., and Lewy, F. H. *Brain*, 1936, *59*, 122-129.
Glaubach, S., and Pick, E. P. *Arch. exp. Path. Pharmak.*, 1933, *173*, 571-579.
Hammouda, M. *J. Physiol.*, 1933, *77*, 319-336.
Hasama, B. *Arch. exp. Path. Pharmak.*, 1929, *146*, 129-161.
Hashimoto, M. *Arch. exp. Path. Pharmak.*, 1915, *78*, 394-444.
Isenschmid, R., and Schnitzler, W. *Arch. exp. Path. Pharmak.*, 1914, *76*, 202-223.
Kahn, R. H. *Arch. Anat. Physiol., Lpz., Physiol. Abt. Suppl.*, 1904, *81* 134.
Keller, A. D. *Amer. J. med. Sci.*, 1933, *185*, 746-748.
Keller, A. D. *J. Neurophysiol.*, 1938, *1*, 543-557.
Lilienthal, J. L., Jr., and Otenasek, F. J. *Johns Hopk. Hosp. Bull.*, 1937, *61*, 101-124.
Magoun, H. W., Harrison, F., Brobeck, J. R., and Ranson, S. W. *J. Neurophysiol.*, 1938, *1*, 101-114.
Magoun, H. W., and Ranson, S. W. *J. Neurophysiol.*, 1938, *1*, 39-44.
Moore, L. M. *Amer. J. Physiol.*, 1918, *46*, 253-274.
Moorhouse, V. H. K. *Amer. J. Physiol.*, 1911, *28*, 223-234.
Pinkston, J. O., Bard, P., and Rioch, D. McK. *Amer. J. Physiol.*, 1934, *109*, 515-531.
Prince, A. L., and Hahn, L. J. *Amer. J. Physiol.*, 1918, *46*, 412-415.
Ranson, S. W., and Clark, G. *Proc. Soc. exp. Biol., N. Y.*, 1938, *39*, 453-455.
Ranson, S. W., Fisher, C., and Ingram, W. R. *Arch. Neurol. Psychiat., Chicago*, 1937, *38*, 445-466.
Ranson, S. W., and Magoun, H. W. *Ergebn. Physiol.*, 1939, *41*, 56-163.
Sachs, E., and Green, P. P. *Amer. J. Physiol.*, 1917, *42*, 603.
Sherrington, C. S. *J. Physiol.*, 1924, *58*, 405-424.
Teague, R. S., and Ranson, S. W. *Amer. J. Physiol.*, 1936, *117*, 562-570.
Thauer, R., and Peters, G. *Pflüg. Arch. ges. Physiol.*, 1937, *239*, 483-514.
Uotila, U. U. *Endocrinology*, 1939, *25*, 605-614.

Behavioral thermoregulation in response to local cooling of the rat brain[1,2,3]

EVELYN SATINOFF

Psychology Department, University of Pennsylvania, Philadelphia, Pennsylvania

SATINOFF, EVELYN. *Behavioral thermoregulation in response to local cooling of the rat brain.* Am. J. Physiol. 206(6): 1389–1394. 1964.—Local cooling of the anterior hypothalamic-preoptic area in rats, at environmental temperatures of 5 and 24 C, caused their rectal temperatures to increase as much as 3.1 C, as well as vigorous shivering. When the animals were allowed to press a bar to turn on a heat lamp directly overhead, they pressed more at both ambient temperatures when their brains were cooled than when they were not, although they worked harder in the cold. They shivered continuously during brain cooling at either temperature. The behavioral and physiological temperature regulations appeared to be complementary, since the same temperature levels were reached whether or not heat could be obtained voluntarily. Central cooling produced, in addition to the usual reflex mechanisms of increased body temperature and shivering, the behavioral motivation for heat.

temperature regulation   hypothalamic cooling   anterior hypothalamus   heat reinforcement

---

MUCH ATTENTION HAS BEEN FOCUSED on the anterior hypothalamic-preoptic region as a central temperature-sensitive device that maintains thermal homeostasis. An extensive series of experiments on chronic cats and monkeys by Ranson and his colleagues (14–16) showed that lesions in this area destroyed the adaptive response to body cooling. Magoun et al. (13) localized the heat-sensitive elements to the region between the optic chiasm and the anterior commissure. Diathermic warming of this area, that resulted in a moderate rise in temperature of the anterior hypothalamus but not of other parts of the brain, evoked in anesthetized cats the heat-loss responses of panting, sweating, and accelerated respiratory rate.

Received for publication 16 October 1963.

[1] This research was supported, in part, by National Science Foundation Grant NSF G 24386, and was conducted during the author's tenure as a Public Health Service Predoctoral Fellow.
[2] This paper is based on a dissertation submitted to the Psychology Department of the University of Pennsylvania in partial fulfillment of the requirements for the Ph.D. degree.
[3] A preliminary report has been published (*Federation Proc.* 22: 283, 1963).

The results of Magoun et al. have since been extended to other animals and to both chronic and unanesthetized preparations; they have been confirmed many times. Hemingway et al. (11) reported that diathermic heating in the anterior hypothalamus of unanesthetized dogs inhibited shivering and peripheral vasodilatation. The work of Magoun and his colleagues was repeated by Beaton et al. (5) and Eliasson and Strom (6) on monkeys and dogs, respectively. Both groups concluded that the thermosensitive area lies mainly between the anterior commissure and the optic chiasm.

More recently, Hardy and his associates (9, 10) conducted a series of carefully controlled studies on the effects of local heating on unanesthetized dogs placed in different environmental temperatures. They found that although the body temperature of the dogs was lowered in response to anterior hypothalamic heating, the physiological responses were to a large degree determined by the environmental temperature. In a cool environment the metabolic rate decreased and there was periodic cessation of shivering. In a neutral environment there was vigorous panting. In a warm environment panting was extreme and the metabolic rate increased slightly. These studies clearly show the interaction of central and peripheral thermal effects.

Local cooling of the anterior hypothalamus also has a profound effect on thermoregulation, producing shivering, increased body temperature, and vasoconstriction at different environmental temperatures. Andersson and his co-workers (1, 3) conducted several important experiments on the effects of cooling the brains of unanesthetized goats. All centrally cooled goats developed a sustained hyperthermia, but without shivering, which lasted up to a week, until cooling was stopped, and then gradually declined. Shivering appeared only in a cold environment or when parts of the body were cooled locally. Peripheral vasoconstriction occurred in all experiments, even in a warm environment, however. Hammel et al. (8) obtained vigorous shivering and peripheral vasoconstriction in dogs kept at a neutral environmental temperature by cooling the same hypothalamic sites where heating had previously inhibited shivering in the cold.

All of the above work has investigated physiological responses to various internal and external thermal loads, e.g., hypothalamic cooling or heating, lesions, high or low ambient temperatures. In the past few years several studies have begun to emphasize behavioral responses to the same sorts of thermal stresses (7, 12, 17–19). Weiss and Laties, in a number of experiments, have demonstrated how behavior contributes to temperature regulation. Typically they measured the latency from the time a rat was placed in a cold environment until it began pressing a lever regularly to receive a burst of heat from an infrared lamp, as well as the total amount of heat obtained by the rat. The willingness with which rats will press the bar under different experimental conditions is an indication of their motivation to keep warm.

The present study has two purposes. The first is to extend to the rat the study of the central mechanism subserving thermoregulation by the method of local cooling of the hypothalamus. The second is to discover whether stimulation of central thermal receptors, in addition to arousing physiological reflex mechanisms, also motivates rats to keep warm.

METHODS

Animals. Nine female rats of the Long-Evans strain, weighing between 250–350 g, were used. They were shaved once or twice a week, as needed, and they were maintained at 80–90% preoperative body weight, since both shaving and food deprivation increase the rats' bar-pressing behavior.

Construction of thermode. Each thermode consisted of 5 cm of 22-gauge stainless steel tubing which, after heating, was bent into a tight U shape. The arms of the thermode extending above the skull were bent away from each other slightly to facilitate connection to polyvinyl tubing. The thermode was insulated with tight-fitting polyethylene tubing from the point where it left the brain to within 1 mm of the tip, a distance of approximately 8 mm. A picture of the thermode is shown in Fig. 1.

Method of implantation. A standard dose of 0.40 ml atropine sulfate U.S.P. (concn. 0.4 mg/ml) was administered intraperitoneally to each rat and 10 min later it was anesthetized with Evipal in a dosage of 140 mg/kg. The rat was then placed in a stereotaxic instrument, its scalp incised, and a large hole drilled over the superior sagittal sinus 1 mm anterior to the bregma. Several smaller holes were drilled in other parts of the skull and jewelers' screws were screwed into them. The thermode was implanted 9 mm anterior to the interaural plane, 7.5–8 mm below the surface of the cortex and as close as possible to the superior sagittal sinus, i.e., approximately 0.5 mm lateral to the midline. The thermode was attached to the skull with dental cement, which also connected to the jewelers' screws, thus making a more secure foundation. After the operation, an intra-

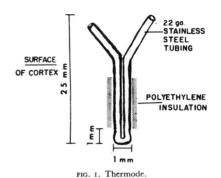

FIG. 1. Thermode.

TABLE 1. *Average rectal temperatures of rats whose brains were cooled in ambient temperature of 24 ± 3 C*

ANIMAL	NO. OF OBSERVATIONS	AVERAGE RECTAL TEMPERATURE °C				MAXIMUM RISE IN SINGLE SESSION °C
		BEFORE COOLING	DURING COOLING		30 MIN AFTER COOLING	
			AFTER 30 MIN	AFTER 60 MIN		
WT 4	2	37.3	38.5	39.1	37.8	1.7
WT 5	2	37.3	38.4	39.3	37.0	2.1
WT 6	3	37.4	39.0	39.1	37.7	1.9
WT 7	1	36.9	39.0	39.2	37.9	2.2
WT 10	4	38.1	40.3	40.5	38.1	3.1
WT 11	2	37.2	38.6	39.3	37.3	2.4
WT 12	2	37.8	38.5	39.6	37.8	1.8

muscular injection of 0.4 ml procaine penicillin (300,000 U/ml) was administered, and the rat was allowed to recuperate from the operation for at least 5 days before any training was begun.

Method of cooling. The brain was cooled by pumping cold fluid through silicone tubing into one arm of the thermode and back to a collection reservoir through the other arm. The tube began at a reservoir of 100% ethyl alcohol which was located in a bath of dry ice and acetone (kept at −60 to −70 C), and ran through a peristaltic pump, from which it led to the inflow arm of the thermode. Since the silicone tubing had an inner diameter of $\frac{1}{16}$ in. and was, therefore, much larger than the 22-gauge stainless steel tubing, it was joined by a connector to a small length of 24-gauge polyvinyl chloride tubing which fit tightly over the thermode. The same type of PVC tubing led from the outflow arm of the thermode back to the reservoir. All this tubing was supported by counterweighted wires passing over pulleys; thus, the rats were able to move freely without undue strain on their heads. The temperature of the liquid passing through the thermode was measured three times. Thirty-six-gauge thermocouple wire was inserted in the connection between the PVC tubing and the inflow arm of the thermode and was connected to a Leeds-Northrop potentiometer. The temperature was found to be 18 ± 2 C, depending on the temperature of the room.

Behavioral techniques. Two identical experimental cages,

similar to those of Weiss and Laties (19), were used. A Plexiglas cylinder, 1 ft high and 9 in. in diameter, was mounted on a grid formed of Plexiglas rods located 1/4 in. apart. This was the test cage. A 3-in. length of Plexiglas rod, 3/8 in. in diameter, was mounted on a microswitch and served as a lever. It was located 1 in. above the grid floor and extended about 1 1/2 in. into the cage. (At first, metal cages and levers were used, but they rapidly became very cold and the rats would not approach the levers.) A 375-w, General Electric, red bulb, infrared lamp was mounted on top of the cage. This lamp was lighted whenever and for as long as the lever was depressed. Its wattage was controlled by a variable transformer, and generally it was kept at 300 w. Each rat had at least three sessions each of at least 3 hr duration with the apparatus housed in a refrigerator at 5 ± 2 C. The refrigerators had windows in their doors so that the rats could be observed. The same cage was used for each rat in all experiments.

Test of cooling technique. The test occurred at room temperature (24 ± 3 C) in the test cage but without the bar, with the rat attached to the cooling apparatus. Rectal temperature was recorded initially and after 15 min. If both readings were the same, cooling was begun. If the second reading was higher, sometimes the animal was kept in the cage until his temperature returned to normal before starting the cooling process; other times, cooling was begun immediately and the higher temperature was taken as the base temperature for that session. The second reading was never lower than the first. After 1 hr, the cooling was terminated, and 30 min later the animal's rectal temperature was taken and it was returned to its home cage. Rectal temperature was also taken after 30 and 60 min of cooling.

Tests for bar pressing for heat during local cooling. These bar-pressing tests were conducted in the test cage in a room at 24 ± 3 C and in a refrigerator at 5 ± 2 C on alternate sessions. At least 2 days elapsed between sessions. Following 10 min of access to the bar, the brain was cooled for 10 min with the rats having access to the bar. This procedure was repeated three times, yielding three 10-min intervals of no cooling followed by 10 min of brain cooling.

Measurements taken. The rats' behaviors were recorded by means of written protocols. A record of the amount of time that the heat lamp was on was obtained from a running time meter and of the pattern of responding from a cumulative recorder. The apparatus controlling the cumulative recorder was arranged so that the recorder pen moved upward steadily as long as the lamp was on. A visual record was thus obtained in addition to the numerical one furnished by the running time meter. Rectal temperatures were measured before and after each session, and irregularly at 10-min intervals during a session. They were obtained from a Yellow Springs telethermometer, model 43, connected to a Yellow Springs no. 402 rectal probe, inserted 5 cm into the anus. The telethermometer is accurate to 0.5 C.

Histology. After the tests were completed, four of the rats were killed by perfusing saline and 10% formalin through the heart, and their brains were removed, embedded in parlodion, and sectioned coronally at 40 mμ. Every fifth section and every section showing signs of the thermode was stained with thionin and mounted on a slide.

RESULTS

One-hour brain cooling at 24 C ambient temperature. All animals in the neutral environment and with no opportunity to bar press for heat showed a marked rise in rectal temperature after 1 hr of brain cooling (Table 1). In three of seven cases at least 75% of the maximum elevation was achieved within 30 min of cooling. Half an hour after the cooling ceased, the rectal temperatures were within a degree of what they had been before the test.

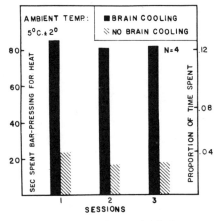

FIG. 2. Amount of time spent bar pressing for heat at 5 ± 2 C.

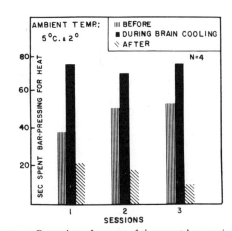

FIG. 3. Comparison of amount of time spent bar pressing for heat before, during, and after brain cooling. Ambient temperature: 5 ± 2 C.

This rise in temperature during cooling and the subsequent drop afterward occurred in all animals on all tests. The maximum increase in temperature in any single session varied from 1.7 to 3.1 C for individual animals. The temperature rises were extremely consistent so that all animals on each test were within 0.3 C of their maximum.

All rats began shivering within 1–5 min after cooling began. Generally they were highly active at first, with slight shivering episodes interpolated between grooming and exploring, but soon they moved near the wall of the cage where they remained huddled in a ball for the rest of the hour. The shivering increased gradually in intensity, then reached a constant level which held throughout most of the session except near the end of the hour when it usually decreased both in intensity and in the frequency of the episodes.

Heat reinforcement in a cold environment. The main results of the heat reinforcement study are shown in Fig. 2. *t* Tests were performed on these data as well as those shown in Figs. 3 and 4, reported later. All the differences were found to be significant at the .005 level with *t*s of 5–30 with 4 df. On the average, the rats depressed the lever, and so kept the heat lamp on for 81 sec out of 10 min during brain cooling and only 17 sec out of 10 min when there was no cooling. In the refrigerator they shivered continuously during brain cooling, even while they held the lever down and received heat. When the cooling was stopped, the rats stopped shivering. At no time was shivering observed at 5 C after a period of brain cooling, although it was always seen at the start of the session.

After 10 min of brain cooling, temperatures were up several degrees, and after 10 min of no cooling they were reduced, but not to the precooling levels. Figure 3 compares the amount of heat received in the first three successive 10-min periods of each session: before, during, and immediately after cooling. The interesting comparison is between the pre- and postcooling periods. Before cooling, rectal temperatures were at normal levels of from 37 to 38 C. At the end of each cooling period, they had risen to higher temperatures ranging from 39 to 40.5 C. During the postcooling periods the rate of bar pressing was greatly suppressed, and the rectal temperatures had dropped to lower temperatures ranging from 38.3 to 39 C. This sequence of rising and falling temperatures was continued during the three alternations of cooling over the entire test session, as measured for one session for each rat. These data are summarized in Table 2.

Heat reinforcement in a neutral environment. In general,

TABLE 2. *Comparison of rectal temperature, bar pressing, and shivering before, during, and after first 10-min period of brain cooling*

	Precooling	Cooling	Postcooling
Rectal temp., °C	37–38	39–40.5	38.3–39
Mean duration of bar pressing, sec	48	81	17
Shivering	slight	vigorous	none

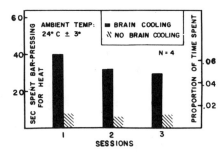

FIG. 4. Amount of time spent bar pressing for heat at 24 ± 3 C.

when a rat was placed in the test cage at a room temperature of 24 C, it pressed the bar a few times and then ignored it, and either explored the cage or sat quietly. When cooling began it became highly active, started to shiver violently, and after a few minutes went to the bar and held it down. Bar pressing was intermittent throughout this interval. During the next 10 min of no cooling the animal sat quietly or groomed. When brain cooling was resumed the animal again went to the bar and pressed it steadily throughout the cooling interval as shown in Fig. 4. The amount of heat for which the animals worked at room temperature during brain cooling, however, was less than in the refrigerator.

Figure 5 presents the cumulative record of bar pressing for a typical animal during two sessions, one in the cold, the other in the neutral environment. It shows clearly the phenomena just described. The two records are closely parallel, except that the rate in the neutral environment is less than in the cold. In the noncooling periods the rate in the neutral environment is essentially zero. The time spent pressing during these periods resulted from perseverance when the cooling was terminated and from, presumably, random depressions of the bar as the animal moved around the cage.

Histology. Four animals' brains were studied histologically. In each case the implant was in the anterior hypothalamic area. The most anterior extent of the thermode in any animal was in the preoptic region and the most posterior extent in the anterior hypothalamic area. All four animals gave good body-temperature responses during 1 hr of brain cooling at 24 C ambient temperature.

DISCUSSION

There are two main findings in the present study, one primarily physiological, the other behavioral. It has been demonstrated that local cooling of the anterior hypothalamus and preoptic area of rats in a neutral environment elicits responses designed to increase heat production. The animal begins to shiver almost immediately after the onset of cooling and shivers more and more vigorously and regularly as cooling progresses. Body temperature increases by as much as 3 C and remains elevated for the duration of the stimulation, up to 1

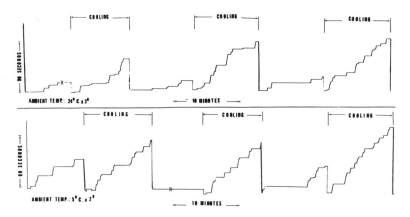

FIG. 5. Cumulative record showing amount of bar pressing in cooling and noncooling intervals for one animal.

hr, but quickly subsides when cooling is stopped. In a cold environment, the body temperature of a normal animal generally rises only a fraction of a degree in an hour, and shivering develops at a much slower and more sporadic rate. The two response patterns seem to be essentially the same, differing only in magnitude. Presumably this difference occurs because it is easier to raise body temperature in a neutral than in a cold environment, in the sense that less heat is lost in the neutral environment and so less energy is needed to effect a given increase in body temperature. In fact, the increase in body temperature of rats cooled in a cold environment is not any greater than it is when they are cooled in a neutral environment. There seems to be a maximum increase in rectal temperatures which cannot be exceeded by increasing the flow of the coolant beyond a certain rate. Some rats achieve this maximum temperature rise within half an hour.

Such a "temperature ceiling" has been noted consistently. Hardy (9) remarked that an increase of 0.5–1.0 C in rectal temperature was the maximum observed with anterior hypothalamic cooling in dogs regardless of the temperature of the environment. Andersson et al. were able to raise the rectal temperatures of goats only 0.5–1.0 C, either by electrical stimulation of the septal area or by local cooling of the anterior hypothalamic-preoptic region (2, 4). These results were interpreted as indicating the importance of peripheral thermal inputs which limit the degree to which central stimulation has an effect.

At least two other possibilities are tenable, however. One reason that the limit of temperature increase in dogs and goats is at least 2 C lower than that obtained in rats may be that the size of the thermode used in rats was much larger relative to the size of the rat's brain than the comparable thermode in the dog or goat brain. Thus, in the rat, a larger number of thermal receptors were excited, causing a larger thermoregulatory response.

Since experiments on local cooling have all used unilateral cooling, the observed limit could also reflect, at least in part, an antagonistic role of the noncooled side of the anterior hypothalamus. Because that side is not receiving a cold signal, it may oppose the temperature rise brought about by central stimulation resulting from cooling of the other side. If bilateral brain cooling produced a greater increase in body temperature than has been found so far, we could credit at least part of the apparent temperature ceiling to action of the opposite anterior hypothalamic area.

The rat is an excellent animal for this work because its temperature can be raised above its basal level by as large an increase as 3 C, and therefore it is easier to examine the effects of experimental manipulation. Also, the increase with fixed cooling is approximately constant and can be repeated at will, as can the shivering. In contrast, Hardy (9) reports that a stimulus that had produced excellent responses on one day with a dog would be ineffective another day. In rats a raised temperature is observed whenever the hypothalamus is cooled, and thus the rat's responses to central stimulation are easier to observe.

The second finding is behavioral: a rat will work to receive a source of external heat when its brain is cooled, even in a neutral environment. Aside from the reflex physiological responses, such as shivering and increased body temperature, rats will make a voluntary response—bar pressing—to increase heat in the environment. The rat's use of an arbitrary learned act in response to local brain cooling proves that the stimulus arouses the appropriate motivation, the urge to keep warm. Turning on the heat lamp is, in fact, an adaptive response to brain cooling, since it reduces the amount of heat lost to the environment. Nevertheless, the same body temperature is attained whether or not the rat can voluntarily increase the external heat. This suggests that the rat regulates the amount of heat received very carefully, since it could press for more heat during the cooling interval without burning its skin.

The almost complete cessation of pressing when the brain cooling was stopped is interesting. The animals were hyperthermic when cooling ceased. Without the strong stimulus of the cooling, they were actually overheated in the cold and so needed to lower their body temperature, rather than obtain more heat. When cooling was started again, their temperatures had dropped, though not to precooling levels. Occasionally, an animal began pressing in earnest toward the end of the noncooling interval. By that time the body temperature had fallen several degrees and so the cold environment was enough to start the rat working again.

At present, there is no way of knowing whether the brain temperature, the body temperature, or some interaction of the two ultimately controls an animal's motivation to work for heat. We do know from this experiment, however, that in response to hypothalamic cooling, rats exhibit not only physiological thermoregulation but also the behavioral urge to keep warm.

The author is indebted to Philip Teitelbaum for his advice and guidance throughout the course of the work.

REFERENCES

1. ANDERSEN, H. T., B. ANDERSSON, AND C. C. GALE. Central control of cold defense mechanisms and the release of "endopyrogen" in the goat. *Acta Physiol. Scand.* 54: 159-174, 1962.
2. ANDERSSON, B. Cold defense reactions elicited by electrical stimulation within the septal area of the brain in goats. *Acta Physiol. Scand.* 41: 90-100, 1957.
3. ANDERSSON, B., C. C. GALE, AND J. W. SUNDSTEN. Effects of chronic central cooling on alimentation and thermoregulation. *Acta Physiol. Scand.* 55: 177-188, 1962.
4. ANDERSSON, B., R. GRANT, AND S. LARSSON. Central control of heat loss mechanisms in the goat. *Acta Physiol. Scand.* 37: 261-280, 1956.
5. BEATON, L. E., W. A. MCKINLEY, C. BERRY, AND S. W. RANSON. Localization of the cerebral center activating heat loss mechanisms in monkeys. *J. Neurophysiol.* 4: 478-485, 1941.
6. ELIASSON, S., AND G. STROM. On the localization in the cat of hypothalamic and cortical structures influencing cutaneous blood flow. *Acta Physiol. Scand.* 20: (Suppl. 70), 113-118, 1950.
7. HAMILTON, C., AND W. SHERIFF, JR. Thermal behavior of the rat before and after feeding. *Proc. Soc. Exptl. Biol. Med.* 102: 746-748, 1959.
8. HAMMEL, H. T., J. W. HARDY, AND M. M. FUSCO. Thermoregulatory responses to hypothalamic cooling in unanesthetized dogs. *Am. J. Physiol.* 198: 481-486, 1960.
9. HARDY, J. D. Physiology of temperature regulation. *Physiol. Rev.* 41: 521-606, 1961.
10. HARDY, J. D., M. M. FUSCO, AND H. T. HAMMEL. Response of conscious dog to local heating of the anterior hypothalamus. *Physiologist* 1: 34, 1958.
11. HEMINGWAY, A., T. W. RASMUSSEN, H. WIKOFF, AND A. T. RASMUSSEN. Effects of heating hypothalamus of dogs by diathermy. *J. Neurophysiol.* 3: 328-338, 1940.
12. LATIES, V. G., AND B. WEISS. Behavior in the cold after acclimatization. *Science* 131: 1891-1892, 1961.
13. MAGOUN, H. W., F. HARRISON, J. R. BROBECK, AND S. W. RANSON. Activation of heat loss mechanisms by local heating of the brain. *J. Neurophysiol.* 1: 101-114, 1938.
14. RANSON, S. W. Regulation of body temperature. *Res. Publ. Assoc. Res. Nervous Mental Disease* xx: 342-399, 1940.
15. RANSON, S. W., C. FISHER, AND W. R. INGRAM. Hypothalamic regulation of temperature in the monkey. *Arch. Neurol. Psychiat.* 38: 445-466, 1937.
16. RANSON, S. W., AND W. R. INGRAM. Hypothalamus and regulation of body temperature. *Proc. Soc. Exptl. Biol. Med.* 32: 1439-1441, 1935.
17. WEISS, B. Pantothenic acid deprivation and thermal behavior of the rat. *Am. J. Clin. Nutr.* 5: 125-128, 1957.
18. WEISS, B., AND V. G. LATIES. Thyroid state and working for heat in the cold. *Am. J. Physiol.* 197: 1028-1034, 1959.
19. WEISS, B., AND V. G. LATIES. Behavioral thermoregulation. *Science* 133: 1338-1344, 1961.

Part III

EVIDENCE FOR MANY THERMOSTATS

Editor's Comments
on Papers 9 Through 14

9   KELLER
    Observations on the Localization in the Brain-stem of Mechanisms Controlling Body Temperature

10 SIMON, RAUTENBERG, and JESSEN
    Initiation of Shivering in Unanaesthetized Dogs by Local Cooling Within the Vertebral Canal

11 CHAMBERS et al.
    Thermoregulatory Responses of Decerebrate and Spinal Cats

12 ROBERTS and MOONEY
    Brain Areas Controlling Thermoregulatory Grooming, Prone Extension, Locomotion, and Tail Vasodilation in Rats

13 CARLISLE and INGRAM
    The Effects of Heating and Cooling the Spinal Cord and Hypothalamus on Thermoregulatory Behaviour in the Pig

14 CABANAC et al.
    Indifférence a la douleur et confort thermique (2 cas)
    English translation: *Indifference to Pain and Thermal Comfort*

During the time that the hypothalamus was the focus of attention in thermoregulation, several scientists were maintaining that other parts of the brain and spinal cord were also critically involved. In the United States Allen D. Keller (Paper 9) did a series of studies showing that cats and dogs whose brains were transected below the level of the hypothalamus, through the rostral midbrain, panted and thereby kept their body temperatures down in a hot environment. (For a more complete discussion of Keller's careful localization studies, see Keller 1963 and Keller and McClaskey 1964.) In Germany Rudolph Thauer (1935) had shown that rabbits whose spinal cord had been sectioned at a high cervical level slowly regained the ability to keep their body temperatures within normal limits in the cold, but this finding was generally ignored by Ranson and others who thought that it was a metabolic rather than a nervous adaptation. However, in 1965 Eckhart Simon, Walter Rauttenberg, and Claus Jessen (Paper 10), working in the Max Planck Institute in Bad Nauheim (where Thauer was the director, and where work on the spinal cord had been continued) demonstrated unequivocally that unanesthetized dogs could be made to shiver when their spinal cords were cooled. To examine the possibility that the thermosensitive elements might not be in the spinal cord itself but rather in the meninges or other tissues of the vertebral canal, Meurer, Jessen, and Iriki (1967) transected the dorsal roots of the lumbosacral cord. They then warmed the intact thoracic cord (to localize the cooling to the denervated part) and cooled the deafferented part. In such a rhizotomized preparation, cooling the deafferented part of the cord evoked a shivering-like tremor in the hind limbs. This experiment left no doubt that thermosensitive structures exist in the spinal cord itself. Later it was demonstrated that shivering could be obtained in rabbits, dogs, and cats during either spinal or whole-body cooling (for example, Kosaka, Simon, and Thauer 1967; Simon, Klussman, Rautenberg, and Kosaka 1966; Herdman 1978).

There is an apparent paradox here. Spinal animals shiver; yet as we saw in the last section, decerebrate animals (such as those of Bazett and Penfield) do not. One would expect that having more brain should lead to a more effective response rather than eliminating it entirely. This paradox has been beautifully resolved by an experiment of William Chambers and his coworkers (Paper 11). Chambers et al demonstrated that there are levels of inhibition and excitation of thermoregulatory responses throughout the nervous system. There is, for example, a region in the midbrain and upper pons that tonically inhibits lower regions in the lower pons and medulla. If a cut is made above the upper region, no thermoregulatory responses are seen. If the cut is made below that upper inhibitory region and its influence is abolished, the lower areas are themselves capable of organizing and facilitating thermoregulatory responses.

This sort of hierarchical control of thermoregulatory responses is strikingly demonstrated in Paper 12 by Harry Carlisle and David Ingram. In this case, two different thermoregulatory behaviors—operant responding for heat and postural changes—were examined during local thermal stimulation of the spinal cord and preoptic area. When the temperatures of the two areas were made to go in opposite directions, the behavioral responses were also in opposition, that is, the operant response was determined by the temperature of the hypothalamus and the postural response was determined by the temperature of the spinal cord. Thus each area can sense temperature and integrate and initiate effector responses appropriate to its temperature almost independent of what is being sensed in the other area.

The next paper in this section, by Warren Roberts and Richard Mooney, illustrates the essential point that not all thermoregulatory responses are activated together when the preoptic area is thermally stimulated. This paper by itself would have been enough to point to other areas of the brain as necessary for integrating thermal responses. In fact, there had been other reports of the same nature, that is, that after preoptic lesions, some thermoregulatory behaviors are impaired, but others are not. For instance, operant responding for warmth or coolness is left almost intact after preoptic lesions (Lipton 1968; Carlisle 1969; Satinoff and Rutstein 1970), as are grooming and locomotion (Roberts and Martin 1977), two of the major behavioral responses to heat stress in rats. However, in the cold, rats with preoptic lesions build very poor nests (Van Zoeren and Stricker 1977), eat less food, and are less active than normal (Hamilton and Brobeck 1964, 1966). In the heat, sprawling is greatly reduced (Roberts and Martin 1977), and rats do not reduce their food intake nearly as much as do controls (Hamilton and Brobeck 1964).

Paper 14, by Michael Cabanac and his coworkers, suggests that humans have the same nervous differentiation of temperature as do other animals. Their subjects who were indifferent to pain were also indifferent to any pleasurable aspects of thermal sensations. Nevertheless, they could discriminate heat and cold perfectly well and showed reflexive responses that were the same as normal subjects.

REFERENCES

Carlisle, H. J., 1969. The effects of preoptic and anterior hypothalamic lesions on behavioral thermoregulation in the cold. *J. Comp. Physiol. Psychol.* **69**:391-402.

Hamilton, C. L., and Brobeck, J. R., 1964. Food intake and temperature regulation in rats with rostral hypothalamic lesions. *Am. J. Physiol.* **207**:291-297.

Hamilton, C. L., and Brobeck, J. R., 1966. Food intake and activity of rats with rostral hypothalamic lesions. *Proc. Soc. Exp. Biol. Med.* **122**:270-272.

Herdman, S., 1978. Recovery of shivering in spinal cats. *Exp. Neur.* **59**:177-189.

Keller, A. D., 1963. Temperature regulation disturbances in dogs following hypothalamic ablations. *In Temperature: Its Measurement and Control in Science and Industry*, Vol. 3, J. D. Hardy, ed. Reinhold, New York, chap. 49, pp. 571-584.

Keller, A. D., and McClaskey, E. B., 1964. Localization, by the brain slicing method, of the level or levels of the cephalic brainstem upon which effective heat dissipation is dependent. *Am. J. Phys. Med.* **43**:181-213.

Kosaka, M.; Simon, E.; and Thauer, R., 1967. Shivering in intact and spinal rabbits during spinal cord cooling. *Exp.* **23**:385-387.

Lipton, J. M., 1968. Effects of preoptic lesions on heat-escape responding and colonic temperature in the rat. *Physiol. Behav.* **3**:165-169.

Meurer, K. A.; Jessen, C.; and Iriki, M., 1967. Kältezittern während isolierter Kühlung des Rückenmarks nach Durchschneidung der Hinterwurzeln. *Pflüg. Arch. Ges. Physiol.* **293**:236-255.

Roberts, W. W., and Martin, J. R., 1977. Effects of lesions in central thermosensitive areas on thermoregulatory responses in rat. *Physiol. Behav.* **19**:503-511.

Satinoff, E., and Rutstein, J., 1970. Behavioral thermoregulation in rats with anterior hypothalamic lesions. *J. Comp. Physiol. Psychol.* **71**:77-82.

Simon, E.; Klussman, F. W.; Rautenberg, W.; and Kosaka, M., 1966. Kältezittern bei narkotisierten spinalen Hunden. *Pflüg. Arch. Ges. Physiol.* **291**:187-204.

Thauer, R., 1935. Wärmeregulation und Fiebertätigkeit nach operativen Eingriffen am Nervensystem homoiothermer Säugetiere. *Pflüg. Arch. Ges. Physiol.* **236**: 102-147.

VanZoeren, J. G., and Stricker, E. M., 1977. Effects of preoptic, lateral hypothalamic, or dopamine-depleting lesions on behavioral thermoregulation in rats exposed to the cold. *J. Comp. Physiol. Psychol.* **91**:989-999.

9

Copyright © 1933 by Charles B. Slack, Inc.

Reprinted from *Am. J. Med. Sci.* **185**:746-748 (1933)

OBSERVATIONS ON THE LOCALIZATION IN THE BRAIN-STEM OF MECHANISMS CONTROLLING BODY TEMPERATURE*

Allen D. Keller

Department of Physiology and Pharmacology, University of Alabama School of Medicine

The work of Isenschmid and Schnitzler[1] in the rabbit, and of Bazett and Penfield[2] in the cat demonstrated by extirpation experiments the dependence of normal regulation of body temperature upon the hypothalamic region of the brain-stem. It seemed worth while to localize further in a *qualitative* manner the mechanisms concerned, with a view that this procedure would eventually lead to exact anatomic identification.

---

* This investigation was begun in the Yale Physiological Laboratory.[3] It is being continued with the aid of a grant from the Committee on Scientific Research of the American Medical Association.

[1] Isenschmid and Schnitzler: Arch. f. exp. Path. u. Pharm., 1914, **76**, 202.
[2] Bazett and Penfield: Brain, 1922, **14**, 569.
[3] Keller, A. D.: Proc. Am. Phys. Soc., 1932.

Three experimental approaches have been used: (1) Complete transection at various levels; (2) localization of conducting tracts mediating heat regulating functions as they passed from the *tonic* cephalic central mechanisms caudal through the mid-brain, pons and upper medulla; and (3) direct attack upon the hypothalmic region itself. The following factors were studied carefully in the preparations after operation: (1) Their ability to maintain normal rectal temperatures at usual as well as low environmental temperatures; (2) the presence or absence of shivering, panting and sweating; and (3) the time relation between the ear vessels and rectal temperature in returning to normal after overheating.

Certain necessary criteria for evaluating material have been adhered to: (1) In semi-acute preparations conclusions were drawn only from lesions that did not eliminate the mechanisms studied and, further, it was made certain that these mechanisms were present at the time of termination; (2) conclusions were drawn in regard to deprivation of paralytic symptoms only when these symptoms persisted as the animals became chronic. (A preparation is considered chronic only after sufficient time has elapsed after operation for the disappearance of edema and organization of débris.)

In all cases the block of tissue containing the lesion or lesions has been sectioned serially, and every third or fifth section stained for fiber tracts. This allows for accurate localization of the gross lesions. In chronic material, cell studies are being made to check the presence or absence of cell groups located adjacent to the gross lesion.

RESULTS. *Complete Transections.* It was necessary to keep midbrain cats in incubators up to about 3 weeks after operation. Gradually they were able to maintain adequate temperatures in unheated cages; however, when placed in an ice-box (45° to 50° F.), the rectal temperature fell progressively. In no instance has shivering been observed even up to 7 weeks after operation. In two preparations with sections through caudal tip of mammillary bodies ventrally, overheating has elicited typical panting. On removal from the hot box the ear vessels constricted to streaks long before rectal temperature returned to normal.

Only semi-acute pontile preparations have been studied. They have exhibited no attempt at heat regulation. Shivering has never been observed on cooling and the maximum respiration rate on overheating has been 120 per minute.

Several acute and semi-acute medullary preparations have maintained higher rectal temperatures than mid-brain and pontile animals when housed in the same incubators. Typical slight shivering has been noted in the preparations when cooled and also when heated. Hyperthermia has likewise been noted, whereas hyperthermia has never been encountered in pontile or mid-brain preparations. It seems possible that in these cases the mechanisms at the level of the obex may be stimulated by lesion hemorrhages, and so forth, since some medullary preparations react as do pontile preparations.

Hypothalamus. A bilateral transverse section just rostral to the chiasm, complete unilateral involvement of the hypothalamus and bilateral involvement of approximately the ventral third of the hypothalamus has been accomplished in the cat and dog without showing any gross impairment in heat regulating powers.

When the complete hypothalamus was involved—the mammillary bodies may be left intact—without gross injury to thalamic nuclei above, the animals were identical to mid-brain preparations in their inability to maintain normal temperature and to shiver. Rapid respiration and vasodilation were constantly present even in spite of low rectal temperatures. Typical panting was readily elicited by such maneuvers as pinching the tail, massage, or by spontaneous movement or urination by the animal.

Semi-acute and chronic cats and dogs having lesions involving a portion of the dorsal half of the hypothalamic gray matter bilaterally exhibited a permanent impairment of the power to maintain a normal body temperature at low extremes in environmental temperatures. These animals otherwise displayed normal health, exhibited vigorous appetites and were frequently pseudo-affective.

Analysis of the foregoing observations demonstrates the morphological separation—at least in part—of the "heat production" mechanism from that of the "heat loss" mechanism, the former being located in the more dorsal portion of the hypothalamus, the latter in the cephalic mid-brain. The separate existence of two such central mechanisms was postulated by Meyer.[1]

Conduction Paths in Mid-brain and Pons. That conducting pathways mediating heat regulating functions pass caudally *both medially and laterally* through the mid-brain and pons, is evidenced by the fact that cats and monkeys (only low pontile lesions have been placed in monkeys) continue to heat regulate after either bilateral section of the medial quarter segments or the lateral quarter segments of the brain-stem at these levels. At the upper level of the mid-brain the tracts seemed to be fairly well concentrated in the medial quarter segments. In the lower mid-brain and pontile levels there is a definite spread to lateral quarter segments.

[1] Quoted by Hasama: Arch. f. exp. Path. u. Pharm., 1929, **146**, 126.

Copyright © 1964 by Birkhauser Verlag, Basel

Reprinted from *Experientia* 21:476-477 (1965)

Initiation of Shivering in Unanaesthetized Dogs by Local Cooling within the Vertebral Canal

E. SIMON, W. RAUTENBERG, and C. JESSEN

W. G. Kerckhoff-Institute of the Max Planck-Gesellschaft, Bad Nauheim and Institute of Physiology of the University of Giessen (Germany), May 26, 1965.

Former observations indicating that cold sensitive structures may exist in the extracerebral deep body tissues have been confirmed by the finding that in dogs with normal or elevated core, brain, and skin temperatures, shivering was induced by cooling within the vertebral cnal (SIMON, RAUTENBERG, THAUER and IRIKI[1,2], RAUTENBERG and SIMON[3], THAUER[4]). This first evidence for the existence of cold sensitive structures in the extracerebral body core was obtained in anaesthetized animals. Since thermoregulatory reactions are known to be impaired by anaesthesia, experiments were carried out to investigate the effect of cooling within the vertebral canal on muscular activity in unanaesthetized dogs.

Method. In dogs weighing 16–27 kg, a U-shaped thermode of polyethylene tubing, about 1.5 mm in external diameter, was chronically implanted into the peridural space of the vertebral canal between the 4th cervical and the 7th lumbar vertebra. Cooling within the vertebral canal was performed in the unanaesthetized resting animal by perfusing the thermode with cold water. The rectal temperature and the temperature in the thoracic vertebral canal were measured with thermocouples. Electromyograms of one foreleg and one hindleg were picked up by needle electrodes. Temperatures and electromyograms were recorded by UV-direct writing galvanometers (Galvomat, CEC). In some cases, the temperature in the vertebral canal at the level of the second cervical vertebra was controlled.

Results. A recording obtained during a cooling period in an unanaesthetized 16 kg dog is shown in the Figure. At an ambient temperature of 23°C cooling within the vertebral canal was performed by perfusing the thermode with water at 22°C. Shivering appeared about 10 sec after the beginning of cooling, when the temperature in the thoracic vertebral canal had fallen below 37.3°C. While this temperature dropped to 34.2°C, shivering became more vigorous. The rectal temperature remained unchanged. Within 47 sec after the termination of cooling shivering ceased, when the temperature in the thoracic vertebral canal had risen to 38.4°C.

Shivering was observed in 25 cooling periods in 5 dogs at ambient air temperatures between 22°C and 30°C. Cooling was performed by perfusing the thermode with water at 18–27°C at a rate of flow between 20 and 40 ml/min. The mean values of the parameters measured in these experiments are given in the Table. The onset of shivering was found to occur 50 sec after the beginning of cooling, when the temperature in the thoracic vertebral canal had fallen by 3.6°C. Shivering ceased 75 sec after the end of cooling, when this temperature had reached 38.2°C.

According to the results of former experiments in anaesthetized animals with the same technique, cooling was found to be restricted to the vertebral canal. Core temperature was not affected by cooling. No significant fall of the temperature in the vertebral canal at the level of the second cervical vertebra – only a few centimetres rostrally from the entrance of the thermode – could be detected in the 12 cooling periods in which this temperature had been measured. HAMMEL, STRØMME, and CORNEW[5] have reported that, in unanaesthetized dogs at indifferent ambient temperatures, shivering was induced by hypo-

Rectal temperatures and temperatures in the vertebral canal during shivering evoked by cooling within the vertebral canal. Mean values of 25 cooling periods in 5 dogs (for details see text)

	Time	Rectal temperature	Temperature in the vertebral canal	
			At 2nd cervical vertebra	Thoracic part
	sec	°C	°C	°C
Start of cooling	0	39.17 $n = 25$	38.82 $n = 12$	38.95 $n = 25$
Start of shivering	49	39.16 $n = 25$	38.83 $n = 12$	35.37 $n = 25$
End of cooling	221	39.13 $n = 25$	38.75 $n = 12$	34.39 $n = 25$
End of shivering	295	39.12 $n = 25$	38.77 $n = 12$	38.22 $n = 25$

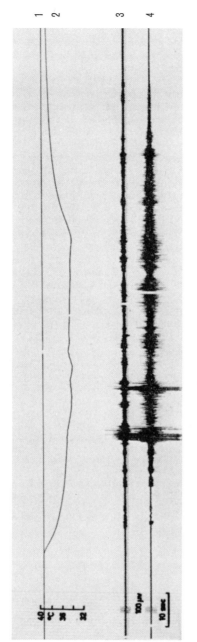

Shivering induced by local cooling within the vertebral canal in an unanaesthetized dog at indifferent ambient temperature. Rectal temperature (1), extradural temperature in the thoracic vertebral canal (2), electromyogram of foreleg (3) and hindleg (4).

thalamic cooling only, when hypothalamic temperature fell to about 37°C. This observation confirms the interpretation that hypothalamic thermosensitive structures were not involved in the initiation of shivering by cooling within the vertebral canal.

The reactions observed in unanaesthetized dogs during local cooling within the vertebral canal prove, therefore, the existence of cold sensitive structures in this area of the body core, which may participate in the physiological shivering mechanism.

Zusammenfassung. Bei unnarkotisierten Hunden konnte durch lokale Kühlung im Wirbelkanal bei indifferenter Umgebungstemperatur Kältezittern ausgelöst werden. Dieser an Tieren mit nicht beeinträchtigter Temperaturregulation erhobene Befund weist darauf hin, dass im Wirbelkanal temperaturempfindliche Strukturen lokalisiert sind, die an der Regelung der Wärmeproduktion mitbeteiligt sein können.

[1] E. SIMON, W. RAUTENBERG, R. THAUER, and M. IRIKI, Naturwissenschaften *50*, 337 (1963).
[2] E. SIMON, W. RAUTENBERG, R. THAUER, and M. IRIKI, Pflügers Arch. ges. Physiol. *281*, 309 (1964).
[3] W. RAUTENBERG and E. SIMON, Pflügers Arch. ges. Physiol. *281*, 332 (1964).
[4] R. THAUER, Naturwissenschaften *51*, 73 (1964).
[5] H. T. HAMMEL, S. STRØMME, and R. W. CORNEW, Life Sci. *12*, 933 (1963).

11

Copyright © 1974 by Academic Press, Inc.
Reprinted from *Exp. Neurol.* 42:282-299 (1974)

Thermoregulatory Responses of Decerebrate and Spinal Cats

W. W. CHAMBERS, M. S. SEIGEL, J. C. LIU, AND C. N. LIU [1]

*Departments of Anatomy, Physiology, and Institute of Neurological Sciences,
University of Pennsylvania, Philadelphia, Pennsylvania, 19174*

Received September 19, 1973

Cooling the spinal cord of the unanesthetized cat elicited shivering, piloerection, and vasoconstriction. A high-level decerebration abolished these effects. Lowering the decerebration to the level of the lower pons or medulla reinstated these responses to spinal cord cooling. In unanesthetized chronic spinal cats, cooling the spinal cord below the level of a T6 transection produced similar thermoregulatory effects limited to the hind limbs, although it was of less intensity and without piloerection. A high-level decerebration abolished shivering in the forelimbs to whole body cooling, while permitting shivering below the level of transection to spinal cord cooling. Lowering the level of decerebration to the lower pons or medulla reinstated shivering, vasoconstriction, and piloerection in the forelimbs. The data suggest that there is a region in the midbrain and upper pontine tegmentum which exerts tonic inhibition on lower regions in the lower pons, medulla, and spnal cord. When released from inhibition these lower regions are capable of facilitating thermoregulatory responses. Such an organization resolves contradictory reports on the abolition of thermoregulation after decerebration and answers the question of why spinal cord cooling produces shivering in spinal preparations but not in decerebrate preparations.

INTRODUCTION

The classical concept of the hypothalamus being the essential thermoregulation center is largely the result of studies after decerebration and spinal transection. Many investigators have reported that a transection in the brain stem at any level between the caudal hypothalamus and midpons abolishes shivering (3–5, 24, 31, 32, 43). Spinal animals fail to shiver below the level of transection (14, 21, 40, 42, 43). However, recent experiments imply that not all thermosensitive cells important in temperature control

---

[1] This work was supported in part by Grants USPHS NS 10464 and USPHS NB 08768 from the U.S. Public Health Service. We gratefully thank Drs. Alexander Beckman and Evelyn Satinoff for their many helpful suggestions on preparing this manuscript.

are confined to the hypothalamus (1, 2, 7, 10, 19, 27–29, 35, 41). The medulla and spinal cord have been shown to be extrahypothalamic sites of thermosensitive elements (10, 22, 33, 47). Thermoregulatory responses are evoked by heating or cooling the spinal cord (10, 20, 27–29), and the isolated spinal cord is capable of initiating shivering (45). Also, in decerebration studies of the early 1930's, Dworkin (18) reported shivering in rabbits with midbrain and medullary transections, and Keller (30, 31) noted slight shivering in dogs with medullary transection. These considerations raise the question of why shivering has been reported in decerebrate and spinal rabbits and dogs (18, 30, 31, 33, 45) but not in decerebrate and spinal cats (3, 5, 14). Unpublished observations in our laboratory that decerebrate cats and monkeys with hemorrhages around the basilar artery, causing ischemia of the pons and mesencephalon, show shivering, piloerection, and vasoconstriction of the foot pads, suggested to us that tonic inhibitory and facilitatory mechanisms for thermoregulation exist in the brain stem. Using both peripheral and spinal cord cooling methods, the present experiments were undertaken to determine by selective levels of decerebration if the brain stem contains inhibitory and facilitatory mechanisms modulating thermoregulatory responses.

MATERIALS AND METHODS

Four normal cats and two with spinal cord transected (T-6) and maintained about 50 days had tubing implanted in the peridural space of the spinal cord. After implantation and testing three normal cats and two with spinal cord chronically transected (one with a second transection at T-3) were decerebrated at several levels using a method similar to that of Bard and Macht (3), and one normal cat was decerebrated at a low level using the anemic technique of Pollock and Davis (38). The four normal decerebrate cats later had the cord acutely transected at either C-2 or around T-3. After all operations, cats were studied by cooling the spinal cord and periphery, as noted in Results.

Tubing was implanted in the peridural space of the spinal cord using a method similar to that of Kosaka et al. (34). In brief, a double-barreled thermode was made by bending polyethylene tubing (PE 60) into a hairpin shape by softening it in hot water. Under ether anesthesia the L-7 spinal process was removed, the U-shaped end of the thermode was inserted into the peridural space and carefully pushed upward extradurally to the thoracic region (T-4, T-5). A recovery period of at least 90 min was allowed.

After testing, a high-level decerebration was carried out using a method similar to that of Bard and Macht (3). During ether anesthesia the trachea was intubated, and the common carotid arteries were tied off. After crani-

284 CHAMBERS ET AL.

otomy, the medial surface of the hemisphere was retracted, and the blunt end of a scalpel was inserted to remove the left occipital and parietal lobes in order to expose the dorsal part of the midbrain. The blunt end of a

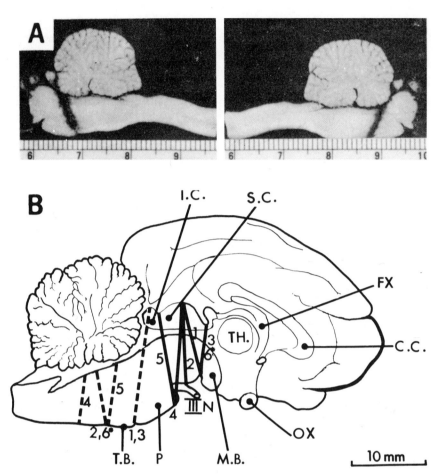

FIG. 1 A. Photographs of the midsagitally sectioned brain stem of cat 1 to show the upper and lower levels of decerebration. Upper: midsuperior colliculi dorsally and just caudal to mamillary bodies ventrally; lower: midinferior colliculi dorsally and just rostral to trapezoid body ventrally. Scale in millimeters. B. Schematic drawing of a midsagital section of the brain of a cat to illustrate the upper (solid line) and lower (broken line) levels of decerebration in six cats. Enlargement: ×2.75. (Cat 6 was anemically decerebrated by tying the common carotid arteries and the basilar artery at the caudal level of the trapezoid body and then scooping out the anemic brain rostral to superior colliculi and caudal to the mamillary bodies.) Abbreviations: C.C., corpus callosum; FX, fornix; I.C., inferior colliculus; M.B. mamilary body; OX, optic chiasm; P, pons; S.C., superior colliculus; T.B., trapezoid body; TH, thalamus; III N, oculomotor nerve.

scalpel was pushed through the brain stem, with the transection usually extending dorsally from within the superior colliculus to emerge ventrally between the mamillary bodies and the rostral edge of the pons (Fig. 1). The rostral bank of the transection was carefully aspirated to form a channel 2–3 mm wide so that the completeness of the brain stem transection could be verified and that if any hemorrhage occurred, it would bleed outward rather than around the brain stem. Blood was prevented from coming into contact with the brain stem by the insertion of loose pledgets of cotton or Gelfoam (Upjohn). The remaining occipital lobe was also removed to allow the verification of the symmetry of the transection. Saline was added to the cranial cavity, which was then loosely covered with cotton. The cat was allowed a recovery period of at least 1 hour before testing. Prior to and during testing neurological examinations were made on all decerebrate cats.

After testing, a low-level decerebration was carried out using no further anesthesia. The cut extended from within or below the inferior colliculus to the trapezoid body (Fig. 1). The cat was allowed a few minutes to recover and then tested. Later, the C-2 spinal process was removed, the spinal cord transected, and the cats tested.

In one normal cat with implanted tube, a low-level decerebration was performed according to the technique of Pollock and Davis (38). Both common carotids and the basilar artery were tied off resulting in an ischemia extending upward from the caudal trapezoid body. To insure nonparticipation of the hypothalamus the forebrain was then scooped out at the level of the superior colliculus, dorsally, and just behind the mamillary bodies ventrally. The cat was allowed to recover, tested, its spinal cord acutely sectioned, and retested as before.

Testing. In cats with implanted tubes, the spinal cord was cooled by circulating water through the polyethylene thermode. The water was injected with a syringe at 5–10 ml/min. For long-term cooling and warming a series of 11–21 injections of 20-ml samples of water were circulated, the temperature of each sample was measured initially. The temperatures were begun at near 37 C, decremented to near 18 C, and incremented to near 37 C. For short-term cooling or warming several 20-ml samples were injected with temperatures from 9.5 to 45 C to determine the latency, intensity, and injected temperature threshold of thermoregulatory responses. In some experiments tests of the effects of spinal cord cooling were followed by experiments on whole-body cooling. Body temperature was decreased either by applying bags of ice to the body surface while immersing a forelimb and hind limb in ice water, or by placing the whole body in ice water. Shivering was measured by EMG, recorded continuously on a Beckman polygraph, and changes in rectal temperature, spinal cord temperature (in some cases), and foot pad temperature (index of vasomotor activity) were measured with thermistors.

Gross Inspection and Histology. The completeness and location of all levels of decerebration were confirmed at autopsy by gross inspection and plotted on tracings of brain stem sagittal sections (Fig. 1). For the anemic decerebration a 1% methylene blue solution was injected into the aorta to determine the level of ischemia. Levels of spinal cord transection and thermode placement were confirmed by gross inspection. Frozen sections were made of chronic spinal cord transections and examined histologically for completeness of lesion.

RESULTS

Normal Cats. In the normal cat with implanted tubes, the most marked, consistently appropriate response to cooling the lower thoracic and lumbosacral spinal cord by circulated cool water was shivering. As the temperature of the injected fluid was decreased to 34.6 ± 0.6 C in cats 1 and 2, twitching and then shivering was observed in the hind quarters. Shivering spread to the forequarters until the entire animal was visibly shivering. With further decreased temperature a related increased intensity of shivering over the entire body was observed and recorded by EMG. At low temperatures large amplitude bursts of shivering were recorded (Fig. 6-A1). Then with increased temperature the intensity of shivering decreased (Fig. 6-A2) until it could not be seen or recorded (Fig. 6-A3). Other responses to spinal cord cooling were less dramatic. Vasoconstriction of the foot pads roughly followed the decrease in temperature of the injected water, but was inconsistent. Slight and inconsistent piloerection was seen over the entire body at injection temperatures of 10–20 C. Little change in rectal temperature was recorded after prolonged cooling of the spinal cord, possibly due to the localized effect of cooling, poor vasoconstriction, poor piloerection, and competing signals from other parts of the body.

Peripheral cooling by decreasing the ambient temperature resulted in a similar pattern of shivering, but was more consistent in eliciting piloerection and vasoconstriction of the foot pads. These thermoregulatory responses in cat 5 at an ambient temperature of 6 C were initiated without a fall in rectal temperature, and after some minutes the temperature rose from 38.7 to 39 C.

High-level Decerebration. In high-level decerebrate cats 1 and 2 (Fig. 1) no thermoregulatory responses were seen after injctions of water at temperatures as low as 10 C. No shivering or piloerection was observed or recorded to spinal cord cooling (Fig. 6-B1, B2, B3). Rectal (Fig. 2-A) and foot pad temperatures continually decreased as if the cats were poikilothermic. Whole body cooling also failed to produce any appropriate

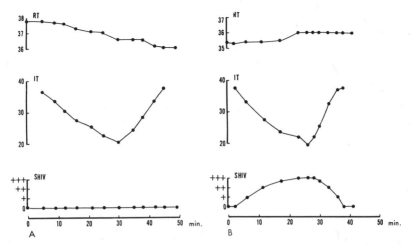

FIG. 2. Cat 2. Graphs showing the effects of spinal cord cooling and warming on rectal temperature and shivering. A. High decerebration. B. Low decerebration. Abbreviations: IT, temperature (°C) of injected water, which was injected at 5–10 ml/min to cool the spinal cord; RT, rectal temperature; SHIV, hind limb shivering, clinical intensity indicated by + signs. Time in minutes. Room temperature in both A and B was 24 ± 0.4 C.

thermoregulatory response even though the rectal temperature of cat 2 was decreased from 36.1 C to 34.2 C over a period of 50 min at an ambient temperature of 24.2 ± 0.5 C.

Low-level Decerebration. In cats 1 and 2 low-level decerebration (Fig. 1) restored the thermoregulatory responses to both spinal cord and whole body cooling (responses which had been abolished by high-level decerebration). Spinal cord cooling produced shivering, piloerection, vasomotor changes, and an increase in rectal temperature. Shivering appeared in the hind quarters of cats 1 and 2 at an injected temperature of 33 ± 1 C. With further decreased temperature, shivering spread to the entire body and increased in intensity (Fig. 2-B) until violent bursts of shivering occurred at 18.8 ± 0.4 C. Then, with increased injected temperature, the intensity of shivering diminished as before (Fig. 6-C1, C2, C3). Thus, shivering was a consistently appropriate thermoregulatory response. Slight and inconsistent piloerection was seen over the entire body when the injected water reached temperatures of 18 to 27.5 C. Vasomotor changes roughly followed the injected temperature, but were as inconsistent as in the normal cat. An inappropriate vasodilatation was found to occur with injected water temperatures of 23.3 ± 1.3 C. Rectal temperature increased 0.7 C in cat 2 at an ambient temperature of 24 ± 0.4 C (Fig. 2-B) and 1 C in cat 1 at an ambient temperature of 24.7 ± 0.3 C during cooling. It remained constant in both cats during warming. Cooling of the whole body produced well-coordinated thermoregulatory responses, although the threshold for these responses was higher than normal and they were unable to maintain the cat's rectal temperature at low ambient temperatures. In cat 6 shivering was seen at a rectal temperature of 37.7 C and its intensity increased only after 54 min of cooling with ice packs at which time the rectal temperature was 32.7 C. Moderate piloerection and good vasoconstriction were also observed. Although low-level decerebrates regulated their rectal temperatures much more effectively than high-levl decerebrates, drastic whole-body cooling overcame the thermoregulatory responses and caused a decrease in rectal temperature. Inadequate postural and vasomotor responses could offer an explanation for this deficiency. Combined whole body and spinal cord cooling interacted in cat 6 to increase the intensity of shivering, but the increased shivering was unable to prevent further decrease in rectal temperature (from 31.8 to 31.5 C).

The restoration of thermoregulatory responses brought about by low-level decerebration suggests that there is a site capable of inhibiting thermoregulation in the pontine tegmentum above a region extending from the inferior colliculus to the trapezoid body (Fig. 1). Thus, loci below this region in the lower pons, medulla, and spinal cord are capable of initiating thermoregulatory responses.

Spinal Cats. The acutely prepared spinal cat showed less intense thermoregulatory responses to cooling the spinal cord than the normal or the low-level decerebrate. With spinal cord cooling, cat 6, which had first been decerebrated at a low level and then transected at C-3, shivered less in the hind quarters than he had when tested after decerebration alone. As before, shivering diminished with warming and was always an appropriate response. Piloerection was abolished, and vasomotor activity was abolished or at times inappropriate (vasodilatation occurred with rapid cooling by injecting water at 9.5 C). There was no change in rectal temperature at ambient temperatures of 27 ± 0.2 C. These results suggest that the transection caused the withdrawal of facilitation of thermoregulatory mechanisms for piloerection, vasomotor activity, and, to a lesser extent, shivering. The spinal shivering mechanism may be moderately facilitated or completely inhibited by supraspinal influences.

The chronically prepared spinal cats showed improved thermoregulatory responses to spinal cord cooling with the most marked change occurring in vasomotor activity. In cats 3 and 4, tested 50 or more days after transection at T-6, the intensity of shivering (Fig. 6D1) was less than that in the low-level decerebrate or normal cats (Fig. 6C1, A1). Shivering was always an appropriate response. It was found primarily in the hind quarters and spread to the forequarters during gradual cooling, possibly due to cerebrospinal fluid or tissue conduction. Piloerection was present above but not below the transection. Little or no change in rectal temperature was observed. With gradual (Fig. 3) or rapid (Fig. 4) spinal cord cooling there was marked vasoconstriction and with warming marked vasodilatation of the hind limbs. The forelimbs showed less response. In Fig. 3 it can be seen that while the hind limb temperature increased with spinal cord warming, the forelimb temperature increased only after a heating pad was applied. However, it should be noted that an inappropriate vasodilatation of the hind limb pads occurred when the temperature of the injected water

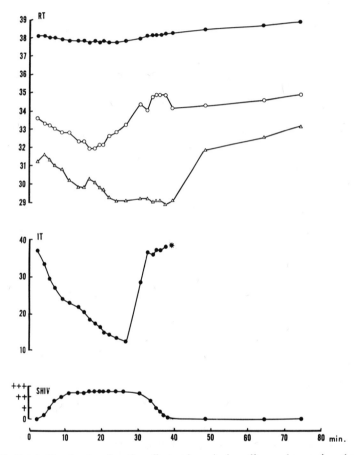

FIG. 3. Cat 4. Graphs showing the effects of gradual cooling and warming the spinal cord below a chronic T-6 transection of 50 days. Abbreviations: same as in Fig. 2, △: left forelimb pad temperature; ○: left hind limb pad temperature. Room temperature was 27 ± 0.2 C throughout the experiment; *: heating pad turned on. Time in minutes.

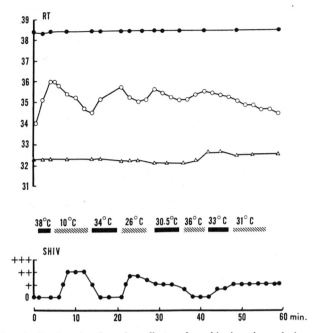

Fig. 4. Cat 4. Graphs showing the effects of rapid changing of the spinal cord temperature below a T-6 transection of 50 days. Bars indicate duration and temperature of injected water. Abbreviation same as in Figs. 2 and 3. Room temperature 27 ± 0.2 C throughout the experiment.

reached 17.2 C and continued through 12.8 C. The results all suggest that thermoregulatory responses are profoundly influenced by supraspinal sources and can be both inhibited and facilitated. The improved responses in the chronic spinal animal suggest a recovery from spinal shock similar to the reflex recovery noted by Chambers et al. (12) in the spinal cat.

To whole-body cooling by immersion in ice water the chronic spinal cat showed marked shivering in the forequarters but only tremors or slight shivering in the hind quarters. This stimulus compared with spinal cord cooling may not be sufficiently fast or direct to elicit a strong shivering response.

Decerebrated Spinal Cats. High-level decerebration of chronically prepared spinal cats 3 and 4 abolished thermoregulatory responses above the level of transection while not affecting responses below it to both spinal cord and whole body cooling. Spinal cord cooling of cat 3 elicited appropriate shivering only below the level of transection. Whole body cooling then increased the intensity of this shivering while above the level of transection there was no shivering at all. Cat 4 shivered only below the level of transection to spinal cord cooling. Although slight, vasomotor responses to

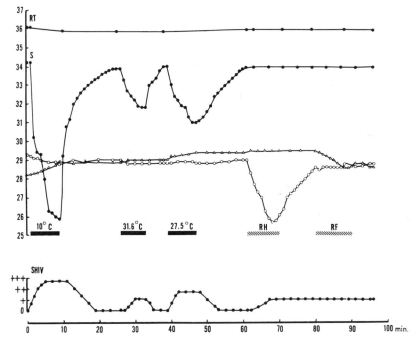

FIG. 5. Cat 4 with a chronic spinal transection at T-6 for 50 days and an acute high decerebration. Graphs showing effects of rapid cooling of the spinal cord below the T-6 transection and the effects of immersion of a hind limb or forelimb in ice water. The epidural space temperature of the spinal cord at T-8 (S) was recorded during spinal cord cooling by injection of water (solid bars) and during immersion of the right hind leg (RH) and the right foreleg (RF) in ice water (dashed bars). Foot pad temperature and shivering are indicated by △ for the left forelimb and by ○ for the left hind limb. Other abbreviations same as in Figs. 2 and 3. Room temperature 26.5 ± 0.5 C.

spinal cord cooling were more appropriate in the hind limbs than in the forelimbs (Fig. 5). Immersion of the right hind limb in ice water induced shivering and caused a marked vasoconstriction of the left hind limb but not in the forelimbs. Similar immersion of the right forelimb induced no shivering and caused only a slight vasoconstriction of the right forelimb. A thermistor placed inside the peridural space of the spinal cord recorded a close proportional relationship between the temperatures of this space and the circulated fluid.

Lowering the level of decerebration in chronically prepared spinal cat 3 restored some shivering in the forequarters to drastic whole-body cooling (rectal temperature 28.5 C) while not affecting thermoregulatory responses of the hind quarters.

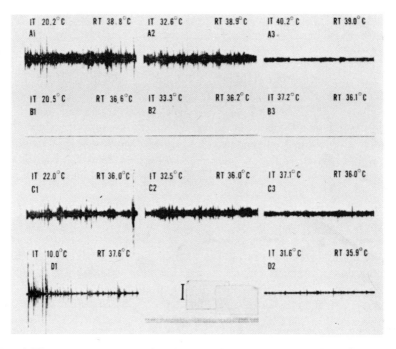

Fig. 6. Electromyograms of a hind limb muscle (quadriceps) to show the patterns of shivering elicited by spinal cord cooling and warming in the following feline preparations: A1-A3 intact; B1-B3 high decerebrate; C1-C3 low decerebrate; D1-D2 spinal. Abbreviations: IT: injection temperature; RT: rectal temperature. A-C cat no. 2 (Fig. 1); D cat no 4. Calibration: Time in seconds; amplitude, 1 mv/cm.

DISCUSSION

Cooling the thoracic and lumbosacral spinal cord of unanesthetized normal, acutely and chronically prepared spinal cats causes corresponding shivering and vasoconstriction, initially limited to the hind quarters, with no or slight rise in rectal temperature. Slight piloerection was noted only in the intact cat. These effects were abolished by heating. Thus, we have confirmed previous reports (20, 27-29, 45) that the spinal cord can influence body temperature. In agreement with most observers (3-5, 24, 31, 32. 43) a high-level decerebration (between caudal hypothalamus and upper pons) abolished these thermoregulatory responses. Lowering the decerebration to the medulla regularly reinstated these effcts. Our results that the effects of cooling and warming the spinal cord can be inhibited and facilitated by neural axis transection, along with our and other (10, 37) results suggesting local thermoregulatory effector mechanisms, suggest that the technique is a potent one for investigating regulation by supraspinal influences. Our results offer an explanation as to why Keller obtained some shivering in his dogs with medullary transection (30, 31) but not from his dogs with mesencephalic or pontine transections (32). Our observations also suggest that the more effective temperature regulatory response in the low than in the high decerebrate rabbit (18) is due to the removal of midbrain and pontine tonic inhibition. This tonic inhibition in the rabbit is probably less than in the cat and dog. The differences in intensity and effectiveness of shivering between spinal, low decerebrate, and intact animals suggest that shivering is normally altered by supraspinal facilitatory as well as inhibitory influences. We propose a modification of the classical concept of thermoregulation such that the hypothalamus can coordinate the activity of lower thermoregulatory regions using the inhibitory and facilitatory influences of the brain stem.

Midbrain and Upper Pons. That a high-level decerebration releases inhibition of thermoregulatory responses and lowering the decerebration restores the responses suggests that a site between the two transections, in the midbrain and upper pontine tegmentum, is capable of tonically inhibiting lower regions. This midbrain inhibition explains why a high-level decerebrate will not respond to spinal cord cooling but a low-level decerebrate or spinal cat will. It is also supported by other reports. Studies of chronically decerebrated cats by Bard and his associates (3, 4) show that "none of the cats was capable of any effective temperature regulation" (3) although they were kept alive and in good condition for months. These results taken together with our results suggest the impairment is not due to irritation, for it is long-lasting, but can be quickly removed by lower decerebration. That the high-level decerebrate cat, chronically prepared, shows some thermoregulatory responses (3) not seen in our acutely prepared cats suggests either a partial recovery of inhibition of the midbrain and pontine tonic inhibitory region, or degeneration of the tonic inhibitory mechanism with time. In further support of an inhibitory region are the findings of Hemingway (23) that electrical stimulation of the midbrain reduces or suppresses shivering in the normal cat. Also, Connor and Crawford (15) reported decreased body temperature both after intercollicular and midpontine decerebrations, but considerable hyperthermia in cats with medial midpontine lesions. They speculated that the different thermogenic responses may be related to the integrity of a shivering pathway thought to descend from the hypothalamus in the extreme lateral regions of the pons (6). It is our belief that by affecting separate pathways these lesions may have preserved the facilitatory influence of the hypothalamus and lower brain stem while destroying the inhibitory influence of the midbrain–midpontine region.

Lower Pons, Medulla, and Spinal Cord. Transection studies suggest that sites in the lower pons, medulla, and spinal cord are capable of initiating thermoregulatory responses, and that the responsiveness is profoundly influenced by the withdrawal of inhibition and facilitation. The present experiments show that these regions, when freed from higher inhibition, respond to local cooling with associated shivering, vasoconstriction, piloerection, and a rise in rectal temperature. Other reports support these findings. Keller (30, 31) reported that medullary preparations shivered when cooled and maintained higher rectal temperatures than did midbrain and pontine animals. Dworkin's previously unexplainable results (18), that rabbits with midbrain and medullary transection shivered but with unequal effectiveness, can now be interpreted. None of his five medbrain and upper pontine decerebrate animals (transections from front of superior colliculus, posterior border of pons, to posterior border of superior colliculus, anterior border of pons) was reported to increase its body temperature while shivering. All nine of his rabbits with lower pontine and medullary transections (from behind inferior colliculus, anterior border of trapezoid body, to 2 mm above calamus scriptorius, lower border of trapezoid body) were reported to increase their body temperature while shivering. We suggest that the difference in effectiveness of shivering between the two groups is due to the removal of midbrain and upper pontine tonic inhibition, and that possibly this inhibition is less in the rabbit than in the cat and dog. Chambers, et al. (11) demonstrated that high-level decerebrate cats with all or most of the midbrain intact lost heat rapidly, never showed shivering, piloerection, vasoconstriction, or fever. In the majority of pontine and medullary cats decerebrated by the anemic method (38) shivering, vasoconstriction, and spontaneous or pyrogenic fevers occurred. Although Constantin and Wang (16) showed that the cerebral ischemia produced by anemic decerebration is not total, our results demonstrate that a high-level transection accompanied by either a low-level transection or a low-level anemic decerebration produces similar thermoregulatory responses, and, therefore, support in this instance the validity of the anemic decerebration technique. Chambers et al. (11) suggested that the tegmentum of the midbrain contains a locus that tends to inhibit shivering, vasoconstriction, and febrile responses, and that an integrative mechanism in the medulla or upper spinal cord, or both, was capable of evoking these effects. Chai and Lin (10) reported that heating the medulla or spinal cord of unanesthetized monkeys produced a decrease in the rectal temperature, vasodilatation, respiratory acceleration, bradycardia, and hypotension. Conversely, cooling the medulla or spinal cord led to an increase in rectal temperature, vasoconstriction, piloerection, decrease in respiratory rate, tachycardia, and hypertension. These changes are similar to those obtained by thermal stimulation of the hypothalamus (40). Since respiratory and cardiovascular changes induced by heating and cooling the medulla in anesthetized cat are not affected by decerebration (8, 9), Chai et al. (10)

suggested that some independent thermosensitive mechanisms exist in the spinal cord or medulla that may share with the hypothalamus in temperature regulation. Our finding that high decerebration abolishes thermoregulatory responses to cooling the lower spinal cord, while lowering the level of decerebration restores these responses, would suggest that shivering, vasoconstriction, and piloerection from cooling the medulla should also be inhibited by a high decerebration. Lipton (36) reported similar changes in rectal temperature and appropriate thermoregulatory behavior with heating and cooling of the medulla of intact animals. The general medullary/rectal temperature relation was not changed by preoptic lesions, leading him to suggest that medullary thermosensitivity can mediate control of physiological and behavioral thermoregulatory activities.

The present experiments suggest that, in addition to the pontine inhibitory effect upon spinal segmental temperature mechanisms, there is an important facilitatory effect from the lower pons and medulla. This facilitation was clearly illustrated when the animal with low-level decerebration subsequently had the spinal cord sectioned; piloerection was abolished, and vasoconstriction and shivering were reduced to spinal cord cooling. The chronic spinal cat showed little shivering to peripheral cooling, and shivered only upon extreme cooling by prolonged immersion in ice water. These findings confirm the observations of others (14, 39, 42, 43) who reported that in response to peripheral cooling spinal animals did not shiver below the level of transection despite a drop of several degrees in rectal temperature. This difference between peripheral and spinal cord cooling in the spinal cat may be due to several factors: (a) The reflex response to peripheral cooling may require more facilitation from supraspinal sources. (b) It may require the integrity of ascending pathways to supraspinal structures. (c) It may be influenced by the rate of temperature change.

Organization of Supraspinal Influences. The organization of potent brain stem inhibitory and facilitatory influences on segmental structures mediating temperature regulation is similar to that of other proposed systems mediating micturition, respiration, and cutaneous reflexes. By transecting the brain at different levels Tang and Ruch (44) attempted to determine regions of the brain stem concerned with controlling the micturition reflex. They reported "at least four levels of the neural axis . . . influence profoundly the excitability of the sacral micturition reflex, namely, *(1)* a cortical inhibitory region, *(2)* a posterior hypothalamic facilitatory area, *(3)* a mesencephalic inhibitory area, *(4)* an anterior pontine facilitatory area" (44). In studies of respiration, Wang, Ngai and Frumin (46) showed that, if the vagi were sectioned, a transection rostral to the upper 2 mm of the pons resulted in eupnea, while a lower transection resulted in apneustic respiration. If the decerebration was made caudal to the trapezoid body the respiration became rhythmic and in some cases appeared entirely normal. In studies of cutaneously elicited reflexes it has been reported (25) and confirmed (13, 17) that regions in the pons and upper medulla exert an inhibitory effect on interneurons between flexion reflex afferents and the motoneurons to the ipsilateral flexor muscles, and the interneurons which inhibit the ipsilateral extensor motoneurons. A lesion at the level of the trapezoid body abolishes bulbar inhibition of the extensor pathway, and a lesion at the level of the obex abolishes inhibition of the flexors.

Although these proposed brain stem systems may interact they do not seem to be coextensive. By heating and cooling the medulla, Chai and Lin (10) have found cardiovascular and respiratory changes as part of a thermoregulatory response. The interaction of shivering with movement has been investigated by Hemingway et al. (23). They reported that skeletal muscles are used in coordinated movements of the skeletal structures or in shivering, but not both at the same time. "When a movement is made there is a rapid suppression of shivering . . . since in the reactions of an animal for fight, flight, or defence, skeletal movements are of more immediate importance than shivering." Thus, it may be that shivering is subordinated to movement, and is suppressed when movements occur. Our findings suggest that these brain stem systems are not all coextensive. A level of decerebration which released shivering, piloerection, and vasomotor responses did not release cutaneous reflexes. A lower decerebration which did release cutaneous reflexes resulted in no further release of thermoregulatory responses. It remains to be seen how micturition, respiration, cutaneous reflexes, and temperature regulation are integrated in the brain stem.

Functional Significance. Traditional concepts of nervous system organization involve Jacksonian levels of integration which serve to amalgamate, adjust, and regulate lower activites for the greatest benefit of the system as a whole (26). Dividing complicated tasks into independent subtasks influenced by higher structures provides the body with an optimum flexibility to respond to a changing environment. However, such an organization has not been applied to thermoregulation. Instead it had been believed that control and effector mechanisms reside only in the hypothalamus, and that other areas only supply information to it. However, the thermoregulatory system may be organized such that the hypothalamus directs the activity of miniature temperature regulation areas, each modulated by the hypothalamus but capable of independent thermosensitive responses. In such a system the hypothalamus would be dominant, for it would amalgamate, adjust, and regulate lower centers; in its absence only a fraction of the capability of the system could be utilized. However, lower centers in the medulla and spinal cord would have some degree of thermosensitivity and a capacity to transduce temperature changes into effector output. Inhibitory regions associated with these centers might form part of the means by which the hypothalamus could exert dominance over them. Regions proposed in this paper would fit into such an organization. Evaluating the relative importance of the constituents of this system would be the purpose of future experiments.

REFERENCES

1. ADAIR, E. R., J. U. CASBY, and J. A. STOLWIJK. 1970. Behavioral temperature regulation in the squirrel monkey: changes induced by shifts in hypothalamic temperature. *J. Comp. Physiol. Psychol.* **72**: 17–77.
2. ANDERSSON, B., C. C. GALE, B. HOKFELT, and B. LARSSON. 1965. Acute and chronic effects of preoptic lesions. *Acta Physiol. Scand.* **65**: 45–60.
3. BARD, P., and M. B. MACHT. 1958. The behavior of chronically decerebrate cats. *Ciba Found. Symp. Neurol. Basis Behav.* 55–75.
4. BARD, P., J. W. WOODS, and R. BLEIER. 1970. The effects of cooling, heating and pyrogen on chronically decerebrate cats, pp. 519–545. *In* "Physiological and Behavioral Temperature Regulation" Hardy, J. D., A. P. Gagge, and J. A. J. Stolwijk [Eds.]. Thomas, Springfield.
5. BAZETT, H. C., and W. G. PENFIELD. 1922. A study of the Sherrington decerebrate animal in the chronic as well as the acute condition. *Brain* **45**: 185–265.
6. BIRZIS, L., and A. HEMINGWAY. 1956. Descending brainstem connections controlling shivering in cat. *J. Neurophysiol.* **19**: 37–43.
7. CARLISLE, H. 1969. Effect of preoptic and anterior hypothalamic lesions on behavioral thermoregulation in the cold. *J. Comp. Physiol. Psychol.* **69**: 391–402.
8. CHAI, C. Y., J. Y. MU, and J. R. BROBECK. 1965. Cardiovascular and respiratory responses from local heating of medulla oblongata. *Amer. J. Physiol.* **209**: 301–306.
9. CHAI, C. Y., and S. C. WANG. 1970. Cardiovascular and respiratory responses to cooling of the medulla oblongata of the cat. *Proc. Soc. Exp. Biol. Med.* **134**: 763–767.
10. CHAI, C. Y., and M. T. LIN. 1972. Effects of heating and cooling the spinal cord and medulla oblongata on thermoregulation in monkeys. *J. Physiol. London* **225**: 297–308.
11. CHAMBERS, W. W., H. KOENIG, and W. F. WINDLE. 1949. Site of action in the central nervous system of a bacterial pyrogen. *Amer. J. Physiol.* **159**: 209–216.
12. CHAMBERS, W. W., C. N. LIU, G. P. MCCOUCH, and E. D'AQUILI. 1966. Descending tracts and spinal shock in the cat. *Brain* **89**: 377–390.
13. CHAMBERS, W. W., C. N. LIU, and G. P. MCCOUCH. 1970. Cutaneous reflexes and pathways affecting them in the monkey, *Macaca mulatta*. *Exp. Neurol.* **28**: 243–256.
14. CLARK, G. 1940. Temperature regulation in chronic cervical cats. *Amer. J. Physiol.* **130**: 712–722.
15. CONNER, J. D., and I. L. CRAWFORD. 1969. Hyperthermia in midpontine lesioned cats. *Brain Res.* **15**: 590–593.
16. CONSTATIN, L. L., and S. C. WANG. 1958. Hypothalamic responses during and following occlusion of cephalic vasculature in the cat. *Amer. J. Physiol.* **193**: 340–344.

17. DOUGHERTY, M., S. SHEA, C. N. LIU, and W. W. CHAMBERS. 1970. Effects of spinal cord lesions on cutaneously elicited reflexes in the decerebrate cat. Tonic bulbospinal and spinobulbar inhibitory systems. *Exp. Neurol.* **26**: 551–570.
18. DWORKIN, S. 1930. Observations on the central control of shivering and of heat regulation in the rabbit. *Amer. J. Physiol.* **93**: 227–244.
19. GALE, C. C., M. MATHEWS, and J. YOUNG. 1970. Behavioral thermoregulatory responses to hypothalamic cooling and warming in baboons. *Physiol. Behav.* **5**: 1–6.
20. GUIEU, J. D., and J. D. HARDY. 1970. Effects of heating and cooling of the spinal cord on preoptic unit activity. *J. Appl. Physiol.* **29**: 675–683.
21. GUTTMAN, L., J. AILVER, and C. H. WYNDHAM. 1958. Thermoregulation in spinal man. *J. Physiol. London* **142**: 406–419.
22. HALES, J. R. S., and C. JESSEN. 1969. Increase of cutaneous moisture loss caused by local heating of the spinal cord in the ox. *J. Physiol. London* **204**: 40–42P.
23. HEMINGWAY, A., P. FORGRACE, and L. BIRZIS. 1954. Shivering suppression by hypothalamic stimulation. *J. Neurophysiol.* **17**: 375–386.
24. HEMINGWAY, A. 1963. Shivering. *Physiol. Rev.* **43**: 397–422.
25. HOLMQUIST, B., and A. LUNDBERG. 1961. Differential supraspinal control of synaptic actions evoked by volleys in the flexion reflex afferents in alpha motoneurons. *Acta Physiol. Scand. Suppl.* **186. 54**: 1–51.
26. INGRAM, W. R. 1960. Central autonomic mechanisms, pp. 951–978. *In* "Handbook of Physiology," Section 1: "Neurophysiology," Vol 2, J. Field [Ed.]. American Physiological Society, Washington.
27. JESSEN, C., and E. T. MAYER. 1971. Spinal cord and hypothalamus as core sensors of temperature in the conscious dog. I. Equivalance of responses. *Arch. Gesamte Physiol.* **324**: 189–204.
28. JESSEN, C., and O. LUDWIG. 1971. Spinal cord and hypothalamus as core sensors of temperature in the conscious dog. II. Addition of signals. *Arch. Gesamte Physiol.* **324**: 205–216.
29. JESSEN, C., and E. SIMON. 1971. Spinal cord and hypothalamus as core sensors of temperature in the conscious dog. III. Identity of functions. *Arch. Gesamte Physiol.* **324**: 217–226.
30. KELLER, A. D., and W. K. HARE. 1932. Heat regulation in medullary and midbrain preparations. *Proc. Soc. Exp. Biol. Med.* **29**: 1067–1068.
31. KELLER, A. D. 1933. Observations on the localization in the brain-stem of mechanisms controlling body temperature. *Amer. J. Med. Sci.* **185**: 746–748.
32. KELLER, A. D. 1935. The separation of the heat loss and heat production mechanisms in chronic preparations. *Amer. J. Physiol.* **113**: 78–79.
33. KOSAKA, M., E. SIMON, and R. THAUER. 1967. Shivering in intact and spinal rabbits during spinal cord cooling. *Experientia* **23**: 385–387.
34. KOSAKA, M., E. SIMON, R. THAUER, and O. WALTHER. 1969. Effect of thermal stimulation of spinal cord on respiratory and cortical activity. *Amer. J. Physiol.* **217**: 858–864.
35. LIPTON, J. M. 1968. Effects of preoptic lesions on heat escape responding and colonic temperature in the rat. *Physiol. Behav.* **3**: 165–169.
36. LIPTON, J. M. In press. Thermosensitivity of the medulla oblongata in the control of body temperature. *Amer. J. Physiol.*
37. MEURER, K. A., C. JESSEN, and M. IKIRI. 1967. Kaltezittern wahrend isolierter kuhlung des ruckenmarks nach durchshneidung der hinterwurzein. *Arch. Gesamte Physiol.* **293**: 236–255.

38. POLLOCK, L. J., and L. DAVIS. 1923. Studies in decerebration. *Arch Neurol. Psychiat.* **10**: 391–398.
39. RICHET, C. 1893. Le frisson comme appareil de regulation thermique. *Arch. Biol.* **5**: 312–326.
40. SATINOFF, E. 1964. Behavioral thermoregulation in response to local cooling of the rat brain. *Amer. J. Physiol.* **206**: 1389–1394.
41. STAINOFF, E., and J. RUTSTEIN. 1970. Behavioral thermoregulation in rats with anterior hypothalamic lesions. *J. Comp. Physiol. Psychol.* **71**: 77–82.
42. SHERRINGTON, C. S. 1923-4. Notes on temperature after spinal transection with some observations on shivering. *J. Physiol. London* **58**: 405–424.
43. STUART, D. G., W. J. FREEMAN, and A. HEMINGWAY. 1962. Effects of decerebration and decortication on shivering in the cat. *Neurology* **12**: 99–107.
44. TANG, P. C., and T. C. RUCH. 1955. Non-neurogenic basis of bladder tonus. *Amer. J. Physiol.* **181**: 249–257.
45. THAUER, R. 1970. Thermosensitivity of the spinal cord, pp. 472–492. *In* "Physiological and Behavioral Temperature Regulation" Hardy, J. D., A. P. Gagge, and J. A. J. Stolwijk [Eds.]. Thomas, Springfield.
46. WANG, S. C., S. H. NGAI, and M. J. FRUMIN. 1957. Organization of central respiratory mechanisms in the brainstem of the cat: genesis of normal respiratory rhythmicity. *Amer. J. Physiol.* **190**: 333–342.
47. WUNNENBERG, W., and K. BRUCK. 1968. Single unit activity evoked by stimulation of cervical cord in the guinea-pig. *Nature* (London) **218**: 1268–1269.

BRAIN AREAS CONTROLLING THERMOREGULATORY GROOMING, PRONE EXTENSION, LOCOMOTION, AND TAIL VASODILATION IN RATS[1]

WARREN W. ROBERTS[2] AND RICHARD D. MOONEY

University of Minnesota

Most general theories and reviews of central thermoregulatory mechanisms have emphasized the preoptic area and anterior hypothalamus with respect to both temperature detection and response control. Numerous studies have shown that localized warming of the region elicits local neuronal activity changes and a variety of peripheral thermoregulatory responses, while localized lesions seriously impair body temperature regulation (Bligh, 1966; Hammel, 1968). However, only a small number of these investigations explored significantly outside of the preoptic and anterior hypothalamic region, and a systematic survey of thermoregulatory mechanisms in the brain as a whole is unavailable for any species. A few reports of thermosensitivity in the midbrain or medulla have been published (Chai, Mu, & Brobeck, 1965; Holmes, Newman, & Wolstencroft, 1960; Lipton, 1971; Nakayama & Hardy, 1969), but their interpretation is limited by use of anesthesia or decerebration, measurement of too few thermoregulatory responses, conflicting or atypical responses, or in the case of single unit recordings, uncertainty regarding the thermoregulatory function of responsive cells (Barker & Carpenter, 1970).

The purpose of the present study was to systematically explore the brain of a single species for thermosensitive zones controlling a variety of heat stress responses using conscious free-moving animals. The rat was selected because of its economy, which permitted a large number of cases, and because of its large repertoire of heat-reduction responses, which include grooming, escape-like locomotion, prone body extension, and peripheral vasodilation (Hainsworth, 1967; Rand, Burton, & Ing, 1965; Swift & Forbes, 1939). Since the rat's central thermoregulatory system has been studied relatively little, it offers an opportunity to test the generality of the classical model based largely on the cat, dog, and goat.

METHOD

Subjects

The subjects were 134 pigmented Long-Evans male rats averaging 10 mo. in age and 600 gm. in weight.

Apparatus

The diathermic warming electrodes were constructed of 4 insulated stainless steel insect pins exposed 2 mm. from the tip, and arranged so that the tips enclosed a 2-mm. cube. They were connected to a socket mounted in cement on the skull.

---

[1] This research was supported by National Institute of Mental Health Grant MH-06901 to W. W. Roberts. Grateful acknowledgement is made to Curtis Olson for assistance in testing the initial animals and to Rebecca Fish for performing the histology.

[2] Requests for reprints should be sent to Warren W. Roberts, Department of Psychology, University of Minnesota, Minneapolis, Minnesota 55455.

The source of the diathermic current was a 2-MHz. Grass LM-2 lesion maker, set for maximum output, with a variable series resistance to permit control of current level with maximum stability. An index of the current level was monitored on a low frequency oscilloscope as a rectified dc voltage across a 10-Ω resistor. An inductance was placed in parallel with the animal to shunt any dc current that might result from rectification in the biological or electronic segments of the circuit. The temperature induced by the diathermy was not measured, but current thresholds for responses were about 30% below the threshold for tissue damage. Observation of the coagulation produced by a similar electrode array in albumin indicated that the diathermic heating was maximal in the center of the array and decreased markedly within .5 mm. outside the zone enclosed by the electrodes.

The test cage was a round stabilimeter cage 27 cm. in diameter and height with 4 microswitches spaced evenly around the base.

Procedure

One week or more after stereotaxic implantation of the warming electrodes, the rats were tested with the diathermic current in the stabilimeter cage. Ambient temperature was 21°–23° C. and humidity was 30%–50%. A 40–60 min. adaptation period in the stabilimeter cage preceded each daily session. Two-minute heating trials alternated with 2-min. control trials without warming, and all trials were separated by 2–5 min. rest intervals. Current levels were low at the start of each session and were raised on successive trials until clear-cut responses were obtained or lesions made. All responses were replicated in 2–7 sessions, and trials were repeated as necessary when responses were weak or equivocal. The experimenter recorded the duration of grooming, prone body extension with the head up, extension with the head on the floor, and the number of counts produced by locomotion in the stabilimeter. The last 40 animals were also tested for vasodilation of the tail at a single high current level, using a restraint cage and a calibrated thermistor taped to the tail. The rats received a 1-hr. adaptation period in the restraint cage for 4 days and were tested after the fourth period. In addition, the last 75 animals were tested with a 60 Hz. current to determine the responses elicited by nonspecific electrical stimulation.

After completion of testing, all animals were perfused with Formalin, and frozen section histology was performed. Without knowledge of elicited responses, the electrodes were localized with a variable microprojector in the de Groot (1959) atlas, Pellegrino and Cushman (1967) atlas, or a locally constructed medullary extension based on Wunscher, Schober, and Werner (1965).

Results

Description of Responses

All 4 of the principal responses of the rat to hyperthermia were elicited by localized central diathermy: grooming, prone body extension, intermittent locomotion, and vasodilation of the tail. They closely resembled responses induced by environmental warming except in the frequency with which they were elicited separately from each other by different electrode placements. The grooming typically began with licking of the forepaws, which were then rubbed over the face and snout. Most of the shorter bouts of grooming consisted entirely of alternation between these 2 elements, but during longer bouts there sometimes appeared licking of the lower abdomen, flanks, hind feet, and rarely the tail. The body extension consisted of relaxed elongation of the body in a prone position with the normally curved spine straightened horizontally, hind legs flexed toward the sides and rear, and forelegs folded beneath the neck. At first the head usually remained in an alert position but later lowered until the chin rested on the floor, often with the eyes partially closed and retracted. The locomotion induced by central diathermy was intermittent and relatively slow. When it was elicited together with relaxation, it was usually performed in a low, stretched out posture, the belly dragging on the floor.

The vasodilation elevated the surface temperature of the tail from about 24° C., slightly above room temperature, to about 33° C., a few degrees below deep body temperature. The temperature rise was slightly greater in the distal $3/4$ of the tail than at the base. The latency on the ventral surface was about 30–45 sec. after the onset of diathermic warming, and the dorsal surface followed by about 20 sec.

Table 1 presents current thresholds and suprathreshold latencies, durations, and percentage of positive trials for grooming, body extension, and activity. Grooming and extension had similar thresholds and latencies, but their durations differed because grooming was performed intermittently, while extension was maintained nearly continuously following its first appearance on a trial. After the diathermy current was turned off, extension was usually terminated within 10–40 sec. Grooming was appreciably more variable from trial to trial than extension or activity, never approaching the near 100%

TABLE 1
Threshold, Latency, Duration, and Probability of Thermoregulatory Behavior During Warming and Control Trials

Response	Threshold (in ma.)	Latency (in sec.)[a,b]		Duration (in sec.)[a]		Percentage of trials[a]	
		Warming	Control	Warming	Control	Warming	Control
Grooming	17.1	57.0	107.0	24.9	4.3	68*	16
Extension	16.6	51.0	118.0	55.1*	6.7	86	9
Locomotion	20.4*	—	—	—	—	86[c]	13

[a] All suprathreshold 2-min. trials for effective electrodes. Control trials alternated with suprathreshold warming trials.
[b] Based on 6 rats displaying grooming and 12 extension. Negative trials included as 120 sec.
[c] Percentage of warming trials exceeding adjacent control trials.
* Different from other values in column, $p < .05$.

probability of the latter 2 responses at high current levels.

The great majority of warming electrodes elicited only 1 or 2 of the 3 behavioral responses, rather than the complete set evidenced under conditions of natural environmental warming. Of the 74 electrodes that reliably induced grooming, extension, or locomotion, only 2 (2.8%) elicited all 3 responses. Nineteen electrodes evoked pure grooming, 17 pure extension, 19 pure locomotion, 11 extension plus locomotion, and 6 grooming plus locomotion. Figure 1 illustrates this dissociation of responses. Rat 069 displayed a strong medium-threshold increase in grooming without any change in extension or activity. Rat 938 evidenced a very low threshold extension response, and a high threshold activity response that emerged only at the highest current level, which approached the lesion threshold.

Nonspecific electrical stimulation of the warming electrodes using 60-Hz. current produced a variety of nonthermoregulatory responses that never appeared during diathermy, including turning, rolling, eye clo-

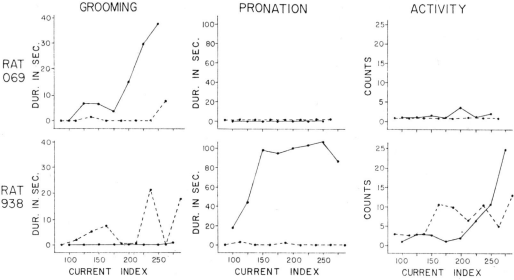

FIGURE 1. Behavioral responses induced by localized diathermic warming in 2 rats. (Rat 069's electrodes were located in the ventromedial hypothalamus, and those of 938 in the preoptic area. Solid lines represent trials with localized warming, dashed lines alternating control trials without warming. Each point is the mean of 4-7 trials.)

sure, and standing. Locomotor activity, usually combined with stereotyped motor responses, was common, but prone extension was absent. Grooming was elicited by only a few electrodes, most of which were located laterally in somatosensory structures and were unresponsive to diathermy.

Anatomical Distribution of Responses

The sagittal diagrams of Figure 2 show the location of midline electrodes. Prone extension was obtained only from the classical thermoregulatory zone in the preoptic and anterior hypothalamic region. Grooming was elicited from a more posterior zone that included the posterior hypothalamus and the anterior ventral midbrain. Locomotor activity was induced most consistently and effectively from 3 zones: the septal area, the ventral midbrain, and the dorsal medulla in the vicinity of the nucleus praepositus hypoglossi. In addition, a small proportion of the electrodes in the relaxation and grooming zones also produced some locomotor activity, usually weak or high threshold. Vasodilation of the tail was much more widely distributed than the 3 behavioral responses, being obtained from all of the zones where behavioral responses were elicited. Figure 3 presents histology for diathermy electrodes eliciting the lowest threshold extension and grooming responses and medium threshold locomotor activity.

In Figure 4, most of the laterally placed electrodes are plotted on frontal diagrams, together with midline electrodes. In general, lateral placements were less effective than midline placements, although grooming was obtained well into the lateral hypothalamus, and a few weak grooming and locomotion responses were evoked from the lateral medulla.

Within the anterior thermoregulatory zone extending from the septal area to the midbrain, thresholds were uniformly low, averaging 15–16 ma. for each millimeter division of the continuum except for the most anterior and posterior placements, which averaged 18–25 ma. The medullary thresholds averaged 21.1 ma., which was significantly higher than the average for the anterior zone ($p = .001$), but this was largely due to the predominance of activity responses in the medulla. When activity thresholds alone were compared between the 2 regions, there was only a small nonsignificant difference favoring the anterior zone.

DISCUSSION

Localized diathermic warming of the brain of the rat induced all of the principal heat-reduction responses of the species: grooming or saliva spreading, prone body extension, escape-like locomotion, and vasodilation of the tail. The behavioral responses of grooming, pronation, and locomotion were seldom elicited together, despite the general similarity of their initiating stimulus and adaptive function, but were obtained from differentiated, though sometimes partially overlapping, zones.

This functional and anatomical dissociation parallels in more pronounced form the partial separation of thermosensitive zones controlling grooming and postural relaxation in the opossum (Roberts, Bergquist, & Robinson, 1969). In both species, different components of the hyperthermic response repertoire are largely or completely controlled by separate channels from central receptor to peripheral effector. A similar functional dissociation exists in the estrus control mechanisms of the female rat, where the running activity and hypophagia of natural estrus can be separately induced by local implants of estradiol in the medial preoptic region or ventromedial hypothalamus, respectively (Wade & Zucker, 1970). Thus, in 2 major adaptive systems, the synchronized activation of different components of complex response patterns does not require integrative "centers" or neural interconnections, since the physical or chemical activating stimulus is widely enough disseminated in the brain and/or periphery to act on separate mechanisms operating in parallel. Whether similar parallel activation exists in other systems controlled by humoral factors, such as male mating, aggressive, maternal, or alimentary behavior, is unknown. However, the finding that electrical stimulation of the hypothalamus elicits similarly fractionated aggressive, male mating, and grooming responses in opossums and

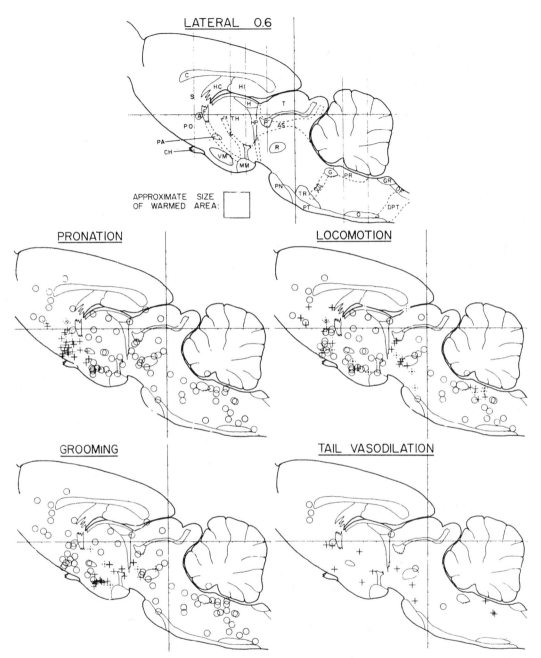

FIGURE 2. Location of warming electrodes eliciting different thermoregulatory responses in midline region of rat brain projected on sagittal diagrams located .6 mm. lateral. (Crosses represent centers of effective electrode arrays, circles ineffective arrays. Heavy crosses indicate lowest thresholds, light crosses medium thresholds, dashed crosses highest thresholds and/or weak responses. Fine vertical lines in upper diagram mark planes of frontal sections of Figure 4. Structures not in plane of section are shown in the upper diagram by dashed lines. Some animals were omitted from some diagrams because quantitative measures of particular responses were equivocal or unavailable. Diagram reconstructed from atlases of de Groot (1959) and Pellegrino and Cushman (1967) with the aid of a sagittal histological series. Abbreviations: A = commissura anterior; AS = aqueductus Sylvius; C = corpus callosum; CH = chiasma opticum; DF = funiculi dorsalis; DPT = decussatio pyramidum; F = for-

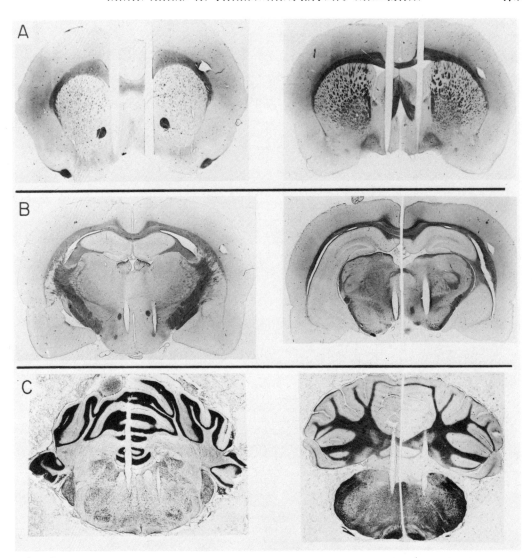

FIGURE 3. Diathermy electrodes inducing pure behavioral responses. (A. Lowest threshold prone body extension; B. lowest threshold grooming; C. medium threshold locomotion elicited from medulla. Anterior electrodes on left, posterior electrodes on right. Divided photos are composites of 2 neighboring sections.)

cats (Roberts, 1970; Roberts, Steinberg, & Means, 1967) indicates that integrative interconnections for these behaviors are absent from the hypothalamus or its efferents, although they may exist in structures afferent to it.

Because of the considerable functional differentiation of thermoreceptors in the rat, localized stimulation or lesion studies can no longer assume that measurement of body temperature alone or a single heat-loss response will reflect the degree of activation

nix; G = genu nervi facialis; GR = n. gracilis; H = habenula; HC = commissura hippocampi ventralis; HI = hippocampus; HP = tractus habenulo-interpeduncularis; M = tractus mamillothalamicus; MM = n. mamillaris; O = n. olivaris inferior; P. = commissura posterior; PA = n. paraventricularis; PN = pons; PO = area preoptica; PR = n. praepositus hypoglossi; PT = tractus pyramidalis; R = n. ruber; S = area septalis; T = tectum; TH = thalamus; TR = corpus trapezoideum; VM = n. ventromedialis hypothalami; VII = nervi facialis descendens.)

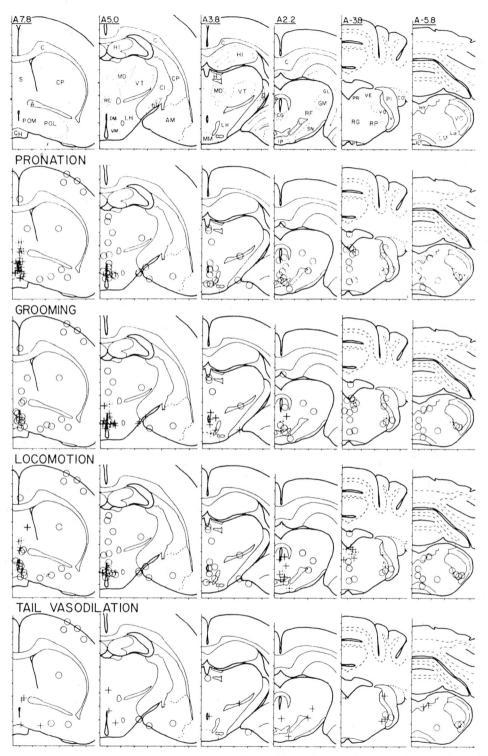

FIGURE 4. Location on frontal diagrams of medial and lateral electrode arrays that overlapped frontal planes indicated in Figure 1, upper diagram (Symbols as in Figure 1. Diagrams based on atlases of

or loss of all thermoregulatory responses. Only by recording all or most responses can systemwide effects be differentiated from alterations in specific components.

Hammel (1968, 1970) has proposed that the functional setpoint for thermoregulation is not determined by the dynamic balance between opposing heat-loss and heat-gain systems, but by a specific neural set-point mechanism localized in the preoptic and anterior hypothalamic region that compares activity in temperature-sensitive and temperature-insensitive neurons. Peripheral thermoreceptors are presumed to elicit regulatory responses indirectly by altering activity in 1 of the 2 types of neurons, which in turn shifts the set-point so as to induce appropriate responses. While this hypothesis is not ruled out by the present evidence for functional and anatomical differentiation in the rat's central thermoregulatory system, such a mechanism would have to be appreciably more complex than originally proposed. The distribution of the temperature-insensitive reference neurons would have to extend considerably beyond the preoptic and anterior hypothalamic region into the septal area, posterior hypothalamus, and midbrain to influence all responses. In addition, recent evidence obtained in our laboratory that peripheral warmth receptors induce grooming and locomotion, but not body extension, (Roberts & Martin, in press) would require that the sensory input be limited to the brain areas where local warming induced grooming and activity, avoiding the classical preoptic and anterior hypothalamic zone that controls extension. Thus, 2 different central set-points would co-exist during early exposure to ambient heat, when input from superficial peripheral receptors would lower the set-point for grooming and activity, but leave that for extension unchanged.

The failure of diathermic warming to elicit any of the nonthermoregulatory responses elicited by nonspecific 60-100 Hz. electrical stimulation through the same electrodes or much smaller electrodes in earlier studies (e.g. Woodworth, 1971) indicates that the thermal stimulus was not exciting all neurons nonspecifically. Together with the consistent effectiveness of diathermy in inducing thermoregulatory responses, this strongly suggests that the excitation produced by the diathermy was largely limited to the central temperature receptors of the thermoregulatory system. It is possible that the location of the thermal and electrical excitatory fields of a given electrode differed slightly, but the electrodes were too widely distributed throughout the brain for this small difference to explain the very large and consistent difference in responses. In addition, the wide range of intensities over which both types of stimulation were tested makes explanations in terms of different sizes of excitatory fields or different degrees of effectiveness unlikely.

The principal thermosensitive zone controlling heat reduction responses in the rat extends from the septal area through the preoptic region and hypothalamus to the ventral midbrain. This is in general agreement with single-unit recording studies that have found thermosensitive neurons in these structures in other species (Bligh, 1966; Edinger & Eisenman, 1970; Hammel, 1968; Nakayama & Hardy, 1969), and in addition, confirms that such neurons exert effective control over the major heat reduction responses of the rat. The relatively uniform temperature sensitivity from preoptic area to anterior midbrain indicates that the pre-

---

de Groot (1959) and Pellegrino and Cushman (1967) and an unpublished atlas of the medulla constructed with the aid of Wunscher, Schober, and Werner (1965). Abbreviations not in Figure 1: AM = amygdala; CI = capsula interna; CP = n. caudatus putamen; DM = n. dorsomedialis hypothalami; GL = corpus geniculatum laterale; GM = corpus geniculatum mediale; HY = n. nervi hypoglossi; IP = n. interpeduncularis; LH = area lateralis hypothalami; LM = n. reticularis lateralis magnocellularis; LP = n. reticularis lateralis parvocellularis; MD = n. mediodorsalis thalami; OT = tractus opticus; PI = pedunculus cerebellaris inferior; POL = area preopticus lateralis; POM = area preopticus medialis; RE = n. reuniens; RF = formatio reticularis mesencephali; RG = n. reticularis gigantocellularis; RP = n. reticularis parvocellularis; SN = substantia nigra; V = tractus spinalis nervi trigemini; VC = n. caudalis tractus spinalis nervi trigemini; VE = n. vestibularis; VO = n. oralis tractus spinalis nervi trigemini; VT = complexus ventralis thalami.)

optic and anterior hypothalamic portion does not have the special sensitivity and importance in the rat that it appears to have in the cat and dog (Hemingway, Rasmussen, Wikoff, & Rasmussen, 1940; Magoun, Harrison, Brobeck, & Ranson, 1938). This is consistent with, and possibly due to, the anatomically differentiated control of different responses within the zone. The adaptive function of the differentiated control is unclear, but may be related to selective activation of different responses by different inputs (such as peripheral thermoreceptors, which induce grooming and locomotion without body extension; Roberts & Martin, in press) or nonthermoregulatory factors (such as circadian cycles, thirst, feeding, maternal condition, etc., that differentially activate or inhibit the 3 behaviors). A second possibility is that the more posterior location of the grooming area may make it less subject to direct respiratory cooling from the nasal air passage, thus insuring the persistence of grooming when hyperthermia is generated endogenously by exercise or a large meal (Abrams & Hammel, 1964; Thompson & Stevenson, 1965) at moderate ambient temperatures.

A second thermosensitive zone controlling activity was found in the dorsomedial medulla in the vicinity of the nucleus praepositus hypoglossi. A few weak or high threshold cases of activity, grooming, and tail dilation were also found in the lateral medulla in the vicinity of the trigeminal nuclei and tract. Holmes et al. (1960) and Chai et al. (1965) have elicited respiratory and circulatory changes during diathermic warming of somewhat similar areas in the medulla in anesthetized or decerebrate cats, but the significance of the responses is unclear because they vary with type of anesthesia and differ from those induced by environmental or preoptic warming. The present finding that activity, grooming, and tail dilation can be induced in conscious rats by medullary warming indicates that bulbar thermoreceptors are capable of inducing most of the normal heat-reduction responses in this species, albeit less effectively than the larger and more sensitive rostral zone.

An additional process of thermoregulation that was not investigated in the present study is the activation of learned responses, such as bar pressing, that produce cooling of thermoreceptors by means of a draft of cool air or a fine water spray (Corbit, 1970). Murgatroyd and Hardy (1970) and Corbit (1970) induced some bar pressing for cool stimuli with warming of the anterior hypothalamic-preoptic region of the rat, but the learned behavior had a higher threshold than other thermoregulatory responses, and was somewhat unreliable. Whether the other thermosensitive areas found in the present study are effective requires direct investigation.

The most consistent temperature-reducing behavioral response in studies of central warming in the opossum, rat, ground squirrel, cat, and dog (Hemingway et al., 1940; Roberts & Robinson, 1969; Roberts et al., 1969; Williams & Heath, 1970) is relaxation in a prone or lateral position. Natural environmental warming also induces relaxation in many mammalian orders (Robinson & Lee, 1946), reaching the extreme of torpidity and estivation in certain species (Hudson & Bartholomew, 1964). In a recent comparative study of responses to environmental warming in rats, guinea pigs, hamsters, mice, and gerbils (Roberts, Mooney, & Martin, in press), relaxation was displayed by all species. However, activity failed to increase in tests with mice, and grooming significantly declined in tests with mice and hamsters (although evaporative water loss increased, probably because of drooling or naso-oral breathing, Schmidt-Nielsen, Bretz, & Taylor, 1970). Thus, postural relaxation, which tends to be underemphasized in most discussions of thermoregulatory responses, is probably one of the most prevalent and important responses to hyperthermia across the mammalian class.

It is frequently assumed that the optimal environmental temperature at which thermoregulatory responses are minimal is the value at which metabolism is lowest, often called the "zone of thermal neutrality" (Davson, 1964). However, the finding that postural relaxation is a prominent response to central warming in a variety of species indicates that reduced metabolism is itself a thermoregulatory response to hyperthermia. Since the threshold for relaxation is lower in many species than thresholds for metabolism increasing hyperthermic responses, such as grooming, locomotion, and panting (Robinson & Lee, 1946), the ambient temperature of minimal metabolism is probably somewhat higher than the true thermally neutral temperature at which thermoregulatory responses are mobilized least. For example, the thermally neutral temperature for the rat is usually given as 28°–32° C. (Altman & Dittmer, 1966; Gelineo, 1964), but Corbit (1970) and Murgatroyd and Hardy (1970) have found that rats will press a bar to obtain cool air at that level, and minimize their responding for both cool and warm stimulation between 21°–24° C.

REFERENCES

Abrams, R., & Hammel, H. T. Hypothalamic temperature in unanesthetized albino rats during feeding and sleeping. *American Journal of Physiology*, 1964, **206**, 641–646.

Altman, P. L., & Dittmer, D. S. *Environmental biology*. Bethesda, Md.: Federation of American Societies for Experimental Biology, 1966.

Barker, J. L., & Carpenter, D. O. Thermosensitivity of neurons in the sensorimotor cortex of the cat. *Science*, 1970, **169**, 597–598.

Bligh, J. The thermosensitivity of the hypothalamus and thermoregulation in mammals. *Biological Reviews*, 1966, **41**, 317–367.

Chai, C. Y., Mu, J. Y., & Brobeck, J. R. Cardiovascular and respiratory responses from local heating of medulla oblongata. *American Journal of Physiology*, 1965, **209**, 301–306.

Corbit, J. D. Behavioral regulation of body temperature. In J. D. Hardy, A. P. Gagge, & J. A. J. Stolwijk (Eds.), *Physiological and behavioral temperature regulation*. Springfield, Ill.: C C Thomas, 1970.

Davson, H. *General physioogy*. London: Churchill, 1964.

de Groot, J. The rat forebrain in stereotaxic coordinates. *Verhandelingen der Koninklijke Nederlandsche Akademie van Wetenschappen (Natuurkunde) Afd. Natuurk.*, 1959, **52**, 1–40.

Edinger, H. M., & Eisenman, J. S. Thermosensitive neurons in tuberal and posterior hypothalamus of cats. *American Journal of Physiology*, 1970, **219**, 1098–1103.

Gelineo, S. Organ systems in adaptation: The temperature regulating system. In D. B. Dill (Ed.), *Handbook of physiology*. Sect. 4. *Adaptation to the environment*. Washington, D. C.: American Physiological Society, 1964.

Hainsworth, F. R. Saliva spreading, activity, and body temperature regulation in the rat. *American Journal of Physiology*, 1967, **212**, 1288–1292.

Hammel, H. T. Regulation of internal body temperature. *Annual Review of Physiology*, 1968, **30**, 641–710.

Hammel, H. T. Concept of the adjustable set temperature. In J. D. Hardy, A. P. Gagge, & J. A. J. Stolwijk (Eds.), *Physiological and behavioral temperature regulation*. Springfield, Ill.: C C Thomas, 1970.

Hemingway, A., Rasmussen, T., Wikoff, H., & Rasmussen, A. T. Effects of heating hypothalamus of dogs by diathermy. *Journal of Neurophysiology*, 1940, **3**, 329–338.

Holmes, R. L., Newman, P. P., & Wolstencroft, J. H. A heat-sensitive region in the medulla. *Journal of Physiology*, 1960, **152**, 93–98.

Hudson, J. W., & Bartholomew, G. A. Terrestrial animals in dry heat: Estivators. In D. B. Dill, E. F. Adolph, & C. G. Wilber (Eds.), *Handbook of physiology*. Sect. 4. *Adaptation to the environment*. Washington, D. C.: American Physiological Society, 1964.

Lipton, J. M. Thermal stimulation of the medulla alters behavioral temperature regulation. *Brain Research*, 1971, **26**, 439–442.

Magoun, H. W., Harrison, F., Brobeck, J. R., & Ranson, S. W. Activation of heat loss mechanisms by local heating of the brain. *Journal of Neurophysiology*, 1938, **1**, 101–114.

Murgatroyd, D., & Hardy, J. D. Central and peripheral temperatures in behavioral thermoregulation of the rat. In J. D. Hardy, A. P. Gagge, & J. A. J. Stolwijk (Eds.), *Physiological and behavioral temperature regulation*. Springfield, Ill.: C C Thomas, 1970.

Nakayama, T., & Hardy, J. D. Unit responses in the rabbit's brain stem to changes in brain and cutaneous temperature. *Journal of Applied Physiology*, 1969, **27**, 848–857.

Pellegrino, L. J., & Cushman, A. J. *A stereotaxic atlas of the rat brain*. New York: Appleton-Century-Crofts, 1967.

Rand, R. P., Burton, A. C., & Ing, T. The tail of the rat in temperature regulation and acclimatization. *Canadian Journal of Physiology and Pharmacology*, 1965, **43**, 257–267.

Roberts, W. W. Hypothalamic mechanisms for motivational and species-typical behavior. In R. E. Whalen, R. F. Thompson, M. Verzeano, & N. M. Weinberger (Eds.), *The neural control of behavior*. New York: Academic Press, 1970.

Roberts, W. W., Bergquist, E. H., & Robinson, T. C. L. Thermoregulatory grooming and sleep-like relaxation induced by local warming of preoptic area and anterior hypothalamus in opossum. *Journal of Comparative and Physiological Psychology*, 1969, **67**, 182–188.

Roberts, W. W., & Martin, J. R. Peripheral thermoreceptor control of thermoregulatory response of rat. *Journal of Comparative and Physiological Psychology*, in press.

Roberts, W. W., Mooney, R. D., & Martin, J. R. Thermoregulatory behaviors of laboratory rodents. *Journal of Comparative and Physiological Psychology*, in press.

ROBERTS, W. W., & ROBINSON, T. C. L. Relaxation and sleep induced by warming of preoptic region and anterior hypothalamus in cats. *Experimental Neurology*, 1969, **25**, 282–294.

ROBERTS, W. W., STEINBERG, M. L., & MEANS, L. W. Hypothalamic mechanisms for sexual, aggressive, and other motivational behaviors in the opossum, *Didelphis virginiana*. *Journal of Comparative and Physiological Psychology*, 1967, **64**, 1–15.

ROBINSON, K. W., & LEE, D. H. K. Animal behavior and heat regulation in hot atmospheres. *Papers from the Department of Physiology, University of Queensland*. 1946, **1**, 1–8.

SCHMIDT-NIELSEN, K., BRETZ, W. L., & TAYLOR, C. R. Panting in dogs: Unidirectional airflow over evaporative surfaces. *Science*, 1970, **169**, 1102–1104.

SWIFT, R. W., & FORBES, R. M. The heat production of the fasting rat in relation to the environmental temperature. *Journal of Nutrition*, 1939, **18**, 307–318.

THOMPSON, G. E., & STEVENSON, J. A. F. The temperature response of the male rat to treadmill exercise, and the effect of anterior hypothalamic lesions. *Canadian Journal of Physiology and Pharmacology*, 1965, **43**, 279–287.

WADE, G. N., & ZUCKER, I. Modulation of food intake and locomotor activity in female rats by diencephalic hormone implants. *Journal of Comparative and Physiological Psychology*, 1970, **72**, 328–336.

WILLIAMS, B. A., & HEATH, J. E. Responses to preoptic heating and cooling in a hibernator, *Citellus tridecemlineatus*. *American Journal of Physiology*, 1970, **218**, 1654–1660.

WOODWORTH, C. H. Attack elicited in rats by electrical stimulation of the lateral hypothalamus. *Physiology and Behavior*, 1971, **6**, 345–353.

WUNSCHER, W., SCHOBER, W., & WERNER, L. *Architektonischer Atlas vom Hirnstamm der Ratte*. Leipzig: Hirzel, 1965.

(Received January 27, 1973)

THE EFFECTS OF HEATING AND COOLING THE SPINAL CORD AND HYPOTHALAMUS ON THERMOREGULATORY BEHAVIOUR IN THE PIG

BY H. J. CARLISLE* AND D. L. INGRAM

From the A.R.C. Institute of Animal Physiology, Babraham, Cambridge

(Received 16 November 1972)

SUMMARY

1. The effects of warming and cooling the spinal cord and hypothalamus on operant thermoregulatory behaviour and posture have been studied in the pig at neutral and cold ambient temperatures.

2. Cooling the spinal cord increased and warming decreased the rate of obtaining thermal reinforcement. The response to cooling began with the onset of the stimulus and persisted for up to 5 min followed by a diminution in rate during the remaining 15 min of cooling. The peak of this 'on' response was greater the lower the ambient temperature. The response to heating was a small reduction in rate of reinforcement.

3. The 'on' response to cooling the spinal cord was related to changes in temperature of only the cervical region of the cord.

4. Cooling the hypothalamus led to an increase in the rate of obtaining heat and this increase was sustained during the 20 min of central cooling. Termination of cooling was followed by a marked depression in rate. Heating the hypothalamus had only a weak inhibitory effect on rate of reinforcement.

5. While working for external heat during periods when thermodes over the spine and in the hypothalamus were not being cooled, pigs lay in 'cold defensive' prone positions 25 % of the time and lay on their sides 75 % of the time. During cooling of the spinal cord the time spent in the prone position was 95 % at 5 and 15° C ambients and 71 % at a 25° C ambient. During cooling of the hypothalamus the prone posture was adopted 50 % of the time.

6. When the temperatures of the spinal cord and of the hypothalamus were changed in opposite directions, the operant response was determined by the temperature of the hypothalamus while the postural re-

* On leave from: Department of Psychology, University of California, Santa Barbara, California 93106, U.S.A.

sponse was most frequently determined by the temperature of the spinal cord.

INTRODUCTION

The importance of the hypothalamus in thermoregulatory function is well known. In recent years considerable evidence has accumulated to show that the spinal cord is also important (Klussmann & Pierau, 1972; Thauer, 1970). In the studies of Jessen and co-workers, heat production and evaporative heat loss responses elicited by cooling or warming of the spinal cord of the dog were very similar to responses elicited by comparable treatment of the hypothalamus and it was suggested that the cord and hypothalamus are basically equivalent core sensors of temperature (Jessen & Mayer, 1971; Jessen & Ludwig, 1971; Jessen & Simon, 1971). The temperature of the spinal cord in the pig is known to influence respiratory frequency (Ingram & Legge, 1972a), blood flow (Ingram & Legge, 1971) and oxygen consumption (Carlisle & Ingram, 1973), and to interact with thermal stimuli applied to the hypothalamus and scrotum (Ingram & Legge, 1972b).

In the studies cited above, the responses measured have been physiological indices of thermoregulation such as blood flow and evaporative heat loss. The behaviour of an animal is also an important variable in thermoregulation (Corbit, 1970), and its contribution has been assessed during cooling and warming of the hypothalamus in the pig (Baldwin & Ingram, 1967) and a variety of other species (Cabanac, 1972). In the present study, thermoregulatory behaviour was used to assess the effect of spinal as well as hypothalamic thermal stimulation. Postural changes have been demonstrated during cooling of the spinal cord in the pigeon (Rautenberg, 1969), and thus performance of a complex temperature-motivated behavioural response might be expected.

METHODS

Animals

Eighteen pigs of the Large White breed aged between 8 and 14 weeks and weighing between 12 and 19 kg were used. Thermodes were inserted into the vertebral canal of ten of these, the hypothalamus of two, and both the spinal cord and hypothalamus of three. An additional three animals were used in acute control experiments. The animals were housed separately in standard pig pens and fed twice daily.

Thermodes and temperature measurement

The thermodes were similar to those used in previous studies (Baldwin & Ingram, 1967; Ingram & Legge, 1971). The thermodes over the spinal cord consisted of a double loop of 2 mm o.d. polyethylene tubing. The thermodes inserted into the hypothalamus consisted of two stainless-steel tubes ending in a disk-shaped tank 5 mm in diameter. The temperature of the spinal cord was sensed by a thermojunction inserted into a blind-ended polyethylene catheter positioned alongside the thermode. Hypothalamic temperature was sensed by a bead thermistor affixed close to the disk of the thermode. The rate of flow through the thermode of cold alcohol, or warm water maintained in a constant-temperature bath, was varied by adjusting inlet pressure or outlet vacuum to provide the desired temperature as measured by the thermistor of thermojunction in proximity to the thermode.

Anaesthesia and surgery

Phencyclidine hydrochloride (Sernylan, Parke-Davis) was given first followed by halothane (Fluothane) delivered through an endotracheal tube from a Boyle's apparatus. All surgical procedures were carried out under sterile conditions. The thermodes over the spinal cord were inserted into the epidural space of the vertebral canal at the level of T10 and pushed rostrally to the level of C1 with the position checked radiographically. A blind-ended catheter which accepted a thermocouple was also inserted with the thermode. The thermodes in the hypothalamus were positioned within 1 mm of the mid line between the optic chiasma and the anterior commissure. A stereotaxic instrument was used to guide the thermode, and its position was verified by multiple radiographs.

Apparatus for heat reinforcement

The apparatus was similar to that described previously (Baldwin & Ingram, 1967). The pig was lightly restrained in a metal frame stall. A Perspex lever which reached to within a few cm of the floor was positioned in front of the pig. The pig could press the lever with its snout while standing, sitting or lying down. Pressure on the lever closed a microswitch and activated 12 red-bulb infra-red heat lamps arranged within a 60 × 40 cm area and mounted 1 m above the floor. The power dissipated in the lamps was 3 kW. Switch closure activated the lamps for 5 sec, and each such activation is termed a 'reinforcement'. The apparatus was placed in a temperature-controlled room (accuracy $\pm 0.5°$ C) with all measuring and control equipment located outside. The animals could be observed by means of a window and mirror arrangement.

Experimental procedure

The animals were provided with a harness by means of which they could be lightly restrained in the stall. They could stand, sit, or lie down, but were not able to turn round. Several sessions of at least an hour were given at a neutral ambient temperature (25° C) to permit the pig to become accustomed to the conditions followed by training at 5° C. Two sessions of 4 hr each at 5° C were sufficient to ensure that the animals were operating the lever and heat lamps proficiently. Surgery was then performed, and 3 or 4 days allowed for recuperation. Experiments were carried out within 2 weeks after recuperation from surgery to assure that the position of the spinal thermode did not change appreciably as the animal grew.

The pigs were tested at ambient temperatures of 5, 15 and 25° C. The standard procedure was either to cool or heat the spinal cord or hypothalamus for 20 min after base line rates of reinforcement had been obtained. Thermode temperatures of 20 or 43° C were standard, but in some experiments cooling was to 10 or 30 as well as 20° C.

The number of reinforcements obtained was noted at 5 min intervals, and in most experiments the predominant posture of the animal was also noted for each 5 min period. Three postures were noted: standing, sitting prone, or lying on the side. Positions intermediate between the prone and lateral posture were also observed. In the prone posture, the pig typically lay on its ventral surface with extremities retracted.

The positions of the spinal thermodes were determined radiographically at the end of the first and second weeks of testing. In some control experiments the position of the spinal thermode was purposely moved caudally and the animal again tested to determine the effect of thermode position on the rate at which heat reinforcement was obtained.

RESULTS

Effect of spinal cord temperature on rate of heat reinforcement

Fig. 1 shows the mean number of reinforcements obtained per 5 min period at three ambient temperatures when the spinal cord was heated or cooled. No attempt was made to heat the cord at an ambient temperature of 25° C since the rate of obtaining reinforcement was already low in the absence of any treatment. The data in Fig. 1 are based on the following number of experiments: cooling the spinal cord at 5° C (eight pigs and twenty-four experiments), 15° C (nine pigs and thirty-three experiments) and 25° C (eight pigs and seventeen experiments); warming the spinal cord at 5° C (nine pigs and sixteen experiments) and 15° C (seven pigs and thirteen experiments). The dashed line in each panel indicates the base line rate of reinforcement obtained from the mean of the 20 min period before thermal stimulation. A number of features are evident in the Figure: (a) the base line rate of reinforcement was inversely related to ambient temperature; (b) the onset of cooling the thermode induced an increase in the rate of obtainng heat and an analysis of variance revealed that this was significant at all ambient temperatures ($P < 0.01$); (c) reinforcement rate diminished but was above base line during the last 15 min of cord cooling at all ambient temperatures; (d) heating the spinal cord resulted in a reduction in rate of obtaining heat although the effect was not apparent during the first 5 min of heating.

The mean rate of obtaining reinforcements during the 20 min periods of heating or cooling was plotted as a function of ambient temperature (Fig. 2). Analysis of variance revealed that heating led to a significant decrease and cooling an increase in rate of reinforcement ($P < 0.05$).

In twenty-seven experiments with five animals at ambients of 5 and 15° C, the thermode over the spine was cooled to 10 or 30° C. The response was of the same form as when the thermode was cooled to 20° C (i.e. an 'on' response followed by a diminution in rate).

Effect of hypothalamic temperature on rate of heat reinforcement

Fig. 1F shows the rate of obtaining heat reinforcement during cooling of the thermode in the hypothalamus to 20° C at three ambient temperatures. The figures are based on six experiments with three pigs at 5° C, ten experiments with five pigs at 15° C, and seven experiments with four pigs at 25° C. The pigs obtained more heat reinforcements when the thermode was cooled. The magnitude of this effect was greater for lower ambient temperatures, and the increased rate was maintained during the entire 20 min period of cooling at ambient temperatures of 5 and 15° C. At an ambient temperature of 25° C there was typically a latency of a few

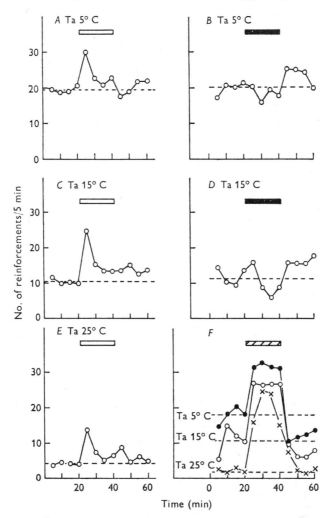

Fig. 1. Mean number of reinforcements per 5 min period before, during and after cooling of the spinal thermode to 20° C (open bar) at three ambient temperatures (A, C, E), heating of the spinal thermode to 43° C (black bar) at two ambient temperatures (B, D), and cooling of the hypothalamic thermode to 20° C (hatched bar) at three ambient temperatures (F). The dashed lines indicate the base line rate of reinforcement.

minutes before the pigs began to respond at a high rate and then the rate of reinforcement diminished during the last 5 min of the period. At ambient temperatures of 5 and 15° C, the reinforcement rate was greatly reduced after cooling was stopped. The values obtained in these experiments are plotted in Fig. 2 as a function of ambient temperature.

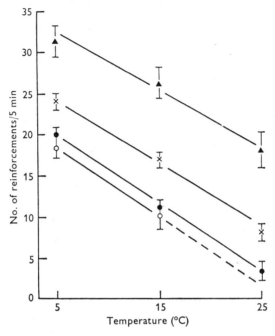

Fig. 2. Mean number of reinforcements per 5 min period averaged over 20 min before thermal stimulation (●), during cooling of the spinal cord to 20° C (×), heating of the spinal cord to 43° C (○), and cooling of the hypothalamus to 20° C (▲) as a function of ambient temperature. Vertical lines indicate S.E. of mean, and the dashed line is an extrapolation.

Effect of spinal and hypothalamic temperature on postural adjustments

The typical posture of the pig while working for radiant heat was to lie on one lateral surface with limbs extended. The response lever was easily flicked with the snout from this position. After a variable period of time, the surface exposed to the heat lamps was reversed. At ambient temperatures of 5, 15 and 25° C, the pigs spent an average of 30, 22 and 18 % of the time respectively, in a prone position. The remainder of the time was spent lying on the side; standing was rarely observed. During cooling of the spinal cord, the pigs lay in a 'cold-defensive' prone position with extremities retraced under the body. This position was taken immediately after the onset of the stimulus, maintained during the remainder of the cooling period, and frequently for 5 min thereafter. At ambients of 5 and 15° C, the prone position was adopted 95 % of the time. At an ambient of 25° C, the prone posture was adopted 71 % of the time. When the hypothalamus was cooled, the pigs assumed the prone position about 50 % of the time regardless of ambient temperature, but the extremities were frequently extended even when the prone position was adopted. The remaining 50 % of the time was spent on a lateral surface (Table 1).

The interaction of spinal and hypothalamic temperature

In three pigs bearing thermodes both in the hypothalamus and over the spinal cord, the temperature of each thermode was opposed (20° C vs. 43° C) in twelve experiments at ambients of 5 and 15° C. Ambient temperature had little effect on the response and therefore the results were averaged (Fig. 3). The rate of obtaining radiant heat was related to the

TABLE 1. Mean percentage of 5 min periods spent in a cold-defensive prone posture

Ambient temperature (°C)	Control	Cool spine (20° C)	Cool hypothalamus (20° C)
5	30	94	48
15	12	96	53
25	18	71	45

temperature of the thermode in the hypothalamus. In ten of these experiments, the posture of the animal was noted, and found to relate to the temperature of the thermode in the hypothalamus in three experiments and to the temperature of the thermode over the spinal cord in six experiments. In one experiment the posture was intermediate.

In four experiments with two animals, the response obtained from simultaneously cooling the thermodes in the hypothalamus and over the spinal cord to 20° C was no greater than that obtained from cooling the hypothalamus alone.

Two examples of the interaction of spinal and hypothalamic temperature are shown in Figs. 4 and 5. In the experiment illustrated in Fig. 4 the hypothalamus was cooled to 20° C for 40 min and during this period the thermode over the cord was warmed to 43° C for 20 min. Reinforcement rate fell to about 25 reinforcements per 5 min during the last 10 min of warming the spinal cord and the first 5 min after termination of warming. This was an instance in which the temperature of the spinal cord appeared to oppose the influence of hypothalamic temperature in determining the rate of reinforcement. The results of another experiment in which the thermode in the hypothalamus was warmed and the thermode over the cord cooled are shown in Fig. 5. The typical 'on' response to cooling the spinal cord was greatly diminished when the hypothalamus was warm. The marked increase in rate of obtaining reinforcements at the end of the test was unusual and more typically seen immediately after warming the cord had been stopped.

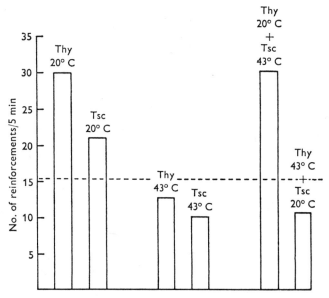

Fig. 3. Mean number of enforcements per 5 min period during cooling of hypothalamus (Thy) or spinal cord (Tsc) and during heating of the hypothalamus or spinal cord separately, and when the temperature of each thermode was simultaneously opposed. Dashed line indicates control rate of reinforcement.

Effect of thermode position over the spine of reinforcement rate

In five pigs the thermode over the cord was moved caudally and the pig's responses tested at an ambient temperature of 15° C. The 'on' response was present in two pigs when the thermode tip was moved to C5–C6, but was greatly reduced in one pig and absent in two others when the thermode tip was at C7–T1.

Effect of cooling or heating the thermode over the spine on tissue temperatures

In three experiments which were completed under general anaesthesia, the spinal cord was exposed and thermojunctions inserted in several sites. A thermode was also implanted over the spinal cord as in previous studies. No effect of either heating or cooling the spine was detected in hypo-

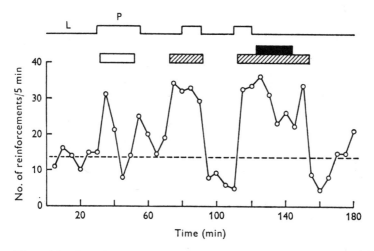

Fig. 4. Number of reinforcements per 5 min when the spinal thermode was cooled to 20° C (open bar), the hypothalamic thermode was cooled to 20° C (hatched bar), and when the spinal thermode was warmed to 43° C (black bar) during cooling of the hypothalamus. L indicates that the animal was lying on a lateral surface and P indicates a prone posture. The dashed line indicates the base line rate of reinforcement. Ambient temperature 5° C.

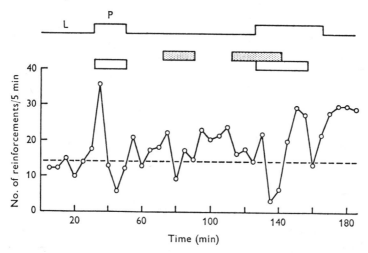

Fig. 5. Number of reinforcements per 5 min when the spinal thermode was cooled to 20° C (open bar) and the hypothalamic thermode warmed to 43° C (stippled bar). L indicates that the animal was lying on a lateral surface and P indicates a prone posture. The dashed line indicates the base line rate of reinforcement. Ambient temperature 5° C.

thalamic temperature. Large variations in intraspinal temperature were however noted during both warming and cooling. When the thermode temperature as judged by a thermojunction threaded down the blind-ended catheter was 10° C the tissue close to the thermode was 12–14° C and on the opposite side of the cord it was 28° C. When the thermode was cooled to 20° C the tissue near to the thermode was 22–26° C and on the far side it was 30–31° C. The temperature in the hypothalamus was 36–38° C while these measurements were being made. Temperature differences within the cord of up to 3° C were noted when the thermode was warmed to 43° C. From measurements of the temperatures in the inlet and outlet of the thermode and the rate of perfusion it was estimated that heat was removed or added at a rate of 2·5–5 W, which would change body temperature by 0·1° C or less during 20 min.

DISCUSSION

The behavioural responses to cooling the spinal cord and the hypothalamus are similar in that the number of heat reinforcements and postural adjustments characteristic of cold defence both increase in frequency. But there are major differences in the magnitude of each response. Cooling the spinal cord elicited a brief but substantial increase in rate of obtaining heat while cooling the hypothalamus elicited a substantial and sustained increase in the demand for external heat, especially when ambient temperature was low. Termination of the cold stimulus was followed by a decrease in the demand for external heat when the hypothalamus had been cooled, but not when the spinal cord had been cooled. On the other hand, cold-defensive postural adjustments were reliably elicited by cooling the spinal cord but not by cooling the hypothalamus. The pigs remained on a lateral surface as frequently as they assumed a prone position during cooling of the hypothalamus, and even when lying prone the limbs were often extended. These observations indicate that the control systems for temperature-dependent posture and motivation for obtaining heat are not identical, the latter being more strongly influenced by hypothalamic stimulation and the former by stimulation of the spinal cord. This generalization was especially evident when the temperatures of the two regions were opposed. The operant response was then determined by the temperature of the hypothalamus while the postural response most frequently related to the temperature of the spinal cord. These results cannot be attributed to ineffective spinal stimulation since comparable stimulation elicits appropriate and reliable changes in oxygen consumption, blood flow, and respiration in the pig (Carlisle & Ingram, 1973; Ingram & Legge, 1971, 1972a).

Heating either the spinal cord or the hypothalamus decreased the rate of obtaining heat reinforcement. The change in response frequency appeared to be weaker than that observed when the thermodes were cooled, and similar observations have been made on the basis of previous work (Baldwin & Ingram, 1967). In contrast, some species (e.g. the ox) show a greater response to heating than to cooling (Jessen, McLean, Calvert & Findlay, 1972), and a preliminary report by Cabanac (1972) noted no effect of cooling the spine on operant response rate for cool air reinforcement in the dog, while warming the cord increased the rate at ambient temperatures as low as 10° C.

It is not clear why cooling the spinal cord elicited only a brief increase in rate of obtaining heat. This form of a response suggests that the rate of change of the stimulus may be an important parameter. In this connexion, Simon & Iriki (1971) have shown that a dynamic component is obtained from somewhat less than half the warm- and cold-sensitive units identified in the anterolateral column of the spinal cord of the cat, while Wünnenberg & Brück (1970) noted an effect of rate of change in spinal units which responded to warming the cord between C5–T2 in the guinea-pig. This evidence for a dynamic component in temperature-sensitive spinal units is of some interest, but it is not clear what significance this may have in large mammals like the pig where rapid rates of change would not be expected to occur in core structures.

A problem in the present study is the difficulty of comparing thermal stimulation of the spinal cord and hypothalamus. The surface areas of the two types of thermode are very different, and the change in temperature with distance from the thermode is poorly understood since this depends on the vascular supply to the region and position of the temperature-sensing element with respect to the thermode. In the acute experiments in this study, marked temperature differences were observed within the cord and it seems likely that large temperature gradients exist in and around the spinal cord during thermal stimulation. This difficulty is compounded by the lack of knowledge of the specific site of thermosensitive cells within the spinal cord. These reservations concerning the specification of the stimulus must be taken into account when comparing the effects of thermal stimulation of the hypothalamus and spinal cord. Nevertheless, the pre-eminence of the hypothalamus in the control of an operant response and of the spinal cord in the control of a postural response must reflect more substance than can be attributed to an ambiguous stimulus.

We wish to thank Mr D. E. Walters of the A.R.C. Statistics Group in Cambridge for his help with the statistical analysis.

H.J.C. was supported by Special Research Fellowship MH-51485 from the U.S. Public Health Service.

REFERENCES

BALDWIN, B. A. & INGRAM, D. L. (1967). The effect of heating and cooling the hypothalamus on behavioural thermoregulation in the pig. *J. Physiol.* **191**, 275–392.

CABANAC, M. (1972). Thermoregulatory behavior. In *Essays on Temperature Regulation*, ed. BLIGH, J. & MOORE, R., pp. 19–36. Amsterdam: North Holland Publishing Co.

CARLISLE, H. J. & INGRAM, D. L. (1973). The influence of body core temperature and peripheral temperatures on oxygen consumption in the pig. *J. Physiol.* **231**, 341–352.

CORBIT, J. D. (1970). Behavioral regulation of body temperature. In *Physiological and Behavioural Temperature Regulation*, ed. HARDY, J. D., GAGGE, A. P. & STOLWIJK, J. A. J., chap. 53, pp. 777–801. Illinois: C. Thomas.

INGRAM, D. L. & LEGGE, K. F. (1971). The influence of deep body temperature and skin temperature on peripheral blood flow in the pig. *J. Physiol.* **215**, 693–707.

INGRAM, D. L. & LEGGE, K. F. (1972a). The influence of deep body temperature and skin temperature on respiratory frequency in the pig. *J. Physiol.* **220**, 283–296.

INGRAM, D. L. & LEGGE, K. F. (1972b). The influence of deep body and skin temperatures on thermoregulatory responses to heating of the scrotum in pigs. *J. Physiol.* **224**, 477–487.

JESSEN, C. & LUDWIG, O. (1971). Spinal cord and hypothalamus as core sensors of temperature in the conscious dog. II. Addition of signals. *Pflügers Arch. ges. Physiol.* **324**, 205–216.

JESSEN, C., MCLEAN, J. A., CALVERT, D. T. & FINDLAY, J. D. (1972). Balanced and unbalanced temperature signals generated in spinal cord of the ox. *Am. J. Physiol.* **222**, 1343–1347.

JESSEN, C. & MAYER, E. T. (1971). Spinal cord and hypothalamus as core sensors of temperature in the conscious dog. I. Equivalence of responses. *Pflügers Arch. ges. Physiol.* **324**, 189–204.

JESSEN, C. & SIMON, E. (1971). Spinal cord and hypothalamus as core sensors of temperature in the conscious dog. III. Identity of functions. *Pflügers Arch. ges. Physiol.* **324**, 216–226.

KLUSSMANN, F. W. & PIERAU, F. K. (1972). Extrahypothalamic deep body thermosensitivity. In *Essays in Temperature Regulation*, ed. BLIGH, J. & MOORE, R., pp. 87–104. Amsterdam: North Holland Publishing Co.

RAUTENBERG, W. (1969). Die Bedeutung der Zentrelnervosen thermosensitivität für die temperatur regulation der Taube. *Z. vergl. Physiol.* **62**, 235–266.

SIMON, E. & IRIKI, M. (1971). Sensory transmission of spinal heat and cold sensitivity in ascending spinal neurons. *Pflügers Arch. ges. Physiol.* **328**, 103–120.

THAUER, R. (1970). Thermosensitivity of the spinal cord. In *Physiological and Behavioral Temperature Regulation*, ed. HARDY, J. D., GAGGE, A. P. & STOLWIJK, J. A. J. ch. 33, pp. 472–492. Illinois: C. Thomas.

WÜNNENBERG, W. & BRÜCK, K. (1970). Studies on the ascending pathways from the thermosensitive region of the spinal cord. *Pflügers Arch. ges. Physiol.* **321**, 233–241.

INDIFFÉRENCE A LA DOULEUR ET CONFORT THERMIQUE (2 cas)

Par M. CABANAC[*], P. RAMEL[**], R. DUCLAUX[*] et M. JOLI[**] (Lyon)

Les stimulations sensorielles thermiques et gustatives donnent naissance à des sensations composées de deux éléments : l'un discriminatif est lié aux caractéristiques physiques du stimulus, l'autre affectif dépend de l'état du milieu intérieur (3, 4). Tout stimulus utile est agréable, tout stimulus menaçant l'homéostasie est désagréable. On peut se demander si ce phénomène affectif est spécifique pour chaque sensation ou s'il est au contraire commun à toutes les sensibilités.

Dans l'indifférence congénitale à la douleur décrite par Dearborn (1932) [Voir aussi Baxter et coll. ; Girard et coll.], les sujets ne perçoivent pas comme désagréables les stimulations douloureuses, mais sont entièrement normaux par ailleurs. On peut postuler que ce syndrome consiste en l'absence de l'élément affectif pour la sensation douloureuse cependant perçue, il est donc intéressant d'explorer dans une telle hypothèse la perception affective des autres sensations.

Nous avons observé deux cas d'indifférence à la douleur, l'une congénitale, l'autre acquise, pour lesquels ont été explorés la sensation thermique cutanée, le confort thermique, et chez l'un des sujets la sensation gustative sucrée et salée.

La méthode d'exploration a consisté à reprendre chez ces deux sujets les expériences faites chez des sujets normaux (2) et à comparer les résultats. L'immersion dans un bain à 33° C n'est pas désagréable, le séjour prolongé est suivi d'une baisse progressive des températures internes (36,5°) et les sujets normaux se plaignent d'inconfort de froid et frissonnent. L'immersion et le maintien dans un bain à 38° C sont suivis d'une élévation des températures internes et d'un inconfort de chaud. La température interne est mesurée en moyen de sondes à thermistance, l'une rectale, l'autre œsophagienne. Les températures interne et cutanée étant connues on porte des stimulations thermiques à la main gauche du sujet à l'aide d'un récipient contenant de l'eau agitée à température connue homogène pendant trente secondes. A l'issue de cette période on demande au sujet de donner une note (+ 2 à — 2) correspondant au plaisir ou au déplaisir de la sensation évoquée par le stimulus porté sur le territoire de la main gauche. La réponse du sujet normal apparaît sur la figure 1. La sensation évoquée par le stimulus thermique rencontre deux seuils de douleur, l'un à 15° C, l'autre à 45° C.

Entre 15° C et 45° C la sensation évoquée peut être agréable ou désagréable et les sujets normaux utilisent toute l'échelle affective de — 2 à + 2.

Première observation d'indifférence acquise à la douleur. — Le malade est peu concerné par les ambiances thermiques des milieux qui l'environnent et complètement indifférent à toute stimulation douloureuse : piqûre, pincement tendineux.

Immergé dans un bain frais il y demeure quarante-deux minutes sans que sa température interne s'abaisse de plus de 0,2° C (minimum 36,8° C). Il ne signale aucun inconfort et ne frissonne pas. Au réchauffement du bain sa température s'élève et après quarante-six minutes atteint 38,5° C sans trace d'inconfort. Les réponses données par le sujet aux stimulations thermiques de sa main gauche est représenté figure 2.

Deuxième observation d'indifférence congénitale à la douleur. — L'immersion dans le bain à 33° C en quatre-vingt-treize minutes fait baisser ses températures internes de 0,8° C (36,15° C) sans inconfort. De l'horripilation sur les bras et le tronc apparaît vers la cinquante-cinquième minute et du frisson vers la soixante-dix-huitième minute. Au réchauffage du bain sa température interne s'élève jusqu'à 38° C en soixante-dix minutes avec sudation du visage et léger inconfort attribué à la sudation. Les réponses aux stimulations thermiques de la main gauche sont représentées figure 3.

L'exploration de la sensation gustative a montré une forte aversion pour le sel (même pour une dilution de 0,25 % normale : 1 g %) et

---

Les demandes de tirés à part doivent être adressées au Pr agr. M. CABANAC, Lab. de physiol. et expl. fonctionnelle, 8, av. Rockefeller, F-69-Lyon (8e).

[*] Laboratoire de Physiologie et Exploration fonctionnelle (Pr J. CHATONNET), C. H. U., Lyon.
[**] Service de Neuropsychiatrie, Hôpital d'instruction des Armées Desgenettes, Lyon.

An English translation follows.

une préférence pour les faibles concentrations de sucre, réponses non modifiées après apport par tubage gastrique de 60 g de saccharose dans 600 ml d'eau.

L'observation de ces deux sujets confirme la nature de l'indifférence à la douleur. Ils maintiennent leur main 30 secondes dans de l'eau à plus de 45° C.

— Le facteur affectif de la sensation thermique semble exister puisque les sujets utilisent, quoique moins largement, la gamme des notes à leur disposition (fig. 2 et 3).

Le plaisir ou le déplaisir de la sensation dans la mesure où il existe n'est pas modifié par l'état thermique interne. Cette absence d'adaptation du plaisir à l'homéostasie les différencie des sujets normaux.

— Le confort thermique était très fruste chez l'un des sujets et complètement inexistant chez l'autre. Or, le confort est lié à la température interne, donc à la thermorégulation (5). Cette fonction semble intacte chez nos deux sujets. Les réactions thermorégulatrices (frisson, sudation) ne semblent donc pas capables à elles seules de créer un inconfort.

— Il s'avère donc que les deux sujets dépourvus d'une interprétation affective de la sensation douloureuse étaient également dépourvus non d'une interprétation affective des stimulations thermiques puisqu'elle existait bien que fruste, mais d'une adaptation à l'homéostasie de l'élément affectif de la sensation. On peut alors se demander si la tonalité affective éveillée par une sensation n'est pas un phénomène non spécifique dont seraient dépourvus les sujets indifférents à la douleur.

BIBLIOGRAPHIE

(1) BAXTER, D. W. et OLSZEWSKI, J. : Congenital insensitivity to pain. *Brain*, 1960, **83**, 381.
(2) CABANAC, M. : Plaisir ou déplaisir de la sensation thermique et homéothermie. *Physiol. Behav.*, 1969, **4**.
(3) CABANAC, M. et CHATONNET, J. : Influence de la température interne sur le caractère affectif d'une sensation thermique cutanée. *J. Physiol.*, Paris, 1964, **56**, 540.
(4) CABANAC, M., MINAIRE, Y. et ADAIR, E. R. : Influence of internal factors on the pleasantness of a gustative sweet sensation. *Communic. Behav. Biol.*, 1968, part. A, **1**.
(5) CHATONNET, J. et CABANAC, M. : The perception of thermal comfort. *Internat. J. Biometeorol.*, 1965, **9**, 183.
(6) DEARBORN, G. : Case of congenital general anesthesia. *J. Nerv. ment. Disease*, 1932, **75**, 612.
(7) GIRARD, P. F. : Communication personnelle.
(8) GIRARD, P. F., MONNET, P., KOHLER, Cl., CONFAVREUX, J., PASQUIER, J. et FRANCOU, M. : L'indifférence congénitale à la douleur. A propos d'une observation. *Lyon Médical*, 1957, **198**, 247.

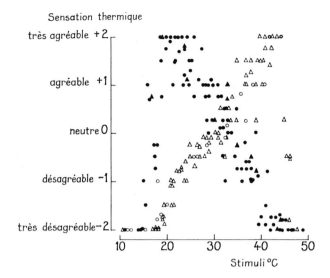

Figure 1. Responses given by a normal subject after 30 seconds of thermal stimulation to the left hand. Each point corresponds to one stimulation: The response describes the value of the affective component of the thermal sensation. The painful stimuli above 45°C, rated –2 by the subject, were interrupted before the 30-second limit.

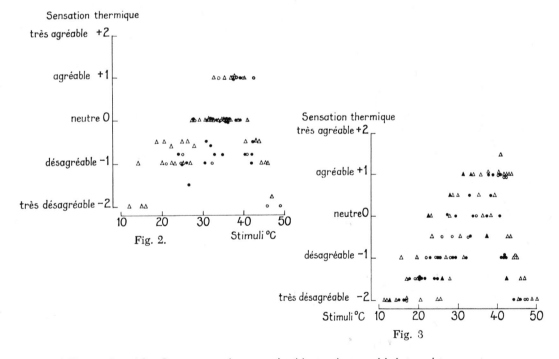

Figures 2 and 3. Contrary to the normal subject, whatever his internal temperature may be, this subject gives the same affective response to thermal stimuli. He tolerates perfectly leaving his hand in water higher than 45°C for 30 seconds.

14

INDIFFERENCE TO PAIN AND THERMAL COMFORT

M. Cabanac, P. Ramel, R. Duclaux, and M. Joli

This article was translated expressly for this Benchmark volume by E. Satinoff, University of Illinois at Urbana-Champaign, from "Indifférence a la douleur et confort thermique," La Presse Med. 77:2053-2054 (1969).

Thermal and gustatory sensory stimulation give rise to sensations composed of two elements: One is discriminative, tied to the physical characteristics of the stimulus; the other is affective, depending on the state of the internal environment (3, 4). Every useful stimulus is agreeable, every stimulus that threatens homeostasis is disagreeable. One can ask if this affective phenomenon is specific for each sensation or if, on the contrary, it is common to all sense modalities.

In the congenital indifference to pain described by Dearborn (1932) (see also Baxter et al; Girard et al), the subjects do not perceive painful stimuli as disagreeable, but otherwise they are entirely normal. One can postulate that this syndrome consists of the absence of the affective element for painful sensations that are nevertheless perceived. Within such a hypothesis, it is then interesting to explore the affective perception of other sensations.

We have observed two cases of indifference to pain, one congenital, the other acquired, in which we have explored cutaneous thermal sensation, thermal comfort, and, in one of the subjects, the gustatory sensations of sweet and salty.

The method of examination has consisted of repeating in the two subjects experiments performed in normal subjects (2) and comparing the results. Immersion in a bath of 33°C is not unpleasant, but a prolonged stay is followed by a progressive lowering of internal temperature (36.5°C), and normal subjects complain of cold discomfort and they shiver. Immersion and maintenance in a bath at 38°C are followed by a rise of internal temperature and heat discomfort. Internal temperature was measured by means of two thermistor probes, one rectal, the other esophageal. The internal and cutaneous temperatures being known, thermal stimulation was given to the left hand of the subject using a vessel containing water stirred to a known, uniform temperature for 30 seconds. At the end of this period, we asked the subject to give a value (+2 to –2) corresponding to the pleasantness or unpleasantness of the sensation evoked by the left-hand stimulus. The response of the normal subject appears in Figure 1. The sensation evoked by the thermal stimulus elicited two pain thresholds, one at 15°C, the other at 45°C. Between 15°C and 45°C, the evoked sensation could be agreeable or disagreeable, and the normal subjects used all the affective gradations from –2 to +2.

First observation of acquired indifference to pain: The patient is little concerned with the thermal environment that surrounds him and is completely indifferent to all painful stimulation: pricking, pinching the tendons.

Immersed in a cool bath for 42 minutes, his internal temperature drops no more than 0.2°C (minimum 36.8°C). He does not signal any discomfort nor does he shiver. Upon reheating the bath, his temperature rises, and after 46 minutes reaches 38.5°C without a trace of discomfort. The subject's responses to thermal stimulation of his left hand are shown in Figure 2.

Second observation of congenital indifference to pain: After 93 minutes' immersion in a bath at 33°C, internal temperatures were lowered by 0.8°C (36.15°C) without discomfort. Piloerection of the arms and trunk appeared towards the fifty-fifth minute and shivering towards the seventy-eighth minute. Upon reheating the bath, the subject's internal temperature rose to 38°C in 70 minutes accompanied by facial sweating and slight discomfort attributed to sweating. The responses to thermal stimulation of the left hand are represented in Figure 3.

Examination of the gustatory sensation has shown a strong aversion to salt (the same for a dilution of 0.25% normal: 1g%) and a preference for the weaker concentrations of sugar, responses not modified after gastric intubation of 60 g of sucrose in 600 ml water.

The observation of these two subjects confirms the nature of indifference to pain. They kept their hand in water hotter than 45°C for 30 seconds. The affective factor of thermal sensation seems to exist because the subjects used, although less extensively, the range of the scale values at their disposal (Figures 2 and 3). The pleasantness or unpleasantness of the sensation, to the extent that it exists, is not modified by the internal thermal state. This absence of adaptation of pleasure to homeostasis differentiates these from normal subjects.

The thermal comfort was very crude in one subject and completely absent in the other. Now, comfort is tied to the internal temperature and thus to thermoregulation (5). This function seems intact in our two subjects. The thermoregulatory reactions (shivering, sweating) therefore do not seem capable by themselves of giving rise to discomfort.

It is established then that the two subjects deprived of an affective interpretation of the painful sensation were equally deprived not of an affective interpretation of the thermal stimulation, since that existed, although crude, but of an adaptation to the homeostasis of the affective element of the sensation. One can then ask if the affective tone evoked by a sensation is not a nonspecific phenomenon of which subjects indifferent to pain would be deprived.

Part IV

FEVER

Editor's Comments
on Papers 15 Through 18

15   BERNARD
       Lectures 20 and 21

16   KING and WOOD
       Studies on the Pathogenesis of Fever. IV. The Site of Action of Leucocytic and Circulating Endogenous Pyrogen

17   VAUGHN, BERNHEIM, and KLUGER
       Fever in the Lizard Dipsosaurus dorsalis

18   SATINOFF, McEWEN, and WILLIAMS
       Behavioral Fever in Newborn Rabbits

Fever is not a pathological process. The current definition of fever is essentially that given by C. Liebermeister in 1875—"Fever is the name of a complex of symptoms based on the fact that *the heat regulation is adjusted to a higher temperature level*" [emphasis added]. This definition implies that in a normal, or afebrile, organism there is a desired, or optimal, or reference, or set level around which body temperature is regulated, and in fever, regulation goes on as usual, only around a higher set level. Some of the experiments that led Liebermeister to this conclusion are contained in Claude Bernard's lectures (Paper 15). Bernard mentions various theories of fever that he considers erroneous and then quickly gets to the essential basic fact about fever—it is an increase in body temperature caused by increased heat production. This clearly distinguishes it from other causes of high body temperatures such as sunstroke in which heat production is not elevated but rather heat-loss mechanisms are inadequate to the task of dissipating sufficient heat to keep body temperature at a normal level.

Bernard, Liebermeister, Jules Le Fèvre (1911), and indeed every early investigator clearly recognized that fever is the result of a bacterial infection and only occurs if special substances (pyrogens) are found in the blood. It was also known that these substances must act on the nervous system. However, it was not until 1948 that Paul Beeson demonstrated that a substance extracted from the polymorphonuclear leucocytes of rabbits was able to raise body temperature when injected intravenously into normal rabbits. This substance was definitely not of bacterial origin but was produced by the white blood cells themselves. Furthermore, there were several differences between this endogenous pyrogen and pyrogens derived from bacteria. The leucocytic pyrogen was heat-labile whereas bacterial pyrogen was not. The onset of fever was more rapid, and no tolerance developed to it (Bennett and Beeson 1953a, b). These studies were a tremendous advance in our understanding of fever. Presumably, the leucocytic pyrogen was carried through the blood stream where it acted somehow on thermoregulatory areas. For some reason, however, it was not until ten years later that M. Kenton King and W. Barry Wood, Jr. (Paper 16), conclusively demonstrated that in fact leucocytic pyrogen caused fever by its action on the central nervous system. We know now that pyrogens act by altering the firing rates of temperature-sensitive cells in the brain in the direction of increasing heat production and decreasing heat loss (for example, Eisenman 1972; Wit and Wang 1968).

During all this time, no attention had been paid to the question of whether ectotherms can develop fevers. Putrid materials had been injected into frogs as early as 1875 (as the excerpt from Claude Bernard notes), but the animals never developed a fever even when they died as the result of the experiment. Bernard was not surprised by these results because, as he notes, "these animals do not have any body heat or temperature regulation system. . . ."

This is one of the few instances where the great French physiologist was wrong. Ectotherms have very good thermoregulatory systems. Their thermoregulatory effector mechanisms are controlled by a combination of brain and other body temperatures just as it is in mammals and birds (see Satinoff and Hendersen 1977 for a review of this area). The point, however, is that those effector mechanisms are *behavioral* and no physiologist ever studied them. If ectotherms are injected with a substance that causes fever in mammals and then maintained at a constant ambient temperature, indeed, they will not develop a fever. They cannot. If, however, they are allowed to choose the environmental temperature that they prefer, they develop beautiful fevers by this behavior, as Paper 17 by Linda Vaughn, Harry Bernheim, and Matthew Kluger was the first to demonstrate. Since 1974 when this paper on iguanas appeared, it has been shown that fish (Reynolds, Casterlin, and Covert 1976), amphibia (Myhre, Cabanac, and Myhre 1977; Casterlin and Reynolds 1977), and crayfish (Casterlin and Reynolds 1977) also develop fevers if given the behavioral opportunity to do so. These experiments all demonstrate that fever is evolutionary a very ancient response.

Infants are another group of organisms that was thought to be insensitive to pyrogens. In fact, infants of some species appear to be able to develop fevers, but a much higher dose of pyrogen is required than is needed in adults (Watson and Kim 1963; Blatteis 1977). However, as Paper 18 by Evelyn Satinoff, Gerald McEwen, Jr., and Bill Williams demonstrates, the behavioral response to pyrogen is much more sensitive than the physiological response, and doses of pyrogen that are too low to cause a reflexive fever will cause a fever if the neonates are allowed to thermoregulate behaviorally. In this respect, infant mammals are very much like reptiles.

REFERENCES

Beeson, P. B., 1948. Temperature-elevating effect of a substance obtained from polymorphonuclear leucocytes. *J. Clin. Invest.* **27**:524.

Bennett, I. L., and Beeson, P. B., 1953*a*. Studies on the pathogenesis of fever. I. The effect of injection of extracts and suspensions of uninfected rabbit tissues upon the body temperature of normal rabbits. *J. Exp. Med.* **98**:477-492.

Bennett, I. L., and Beeson, P. B., 1953*b*. Studies on the pathogenesis of fever. II. Characterization of fever-producing substance from polymorphonuclear leukocytes and from the fluid of sterile exudates. *J. Exp. Med.* **98**:493-508.

Blatteis, C. M., 1977. Comparison of endotoxin and leukocytic pyrogen pyrogenicity in newborn guinea pigs. *J. Appl. Physiol.: Respir. Environ. Exercise Physiol.* **42**:355-361.

Casterlin, M. E., and Reynolds, W. W., 1977*a*. Behavioral fever in anuran amphibian larvae. *Life Sci.* **20**:593-596.

Casterlin, M. E., and Reynolds, W. W., 1977*b*. Behavioral fever in crayfish. *Hydrobiologia* **56**:99-101.

Eisenman, J. S., 1972. Unit activity studies of thermoresponsive neurons. *In Essays on Temperature Regulation*, J. Bligh and R. Moore, eds. North-Holland; Amsterdam, chap. 6, pp. 55-69.

Liebermeister, C., 1975. *Handbuch der Pathologie und Therapie des Fiebers.* Leipzig; Vogelwelt.

Le Fèvre, J., 1911. *Chaleur animale et bioénergetique.* Paris; Masson.

Myhre, K.; Cabanac, M.; and Myhre, G., 1977. Fever and behavioural temperature regulation in the frog *Rana esculenta. Acta Physiol. Scand.* **101**:219-229.

Reynolds, W. W.; Casterlin, M. E.; and Covert, J. B. 1976. Behavioural fever in teleost fishes. *Nature* **259**:41-42.

Satinoff, E., and Hendersen, R., 1977. Thermoregulatory behavior. *In Handbook of Operant Behavior*, W. K. Honig and J. E. R. Staddon, eds. Prentice-Hall, Englewood Cliffs, New Jersey, pp. 153-173.

Watson, D. W., and Kim, Y. B., 1963. Modification of host responses to bacterial endotoxins. I. Specificity of pyrogenic tolerance and the role of hypersensitivity in pyrogenicity, lethality, and skin reactivity. *J. Exp. Med.* **118**:425-446.

Wit, A., and Wang, S. C., 1968. Temperature-sensitive neurons in preoptic/anterior hypothalamic region: Actions of pyrogen and acetylsalicylate. *Am. J. Physiol.* **215**:1160-1169.

15

LECTURES 20 AND 21
Claude Bernard

This excerpt was originally translated for the National Aeronautics and Space Administration by Leo Kanner Associates from pages 405-445 of Leçons sur la chaleur animale: sur les effets de la chaleur et sur la fièvre, *Paris: Bailliere, 1876, 471pp.*

LECTURE TWENTY

Gentlemen:

You all know what we mean by the word "fever"; thus it will not be necessary for us to define it here. We need only recall the principal manifestations of this set of symptoms, which primarily concern changes in heart rate, vascular tension, respiration, and heat production.

We will not examine each of these symptoms specifically. Instead, we should ask first of all whether these modifications of normal phenomena do not include one whose constant occurrence, duration, and predominant intensity give it precedence over all the others, making it the principal phenomenon of fever. This essential modification does in fact exist and will provide the subject for our final lectures. As you may have guessed from the order and nature of the subjects we have studied so far, this change is the modification of temperature.

The very word "fever" (from *febris, fervere*) and the name Πvρεξισ (πvρ, *fire*) given it by the Greeks show that the earliest physicians already regarded heat production as predominant in this symptomatology. However, since Galien who spoke of *calor praeter naturam*, pathologists seem to have paid little attention to febrile heat as a symptom, if not overlooked it completely, both in their research and in the theories they have proposed on the causes and mechanism of fever. Only in the last few years has this change in heat production been restored to its proper significance.

Here, as in so many other problems of physiology, as I have pointed out to you on a number of other occasions, the predominant concepts and special discoveries of each period have turned scientific attention in completely new directions.

At the beginning of the modern experimental period, Harvey's discovery of the circulation of the blood resulted in a pervasive concern with vascular phenomena and the mechanical conditions of blood flow. Fever was considered exclusively in terms of changes in pulse rate, a doctrine that Boerhaave formulated with an aphorism that remained classical for many years: "*Signum pathognomicum omnis febris est pulsus aucta velocitas*" [The pathognomic symptom of all fever is an increased pulse rate].

Later on when pathologic anatomy came into vogue, the sole concern was the specific lesion that caused the fever, and overlooking the symptom itself, physicians primarily sought to determine the organ whose inflammation was the point of departure for this set of reactions. This resulted in the terms "angiotenic fever," "meningogastric fever," "adenomeningeal fever," etc., designations sanctioned by Pinel's nomenclature.

This negligence in the observation of febrile heat and the evaluation of its theoretical importance may be explained still more easily by the inadequacy of available means of examination. Although thermometric measurement was used in ancient times by Sanctorius and especially by Haen, for a long time thereafter application of the hand was considered sufficient for evaluation of heat. This is the most misleading procedure of all, however, since the impressions given by the sense of touch in this case depend both on the temperature of the hand of the observer and on the humidity of the body of the subject observed.

Only over the past twenty years have thermometric measurements been used for fever under strict conditions in France and, it must be acknowledged, especially in Germany and England.[1] Theory and practice have both benefited. However, we will examine this problem from the standpoint of theory alone, giving an account of the advances made in study of the pathologic physiology of fever over the past few years, advances that have been based on more exact research on febrile heat in patients and experiments performed on animals.

However, before dealing directly with the data furnished by temperature measurement, let us review the thermal sensations experienced by the patient. We will then see whether the subjective impressions match the results of objective analysis.

All fever is characterized in its subjective manifestations by three constant periods whose duration varies according to the forms and the circumstances.

A period of chills: This sensation of cold, which may range from simple piloerection to shivering and chattering of the teeth, is of extremely variable duration. In some cases it is so brief and so little accentuated that it seems to be lacking; however, precise observation reveals that it is consistently the first subjective phenomenon of fever: At the same time the skin is observed to be pale, wrinkled, bloodless and cold, the pulse low and rapid. We have described the skin as cold: The thermometer does in fact show the temperature of the surface of the body to be lowered a few degrees. Thus the sensation of cold is not purely subjective. This coldness is real, but it is only peripheral. When introduced into the splanchnic cavities (mouth, anus, vagina) the thermometer shows a rise in internal temperature.

Thus in the first period of fever, we already find a significant phenomenon of temperature increase due to either the production or the distribution of heat. The peripheral capillaries are contracted and anemic, and all the blood flows toward the inside, concentrating the heat in the deep body. We might ask, however, whether this is merely a phenomenon of concentration. Actually, another phenomenon is involved: There is already excessive *production* of heat, a fact that will merely be pointed out now and will be discussed in more detail later. This heat production may be detected by thermometric measurement even before the period of chills, during what is termed the period of "incubation" of the fever, when the patient as yet shows only the mild symptoms of lack of appetite, stiffness, and thirst.

The chills are followed by a period of *heat,* so-called on the basis of the sensations of the subject. The patient feels himself to be heated by sudden flushes, pleasant at first, and then intolerable due to their dryness. The skin is in fact dry and burning:

A thermometer placed in the armpit will read 38 to 40°, and the pulse is heavy and rapid. On the basis of what we have just said in regard to chills, with the addition of the fact that here again, excessive production of heat may be demonstrated, this change in the pathologic pattern can be attributed only to dilation of the peripheral vessels, with the heat no longer being concentrated in the deep body.

Finally, the third period is characterized by heavy sweating, during which the temperature lowers in both the superficial and deep areas of the body. This is the period of return to normal conditions and normal temperature.

This rapid survey shows that the phenomenon of heat production dominates all the manifestations of fever, and we will see that high temperature controls and dominates all other findings. However, we should first try to establish whether heat is actually being produced, or whether there has merely been a change in its distribution or rate of dissipation. This brings us to the theory of fever. We will first examine a few of the explanations for this phenomenon that have been offered. Although these explanations are too exclusive to account for the entire pathologic disturbance, nevertheless they are valuable in that they show the links between some of the attendant pathologic phenomena.

Boerhaave's theory will be reviewed merely for the record. This iatrophysicist saw fever merely as a complex of circulatory phenomena and held that the heat observed was derived from mechanical movements rather than the chemical changes revealed by modern science. According to Boerhaave, the heightened friction of the blood in the arteries caused the rise in temperature. Today we know that this friction of the blood in the arteries actually increases temperature; however, this heat is not produced but merely reinstated since it is movement and work (blood flow rate) which return it to its initial form, that of heat. Ever since our first research on the effect of the nerves on animal heat,[2] we have attributed the phenomenon of fever to heat production actuated by the nervous system, as we will describe later on.

Marey, without completely rejecting the idea of heat production, states that "the rise in temperature occurring during fever consists more in a leveling of the temperature at the various balance points than in heating per se."[3] According to this investigator, during the period of chills, heat is concentrated with the blood, in the body center, followed by an afflux toward the peripheral body during the second period, that of heat. If the temperature remains high despite this peripheral afflux, this is because outside conditions are opposed to cooling since the high sensitivity of the febrile subject to cold leads him to put on extra clothing: "He is supplied with extra covers, not to mention hot drinks and the warm atmosphere of the sickroom."

The theory of Traube[4] is not much different from the preceding: Fever is not seen as an excessive production of heat, but merely the retention and accumulation of normal heat. This theory, whose elements are quite logically combined on their own level and may remain valid for some special forms but without explaining fever in general, may be summarized as follows. Taking the initial stage of fever, chills, as a point of departure, Traube stops with an analysis of the first phenomenon, the most obvious of all, the contraction of the peripheral arteries. This local anemia drives blood back into the deep body, and with it the heat it carries, resulting in elimination of or at least a decrease in external radiation, the immediate consequence being a rise in central temperature. Furthermore, Traube states, since the parts made anemic in this way no longer receive their usual quantity of nutritive fluids, the secretions dry up, the skin and the mucous membranes become dry, and evaporation is checked, a further condition that acts with equal force toward the same end, since evaporation is the very means by which the body expends the most heat.[5]

Everything in this theory is based on incontestable findings, but it is too exclusive. We know that by removing a cold source from the blood in any manner whatever, we increase its temperature, while by limiting peripheral circulation by tying off the ventral aorta in dogs, one obtains a slight increase in temperature.

Some of this theory might be applied to the chill stage in intermittent fever since here the temperature reaches its maximum at the end of this stage and does not undergo any further increase during the period of dry heat, but there is no basis for its application to fever in general, especially those forms in which the heat stage appears to be established at the outset. Here the cutaneous anemia lasts so short a time that it goes virtually unnoticed, to give way to a characteristic redness, a heat that is so sensible to the touch that clinicians have termed it "biting," and a peripheral congestion that causes the body, far from retaining heat, to yield it to the environment for a period of days or weeks (continuous fever, pneumonia, etc.). Moreover, heat is quite obviously being radiated since the bed of a febrile subject is warmer, all other things being equal, than that of a healthy subject. In addition, a thermometer brought within equal distances of the skin of a febrile subject and that of a healthy subject will rise more quickly in the first case than in the second.

Finally, this theory is inadequate even for the period of chills in intermittent fever. This is because, as I pointed out at the beginning of this lecture, vascular contractions and peripheral anemia—chills, in short—are not an essential, causative phenomenon. In all cases where it has been possible to make thermometric measurements, an appreciable increase in temperature has been found for a given time, sometimes even one or two hours, prior to the onset of chills, that is, before any vascular or cutaneous spasm. Before the pulse changes, the thermometer has already risen. Thus the temperature increase is a phenomenon that precedes all others and that seems to be enhanced by them but that cannot be derived from any other phenomenon.

Heat production is therefore the phenomenon we must study, from the standpoint of more exact measurements than those we have mentioned so far, and from the standpoint of its intensity, its sources, and its mechanism.

Heat production may be demonstrated in two ways: a direct method using calorimetry and an indirect method involving the study of organic combustion and the products of this combustion.

For many years attempts were made to determine a site of local combustion in the body that in a pathologic condition, would exaggerate the normal temperature. During the period referred to at the beginning of this lecture as overly concerned with pathologic anatomy, some investigators attempted to link febrile heat directly to the local inflammatory process. For fever in general, this was difficult to demonstrate, at least by experiment, since some cases were caused by hemitis or inflammation of the blood and others by angiocarditis or inflammation of the blood vessels.

Such experiments could be performed, however, in cases where the inflammation was more precisely localized or at least more accessible. Hunter measured the temperature of traumatically inflamed areas, and while finding it higher than that of healthy areas, found the difference to be too slight to account for the rise in temperature of the entire blood supply of the subject. Breschet and Becquerel obtained the same results with more accurate instrumentation.

These experiments have been reperformed and have shown that the blood is actually heated as it passes through an inflamed area, but since this slight heating must be caused by the more intense combustion and organic transformation occurring in such areas, it is not sufficient to account for the overall increase in heat production. Such are the experimental findings of Zimmermann, Weber, John Simon,[6] and Billroth.[7] The latter investigator induced acute vaginitis in dogs by injecting tincture of iodine and found that the temperature of the vagina was equal to or less than that of the rectum, and in any case less than the internal temperature. In wounded men or dogs used for experiments, the temperature of the wound was always found to be lower than that of the rectum, and in some cases even lower than that of the armpit.

In a previous lecture, we showed that when the sympathetic nerve of the neck of a horse is cut on one side, the increase in blood temperature with the fever, which is local at first, ultimately becomes generalized.

Thus fever does not occur through a purely local change in a tissue or surface but through a change in the nervous system. If there is heat production during fever, this heat, like normal heat, arises as part of a general process. Damage to a given organ is only the point of departure for the workings of the nervous system, which govern normal heat production and its pathologic exaggeration.

This is the sense in which calorimetric tests must be understood. It is these tests that we should summarize first of all since we must not forget that we have yet to prove our hypothesis that during fever there is not only a change in the distribution and consequently the dissipation of heat but also a change in its rate of production.

Leyden[8] and Liebermeister[9] have advanced theories along these lines.

These investigators used calorimeters similar to those used in physics experiments. Leyden's apparatus, which was a cylinder filled with water in which a limb or a larger part of the body of the test subject could be immersed,[10] need not be described here. I will merely point out his results, adding, however, that the calorimetric device was designed with sufficient accuracy to yield, if not absolute data, data that at least could be used in comparative evaluation. The problem, as has been mentioned, was to determine whether febrile subjects lose less heat than normal individuals.

These experiments showed that heat loss always increases during fever; this loss may range from one and a half to two times the norm.

Liebermeister placed febrile subjects in a bath at a given temperature and determined the quantity of heat they yielded to the water over a given period of time. Obviously, in this way he overlooked heat losses from the head, which remained out of the water, and losses through pulmonary exhalation. Despite this source of error, plus a few others having the same effect, that is, tending to decrease the result being sought, he still found that heat losses were greater during fever than in normal condition.

Here is how he set up the experiment: "It is not sufficient to place the subject in a bath and to measure the temperature of this bath as the subject enters and leaves it. Such an estimate would be inexact, since water does not naturally remain at a constant temperature. Instead, the temperature of the water must also be measured for a given length of time before and after the bath in order to see how it decreases by radiation over a given period of time. This information may be used to determine the quantity by which this temperature would be lowered and the thermometer reading which would be given at the end of the bath if the subject had not been placed in it.

"The temperature of the water as the subject leaves the bath is noted; one thus knows the temperature rise induced by his presence in the bath, and knowing the weight of the water, it is easy to calculate the number of calories the subject has yielded to the water."[11]

Such is the direct evidence for excessive heat production during fever. However, we still have considerable indirect evidence furnished by observation of the increase waste resulting from organic combustion. These products of organic transformation are found in the expired air and in the urine.

So far, we have said relatively little about respiratory changes during fever. First, let us review a few widely known facts: The respiratory rate becomes much faster in febrile subjects than in normal subjects. Initially, a link was sought between this increase in respiratory rate and the increase in pulse rate. Careful observation has shown that there is no such absolute relationship but that there is a relationship which is much more interesting from our point of view. This is a link between increased respiratory rate and temperature rise: The respiratory rate becomes faster as the temperature increases.

Here then was an initial phenomenon attributable to chemical processes and increased combustion. The next step was to determine, by analysis of the expired air, whether or not a febrile subject produced more carbonic acid than a normal subject. The first experiments along these lines, performed by Senator and Traube, did not appear favorable and yielded negative results.[12] We should hasten to add, however, that the problem did not lie in an erroneous interpretation of observed phenomena. If less carbonic acid was found in the air breathed out by a febrile subject than a normal subject, this result did not allow for the fact that the increased respiratory rate results in the passage of a larger quantity of air through the lungs. Consequently the factor that should be taken into account is not the amount of carbonic acid contained in a given volume of expired air but the amount of carbonic acid produced in a given period of time.

This was, in fact, the result shown by the more recent and more accurate experiments of Leyden (of Koenigsberg) and Liebermeister (of Basel). The latter performed his research by means of a carefully developed assembly whose principal features are given below, based on the description furnished by Weber.[13]

"The assembly consists of a large zinc box approximately 2 meters long, 1.50 meters high and 0.80 meters wide. It is large enough to accommodate a man in either sitting or reclining position, or even in a bath, and alternatively in a bath and on a chair. An opening at one end serves as an inlet for outside air, whose composition is carefully examined at the time of the experiment. At the other end, a rubber tube places the box in communication with the containers in which the carbonic acid is measured, followed by a gasometer which measures the quantity of air passing through the chamber. The rapid flow of air is obtained by an ingenious method involving a current of water which draws the air along as it falls."

Liebermeister's procedure was as follows: Slightly before the onset of a febrile attack, he took the patient's temperature and placed him in a chamber. When the chills stage began, he replaced the container serving for carbonic acid measurement with another container. Every half hour the temperature was measured and the carbonic acid container changed so that it was possible to measure the quantity of carbonic acid exhaled every half hour.

The results of these experiments of Leyden and Liebermeister showed that the proportional carbonic acid concentration of the expired air was decreased: It was to the quantity obtained with a normal subject as 3 is to $3\frac{1}{3}$. However, the absolute quantity of carbonic acid eliminated over a given time was considerably increased, being in a ratio of $1\frac{1}{2}:1$ to the quantity exhaled by a normal subject.

Liebermeister, in experimenting on the various stages of fever, arrived at interesting results relative to the elimination of carbonic acid during the chills stage. We have seen that it is during this stage that the internal temperature increases the most rapidly and to the highest level. We have also seen that the phenomena in this stage have been those that investigators have primarily attempted to explain as a simple concentration of normal heat in the deep body. Thus in the field of pathologic physiology, it was most important to determine the quantity of carbonic acid produced in this particular stage, in the chill stage, to see whether there was actually an increase in combustion and heat production. Here is what Liebermeister found in this regard.

"In the second half-hour of the experiment, while the temperature was rising slowly, carbonic acid production increased by 45%. In the third half-hour, the temperature rose rapidly and carbonic acid production increased 47%. The huge quantity of 34.2 g was eliminated in one half-hour. In the fourth period the temperature was still rising, but more slowly; carbonic acid production decreased to a point only 39% above the norm. Finally, in the last two half-hours, where the temperature was nearly constant at 40 degrees, the elimination of carbonic acid was only 28% above the norm."

Thus these experiments have afforded the most complete possible demonstration of this physicochemical finding and the law linking the various elements: temperature, heat production, combustion, and products of this combustion. The quantity of carbonic acid exhaled increases in direct proportion to the excess heat manifested, and the largest quantities are eliminated at the point when the temperature increase is the most rapid and greatest.

After this demonstration, we might wish to skim over any study of the residues of organic combustion and analysis of the urea contained in the urine. This aspect of the problem should not be overlooked, however. If the excess quantity of carbonic acid indicates the intensity of hydrocarbon combustion, the urea should show the metamorphoses undergone by albuminoid materials. The latter research is important from several standpoints, as will be explained later on.

The initial results obtained on the urea eliminated in the urine were highly contradictory. This was because the experimental conditions were not precisely comparable. It was acknowledged that urea was produced by the combustion of albuminoid materials and that its presence in the urine was in direct proportion to the nitrogen supply. The quantity of urea might vary from one to twice as much, depending on the quantity of albuminoid material—meat—consumed by the subject. Thus it was not very legitimate to set up any sort of comparison between a febrile patient on a diet and a subject eating normally. It was therefore necessary to determine the quantity of urea excreted by a normal subject on a strict diet. Under these circumstances, a mean value of 17 to 18 grams was obtained. However, we should also point out that it is necessary to allow for the size of the individual since as you know, an undernourished subject will use its own body as food, feeding on its tissues, its albuminoid substances, and other materials. Thus rather than comparing a healthy individual on a diet with a febrile individual also on a diet, one must compare the unit of weight of the one with the unit of weight of the other. Although this method is not completely free of error, it is one that should furnish comparable results and was used to determine the statistics that I must cite here: they demonstrate that 1 kg [of body weight] of a healthy individual on a diet produces 0.58 g urea every 24 hours. The same tests performed on febrile subjects have shown that such individuals eliminate 1.50 to 1.89 g/kg every 24 hours, rather than 0.59 to 0.53 g. The ratio of the urea produced by a healthy subject to that produced by a febrile subject is therefore 100:225 to 100:300. In other words, the febrile subject eliminates an average of one and half times the quantity of urea eliminated by a normal subject.[14]

Finally, there is a further method of demonstrating the increase in organic combustion during fever. This consists of analyzing the weight loss undergone by the patient: To some extent, this weight loss represents the sum total of combustion of ternary and quaternary compounds. Here again, the results are decisive. In a series of experiments on animals, O. Weber found that by injecting pus into dogs, he could obtain a greater decrease in weight than in undernourished animals. He obtained similar results in a comparative study of the weight losses of healthy individuals and febrile individuals placed on a diet. Some of his statistics, again with reference to a unit of body weight of the individual, are as follows: The weight loss undergone by a healthy individual on a diet was 23 to 30 g/kg per day, while that undergone by a febrile subject was 30 to 44 g. We can see that the ratio of the former to the latter is 100:120 to 150.

It should be pointed out that the weight loss is not as great as the elimination of urea since the weight loss ratios did not exceed 150:100 while the ratios of urea loss are as high as 300:100. This is an observation that we will undoubtedly have occasion to reexamine later on.

For the time being, it is sufficient to have established that febrile subjects undergo a change in both heat distribution and heat production. However, while this distribution may vary or even be reversed in a given phase of the fever, heat production always proceeds in the same direction, both in the period of chills and in that of dry heat. The period of chills itself, the one period that might be expected to contain changes in distribution alone, is actually that in which there is the most obvious change in heat production, and moreover, this change is still in the same direction, that is, one of increase. This excess heat production has been demonstrated by analyzing all the phases of combustion, by analyzing the physical and chemical results of combustion, that is, heat production, by quantifying carbonic acid and urea content, and finally by determining the weight loss undergone by the subject.

The next problem is to show that in both pathologic and normal states, this increase in heat production, with all its consequences, is the result of a change in the functioning of the autonomic nervous system or the vasomotor nerves, as we indicated to the Academy of Sciences and to the Society of Biology in 1852. Finally, we should add that the development of pathologic and physiologic phenomena is always correlative. Although heat production in cold-blooded animals is latent and invisible to some extent, at the same time, no febrile phenomena can be observed in these animals, a fact that has been known for a long time.[15]

Notes and References

1. As early as 1839 Andral and Gavarret had clearly shown the importance of thermometric measurements in clinical examination. (*Recherches sur la température du corps humain dans les fièvres intermittents [Research on the temperature of the human body during intermittent fever]*, Paris, 1839).
2. De l'influence du grand sympathique sur la chaleur animale (*Comp. rend. de l'Acad. des sc.*, t. XXXIV, p. 472, 1852, et Société de biologie 1852).
3. Marey, *Physiologie de la circulation du sang*. 1863, p. 361.
4. Traube, *Zur Fieberlchre* (Med. Centralzeitlung, 1863 et 1864); Traube et Jochmann, *Zur Theorie des Fiebers* (Deutsche Klinik, 1855).
5. For a history of the work on fever, see H. Hirtz, *Essai sur la fièvre en général [An essay on fever in general]*, thesis, Strasburg, No. 289, 1870; published by Weber, *Des conditions de l'élévation de température dans la fièvre [The circumstances of temperature rise in fever]*, thesis, Paris, 1872.
6. *Deutsche Klinik*. 1862, 1863.
7. *Archives de Langenbeck*, t. VI; see also: *Archiv. gener. de med.* 1865; et Hirtz, op. cit., p. 26.
8. Leyden, Unters. über das Fieber. (*Deutsch. Archiv.*, vol. V, 1869.)
9. Liebermeister, Ueber die quantit. Bestimmung der Waermeproduction in kalten Bade. (*Deutsch. Arch.*, 1868, vol. V, p. 216.)
10. See a complete summary of the research of Leyden and Liebermeister, plus a description of their equipment, in E. Weber, *Des conditions de l'élèvation de température dans la fièvre [Conditions of temperature rise during fever]*, thesis, Paris, 1872.
11. Ed. Weber, op. cit., p. 22.
12. H. Senator, Bieträge zur Lehre von der Eigenwärme und dem Fieber. (*Virchow's Archiv fur patholog. anat.*, 1869, vol. XLV, p. 351.)
13. E. Weber, op. cit., p. 91.
14. See analytical tables in H. Hirtz, op. cit., p. 45.
15. In a recent article on fever in cold-blooded animals ("Fever in cold-blooded animals," *Arch. de W. Pflüger*, June 1875), O. Lassar observed that frogs into which putrid materials had been injected did not show any increase in either body heat or heat emitted by radiation. Even when the animal died as a result of the experiment, no heat was produced, and consequently there was no fever from a temperature standpoint. However, since these animals do not have any body heat or temperature-regulation system, it is not surprising that they do not show these pathologic manifestations, which would basically consist in disturbance of the heat-producing capability and its nervous regulatory system. This is not to say, however, that cold-blooded animals do not experience local inflammations, as has been suggested elsewhere. (See Robert Lator, *Expériences servant a démontrer que le pathologie des animaux à sang froid est exempte de l'acte morbide qui, dans les animaux à sang chaud, a reçu le nom d'inflammation [Experiments showing that the pathology of cold-blooded animals is free of the pathologic phenomenon which has come to be known as "inflammation" in warm-blooded animals]*, Paris, 1843.)

Claude Bernard

LECTURE TWENTY-ONE

Gentlemen:

In the preceding lecture, we saw the importance of heat in fever. No clinician today would hesitate to agree that it is heat that constitutes fever: Increased heat production is synonymous with fever. More importantly, it is primarily after the rise in temperature that the physician can gain a rough idea of the seriousness of certain fevers, by diagnosing the period of its onset; this is the factor that indicates how soon improvement may be expected.[1] There is no need to elaborate on this subject here; all medical treatises over the past few years contain graphic representations of the evolution of temperature during fever, and for the most part, these graphic tracings are totally characteristic.[2]

Today we intend to examine more specifically the effects of the excess heat produced during fever and to compare then with the effects we have produced by applying heat to animals. We will follow the same order of discussion for pathologic heat as for normal body heat: After determining its sources, we will consider its effects.

The main question is this: Are there lesions or functional disturbances that are a result of the increase in temperature of the febrile subject? The significance of this question is easy to understand since if the answer is affirmative, it will immediately furnish a number of therapeutic indications: It will be necessary to combat the heat produced not only by moderating the combustion processes but also by drawing off heat to prevent it from acting as heat, that is, as a physical agent pure and simple.

In our experimental physiological research,[3] we have seen that the increase of temperature to a given level may result in the death of the animal. Today there is no doubt that febrile heat can have the same fatal outcome; to be convinced of this fact, one need only consider the level to which the temperature of the subject can rise in some diseases. In rheumatism, for example, temperatures of 42.7° and 44.1° have been observed.[4] Clinical experience shows that in acute diseases, a continuous rise in temperature presages a fatal outcome. All physicians agree that a temperature can rarely exceed 41.9° for a period of several days without eventually proving fatal, indicating the importance and extreme seriousness of this excess heat.[5]

Thus it is justifiable to compare these rapidly fatal effects of febrile heat with those observed in the animals that I have presented to you in previous lectures. We have produced rigor mortis in the heart, the diaphragm, and the intercostal muscles, and we have caused the death of the animal by fatally damaging one of its essential constituents, the muscle fiber, especially the muscle fiber of the heart.

As it happens, it is precisely muscular damage that predominates in prolonged pyrexia with extremely high temperatures. We have seen that, in our animals killed by heat, the microscope revealed a characteristic state of coagulation of the contents of the muscle fiber. A similar change has been noted in the muscles of subjects having died as a result of severe fever; this phenomenon is known as Zenker's degeneration (vitreous coagulation).

Finally, with equal justification, these effects of febrile heat in cases of extremely high temperature have been compared with the phenomenon known as insolation or sunstroke.[6,7] Under these circumstances, a state of rigidity of the muscle fibers, especially those of the heart and the diaphragm, is observed, which is no different from the states we obtained experimentally in animals.

There can be no doubt that this increase in temperature also affects the nervous system. We have talked about anesthesia by heat: The irritability of the nerves is observed to decrease in response to high temperature. However, we do feel that the alteration of the muscles is the most important factor and at any rate, the best-established one.

These are therefore the effects of extremely high febrile temperature: rapid, precise effects that are easy to interpret. However, there are other effects and although they can be ascribed to the same cause, they cannot be described quite so definitively as yet. These are the general degenerations observed in almost all the tissues of subjects who have died after prolonged fever in typhus or scarlatina. Not only are the muscles of the heart modified but there is also a more or less advanced fatty degeneration of the liver, kidneys, and encephalon. These lesions may also be attributed to a prolonged high temperature. This rise does not go so far as to cause sudden death, or at least extremely rapid death, by the process revealed by our experiments; but by its prolonged duration has slowly and profoundly altered the nutrition and consequently the composition of the anatomical components of the tissues.

This interpretation is not accepted by all investigators who have studied this difficult problem. To give an example of the various theories that have been proposed in this regard, we must retrace a few steps and return to the findings I pointed out to you in the previous lecture.

You may recall that in order to demonstrate the production of excess heat during fever, we analyzed the products of combustion exhaled and excreted by the patient and observed the weight loss occurring over the same period of time. In citing the figures obtained by the various observers, we noted that the weight loss, in contrast to the elimination of urea, is not as great in febrile subjects on a diet as in normal individuals also on a diet.

To Senator,[8] this lack of agreement between the curve of urea production and the curve of weight loss has suggested quite specific views on the nature of febrile combustion. We have previously seen that Lussana and Ambrosolli[9] considered pathologic heat to be different from normal heat. According to Senator, however, febrile heat does not have its own characteristic sources. From the fact that the weight loss does not increase in the same proportion as urea elimination, he concludes that febrile combustion is only partial. According to his theory, only nitrogenous materials undergo a decomposition that has little or no effect on the hydrocarbons.

However, if the hydrocarbons undergo little or no combustion, or at least if they are not subject to excessive decomposition during fever, there should not be any increase in the exhalation of carbonic acid. However, all experiments have shown this superactivity of gas exchange in the lungs. Senator has therefore analyzed the respired gases and states that he found the carbonic acid to remain in normal proportions. Furthermore, he states that he found the blood of a febrile animal to contain less carbonic acid than that of a healthy animal.

Research performed by Liebermeister and by Leyden, as summarized earlier, showed, on the contrary, an increase in carbonic acid production in febrile man.

If Senator had not consistently found as large a quantity of carbonic acid in the expired air as that observed by other investigators, this might be ascribed to the other channels through which this gas can be eliminated, especially the kidneys. An article has recently appeared on this subject.[10] Ewald, of Berlin, was able to ascertain that the urine contains more carbonic acid during fever than during apyrexia. Febrile urine contains an average of 16 to 17% carbonic acid, while normal urine contains barely 12%. In some cases, when it is extremely abundant and heavily diluted, febrile urine contains less carbonic acid than an equal quantity of normal urine. However, if the entire quantity of urine produced over a twenty-four-hour period is taken into account in these cases, one can see that the quantity of carbonic acid eliminated in this way does exceed the norm.

In regard to the decrease in carbonic acid concentration in the venous blood, Senator has sought an explanation for this phenomenon and refutes his own proposal to a certain extent. In effect, he found that in a number of his experiments, the total quantity of carbonic acid eliminated through respiration was higher in febrile subjects than that eliminated by normal individuals over a given period of time. He explains this result, not by overproduction, but by overexcretion. He states that this excretion is encouraged: "by increasing the difference in temperature between the body and the ambient medium (the absorptive capacity of the blood for carbonic acid, like that of any liquid, is known to be in inverse proportion to temperature); febrile acceleration of the circulation and an increase in vascular tension in the lungs could be another condition encouraging the release of carbonic acid; and finally—and this is undoubtedly the most important cause—an increase in respiratory rate would act in the same way."[11]

Thus if all these causes act over a given period of time to accelerate the elimination of carbonic acid from the bloodstream, it is not surprising that analyses of the blood show little or no increase in its carbonic acid concentration. The most significant and least arguable of these causes is obviously the increased respiratory rate. One can easily be convinced of this fact by reference to Viérordt's experiments on pulmonary ventilation. This physiologist observed that by doubling his rate of inspiration, he was able to increase his carbonic acid output by 60%.

Moreover, this increase in elimination does not occur solely through the lungs and the urine but also through the skin, and it is greater in febrile than in healthy subjects. The rise in temperature is obviously the cause of this difference. The research of Neumann seems to show that the increase in temperature and in imperceptible losses occur in parallel.[12]

A little earlier I mentioned the research of Senator, and I attempted to compare the contradictory results obtained by other investigators in order to show you how little is actually known on the process of fever and all its heat phenomena. One can see that the path we must take in attempting to explain these pathologic disturbances is already clearly marked; but there is still much to be done before the ground can be completely cleared, in terms of overcoming each of the problems we encounter as we advance in this research. Moreover, we have seen that in normal conditions the temperature of the tissues increases with the conservation of oxygen, almost all of which passes into the venous blood, which remains red in color. Thus in the pathologic state it is not always necessary for high temperature to coincide with an increase in carbonic acid in the venous blood since this phenomenon can also be observed in the normal state. We will have occasion to return to these findings later on.

Earlier we talked about the profound alterations observed in almost all tissue elements in subjects having died of severe pyrexia, and we stated that we were tempted to consider this fatty degeneration as the result of prolonged high temperatures. There are some experiments that support this viewpoint, and these will be cited later on. For the time being, however, let us review the explanation given by Senator, which is related to the theory of combustion he has proposed.

According to Senator, it is primarily the albuminoids that are burned which undergo excessive degeneration during fever. There is little or no increase in the decomposition of hydrocarbons and fat, and there is soon a relative excess of these fats as a result. This, Senator believes, is the mechanism of the rapid, unhindered fatty metamorphosis of the tissues during the febrile process.

It is not our purpose to evaluate this hypothesis, for the time being; we only wish to point it out.

We know that all the food materials and constituents of the body do not produce an equal quantity of heat during combustion. It makes little difference whether this combustion is direct oxidation, or whether, as we are inclined to believe, it is the result of a series of decompositions whose precise nature is difficult to determine in the current state of organic chemistry: the final result is always roughly the same, and calorimetric testing has made it possible to determine in an exact manner the final result of the combustion of quaternary and ternary compounds. At equal weight, albumin yields only one eighth the heat yielded by fat. Thus combustion of the albuminoids alone, with no involvement of hydrocarbons, will produce little or no increase in temperature. Consequently, the heat observed in fever must arise not from overproduction but from a change in distribution or dissipation.

According to these theories, fever entails changes only in blood distribution and consequently heat distribution. These variations in blood supply undoubtedly produce momentary, more or less local increases in heat production and combustion. On the other hand, we also know that the heat developed and released throughout the duration of the fever is greater than that produced by the same animal under the same conditions over the same period of time but with no fever.

We do not wish to discuss these theories any further, having dwelled on them merely to show the variety of the viewpoints from which this highly complex problem may be considered. We repeat that at present, without going into detail on specifics that will be clarified by further experiments, fever may be considered an exaggeration of normal heat production, that this excess heat is the main hazard of fever, and that this temperature disturbance is the main phenomenon that should be combatted by therapy. Moreover, medical therapy today is primarily concerned with combatting this increase in temperature.[13] For this reason we must give considerable importance to study of the means used for this purpose.

Before setting aside the problem of pathologic physiology, however, we should give some attention to the role of the nervous system in these abnormal heat production phenomena. On the basis of what we have previously seen on the effect of the nervous system on heat production in normal conditions, it is impossible to doubt that fever is a nervous disturbance. The "heat-producing" nerves are involved here, this general term being intended to include the vasoconstrictive and vasodilative nerves and the nerves we have called "heat-producing" and "cold-producing."

However, we should take care to point out at this point that in attempting to determine the role of the nervous system in fever, we are not seeking the *cause* of fever. We will obtain the characteristics of its mechanism in this way, but the cause itself will remain to be determined—that is, the point of departure or initial change which the nervous system only transmits and generalizes throughout the body.

Although this point of departure cannot yet be determined precisely on the basis of available information, there are a few findings that furnish some idea of its mode of action. Earlier I described the effect that the sensory nerves exert by reflex action on the heat-producing nerves, the vasomotor nerves. In another series of lectures, I showed that local trauma is especially likely to result in fever when the sensory nerves are preserved since these nerves place the damaged area in communication with the nerve centers.

We have been led to consider the vasoconstrictive sympathetic nerve as acting as a type of *brake* on heat production, opposing any increase in combustion or degeneration. This nerve "brake" acts by a sort of reflex tonus that is similar to vascular tonicity and muscular tonus itself. Muscular tonus disappears when the motor nerves are cut or when the gray substance of the spinal cord, the seat of reflex action, is destroyed. Now, our experiments and others performed by a number of physiologists over the past few years have shown that the case is exactly the same with tonus, which serves as a brake on organic combustion and heat production.

When the spinal cord of animals such as the dog is cut, the body temperature of the animal increases appreciably, especially if certain favorable conditions, which I have described elsewhere, are met. I have shown that small dogs usually become cold as a result of this operation, while smaller animals such as rabbits always do so.[14]

However, we know that this cutting of the spinal cord in a given region, which paralyzes or excites the vasomotor nerves, considerably dilates the entire peripheral vascular network. This dilation causes a heavy afflux of blood to the surface of the body, where it undergoes considerable losses of heat by radiation. The smaller the animal, the greater the loss of heat.

One of the first questions that must be answered, therefore, is the following: Is the cooling observed in animals whose spinal cord has been cut at the bottom of the cervical region in fact due to enormous losses of heat by radiation? Furthermore, could these losses be so heavy as to overcome the effects of any real increase in heat production?

A number of experiments have been performed to establish this proposal. Here we will mention those of Naunyn and Quincke, controlled by comparative experiments on animals that had not been operated upon.[15]

In the experiments of Naunyn and Quincke, the spinal cord was severed by crushing to prevent any flow of blood. The animal was chloroformed to eliminate any convulsive movements that would have resulted in the release of heat. The cord was usually crushed at the level of the sixth cervical vertebra.

In their initial experiments,[16] these investigators observed a significant drop in temperature after severance of the spinal cord. As we stated a moment ago, they attributed this drop to excessive heat losses. The problem was, therefore, to protect the animals from this heat radiation by placing them in a device that would make it possible for them to conserve their heat. As soon as the animal was enclosed in a medium kept at 26 to 30°, where heat losses were consequently minimal, its body temperature no longer decreased, but instead began to increase after one hour. It continued to increase up to 42 to 43, or even 44°, and the animal then died as a result of its excess heat.

Thus after severing the spinal cord, if heat losses resulting from so-called vasomotor paralysis are prevented, one finds, not a drop in temperature but exactly the reverse, that is, excessive heat production. Thus the severance of the spinal cord has not only destroyed vascular tonicity but also the specific tonus we have compared to a "brake," that is, one opposed to excessive organic combustion.

It may be objected, however, that the experiments in question do not offer any proof as to the role of the spinal cord as a conductor of these temperature-moderating nerves. One might assume that the phenomenon observed in these animals is not the result of nerve severance but of the overall trauma that must have resulted from severance of the spinal cord. The temperature rise would thus be due to traumatic fever developed by the wounds and would not furnish any enlightenment as to the heat-producing nerves along the spinal cord.

This quite natural objection is easily answered by means of a comparative experiment. A dog was subjected to exactly the same wound as that required to sever the spinal cord: an incision was made through the integument, the bones and the meninges, and the animal was enclosed as described above. After several hours, its temperature had increased barely a few tenths of a degree. The spinal cord was then cut, and after one hour the temperature was observed to increase by 2.2°.[17]

Clinical practice offers observations that are quite similar to those we have been able to make experimentally on animals. Earlier I mentioned the case cited by Brodie: This was actually the point of departure for the initial research in this field, his purpose being to determine the role of the nervous system in heat production. The spinal cord of Brodie's patient had been crushed at the bottom of the cervical region[18] resulting in paralysis of all of the muscles of the arms and legs and the trunk, with the exception of the diaphragm. The temperature of this patient during the 42 hours he lived after the accident increased to 43.9°.

Similar observations have been published more recently. In a case of dislocation of the sixth cervical vertebra complicated by fracture, Billroth observed a temperature of 42.2° four hours before death[19]. The same exaggerated heat production is apparent in the observations collected by Fischer, Naunyn, Quincke, and Erb[20,21,22].

All these findings seem to indicate the following conclusion: The temperature-moderating nerves pass through the spinal cord and are cut when it is severed. The "brake" normally opposing heat production is thus eliminated, and the temperature increases considerably due to excessive heat production. In animals exposed to the ambient air, however, it decreases in the long run since this exaggerated heat production is neutralized and counteracted by excessive heat losses resulting from the fact that the vasomotor nerves have been cut at the same time as the heat-producing nerves.

This conclusion, which is that of Naunyn and Quincke, led these investigators to perform a further series of experiments, which we should also consider briefly here. Although the path of the heat-producing nerves follows the spinal cord, it is probable that they leave the spinal cord in turn, in company with the other spinal nerves. As a result, the lower the region in which the spinal cord is severed in experiments such as the above, the less effect this operation will have on the heat-producing nerves, and the smaller the increase in heat production that will result.

Using the precautions described earlier, therefore, the spinal cord of a dog was crushed at different levels from bottom to top in order to perform successive comparative experiments on a single animal. When the spinal cord was crushed at the tenth dorsal vertebra, the temperature increased from 39.1 to 41.1°. When it was then crushed at the sixth cervical vertebra, the temperature increased to 43.7°.[23]

Thus the moderating nerves such as the vasomotor and ordinary motor nerves diverge from the spinal cord one after the other: The higher the region at which the spinal cord is severed or crushed, the greater the number of these nerves that will be cut, and consequently, the greater the increase in heat production. This operation cannot be performed higher than the sixth cervical vertebra since severance at this level eliminates the innervation of the muscles of the chest and diaphragm, and the artifical respiration that must be performed as a result complicates the experimental conditions, in research where it has already been necessary to simplify natural circumstances and to distinguish sharply between essential natural phenomena and those which are secondary and adventitious.

However, there is one serious objection to be made to the preceding explanation. Elsewhere[24] I have shown that the temperature of animals whose spinal cord has been severed does not increase at all but always drops. Now, although this temperature does increase when heat losses are prevented, this phenomenon is in no way applicable to fever, which always results in an increase in temperature when the surface of the body is exposed to the ambient air. Furthermore, I have mentioned previously, and will explain elsewhere in greater detail, the fact that there are two types of nutritional phenomena: those of destruction, degeneration, material disorganization or combustion, and those of organic organization or synthesis. The latter phenomena are controlled by the cold-producing nerves, belonging specifically to the autonomic nervous system, while the combustion phenomena are expressly governed by the vasodilator or heat-producing nerves, which belong specifically to the cerebrospinal system. Now, fever is merely an exaggeration of the activity of these heat-producing nerves, which diverge from the spinal cord, but not a paralysis of the vasodilator nerves.

At any rate, these physiological experiments have furnished some valuable information on the existence and pathways of the heat-producing nerves. This information indicates that fever is nothing more than an exaggeration of a normal phenomenon since we have previously shown that there are reflex actions that produce local exaggerations of heat production in the normal state. However, these findings, with the remarks we have made on them, lead to a number of other, more complex questions that I will merely indicate here.[25] Where are the overall centers of these heat producing activities? Should they be looked for in the medulla oblongata, in the pons, in the encephalic centers themselves, as some clinical observations seem to indicate, or merely in the spinal cord? How are these vasodilative nerve centers related to the vasoconstrictive centers? Although we are not yet able to answer all these questions, at least the future lines of research on fever are clearly marked: It is by studying the normal conditions of the nervous mechanism of heat production in detail that it will be possible to grasp the laws governing pathologic heat production, which can only be considered an exaggeration of normal phenomena. Fever, in short, is only an exaggeration of normal combustion phenomena due to excitation of the nerves governing such phenomena.

Notes and References

1. The Society of Biology has awarded a research grant to Bourneville on temperature as a means of diagnosis. (See Bourneville, *Études cliniques et thermo-*

métriques sur les maladies du système nerveux [*Clinical and thermometric studies on diseases of the nervous system*], Paris, 1873.)
2. Sée, Du diagnostic des fièvres par la température (*Bulletin de thérapeutique*, 1869, p. 145); Botkin (de Saint-Petersbourg), *De la fièvre*. Trad. fr. par A. Georget. Paris, 1872.
3. Ibid., p. 349.
4. R. Macnab, Very high temperature in rheumatism (*The Boston Med. and Surg. Journ.*, 5 mars 1874.)
5. Liebermeister, *Klinische Untersuchungen über das Fieber [Clinical Investigations of Fever]*, Prager Vierteljahrschrift, 1865) believes that in malignant and pernicious fever, it is the heat alone that causes so-called malignant episodes. Wunderlich's statistics support this opinion. Out of 45 cases of exanthematous typhus, he observed temperatures of 42° or more in five cases, and all five cases terminated in death. Out of 20 patients whose temperatures remained between 40° and 41°, there were nine deaths, and these were the patients closest to 41°. (See Hirtz, "Fever," *Diction. de med. et de chirurgie pratiques 19*, 702, 1871.)
6. See Hestres, *Études sur le coup de chaleur, maladie des pays chauds*. These. Paris, 1872.
7. Vallin, Du mechanisme de la mort par la chaleur. (*Arch. génér. de méd.*, janvier 1872.)
8. Senator, *Untersuchungen über den fieberhaften Process und seine Behandlung*. Berlin, 1873.
9. Ibid., p. 392.
10. C. A. Ewald, Ueber den Kohlensaüre-Gehalt des Harns in Fieber (*Arch. de Reichert et Du Bois-Reymond*. 1873, Heft I, p.1.)
11. See the review of Senator op. cit. in *Rev. des sciences medicales 3*, p. 542.
12. Fr. Neumann, *Experimentale Untersuchungen über das Verhalten der insensiblen Ausgabe in Fieber*. (Dissert. inaug. Dorpat, 1873.)
13. See Peter, Des températures élevées excessives dans les maladies (*Gaz. hebd.*, 1872, p. 54); Du Castel, *Des températures élevées dans les maladies*. (Thèse de concours. Paris, 1875.)
14. See my *Leçons sur le système nerveux*, t. II, p. 13.
15. Naunyn, Bieträge zur Fieberlehre (*Arch. de Reichert et Du Bois-Reymond*. 1870).
16. These details have been extracted from E. Weber, *Des conditions de l'elevation de température dans la fièvre [Conditions of temperature rise during fever]*, Paris, 1872.
17. Naunyn and Quincke, in Weber, ibid.
18. *Medico-chirurgical transactions*, 1837.
19. *Langenbeck's Arch.*, 1862.
20. Quincke, *Berlin. klin. Wochenschrift*. 1869, no. 29.
21. Erb, *Deutsches Archiv für klinische Medicin*, 1865.
22. See also Charcot, *Gaz. hebdom.* (November 19, 1869).
23. Naunyn and Quincke, see Weber, op. cit., p. 66.
24. See my *Leçons sur le système nerveux*.
25. See our initial research on the distinction between vasomotor and heat-producing nerves (Experimental research on the vascular and heat-producing nerves of the autonomic nervous system, *Acad. sci. Comptes Rendus*, 55, August 1862).

STUDIES ON THE PATHOGENESIS OF FEVER

IV. THE SITE OF ACTION OF LEUCOCYTIC AND CIRCULATING
ENDOGENOUS PYROGEN*

BY M. KENTON KING,[‡] M.D., AND W. BARRY WOOD, JR., M.D.

*(From the Department of Microbiology, Johns Hopkins University School of Medicine
and School of Hygiene and Public Health, Baltimore)*

(Received for publication, October 2, 1957)

The heat-labile endogenous pyrogen originally extracted from rabbit polymorphonuclear leucocytes by Beeson (1) has subsequently been shown to have many characteristics which distinguish it from heat-stable exogenous pyrogens derived from bacteria (2–4). The circulating endogenous pyrogen which has recently been identified in the blood streams of rabbits (5, 6) and dogs (7–9) following the injection of bacterial endotoxins also has properties quite different from those of bacterial pyrogens (10, 11).

The leucocytic and circulating endogenous pyrogens, on the other hand, are strikingly similar; indeed, they cannot be distinguished on the basis of the biological criteria presently available (7–9).

Intravenously administered bacterial pyrogens regularly produce a leucopenia which precedes the onset of fever (10–14). Subsequently, the febrile response is accompanied by the appearance in the blood of circulating endogenous pyrogen (6–9). Since bacterial endotoxins are known to be injurious to leucocytes (15–17), it has been postulated that the fever which results from their injection is due to endogenous pyrogen released from injured leucocytes and transported *via* the blood stream to the thermoregulatory centers of the brain (6).

The results of the following studies confirm the hypothesis that both leucocytic pyrogen and circulating endogenous pyrogen act directly upon thermoregulatory centers in the central nervous system.

Methods

A single lot of typhoid vaccine (monovalent reference standard NRV-LS No. 1) made from *Salmonella typhosa* V-58 was used as the bacterial pyrogen in all experiments.[1] The methods

* This work was supported by a grant from the Life Insurance Medical Research Fund.

‡ Fellow of the National Foundation for Infantile Paralysis. Present address: Departments of Medicine and Preventive Medicine, Washington University School of Medicine, St. Louis.

[1] Secured through the courtesy of Dr. Geoffrey Edsall of the Army Medical Service Graduate School, Walter Reed Army Medical Center, Washington, D. C. This vaccine had a bacterial count of approximately 500 million per ml. and a nitrogen content of 0.03 mg. per ml. (±10 per cent).

of handling the rabbits and recording their temperatures,[2] and the precautions taken to avoid contamination of solutions and glassware with extraneous bacterial pyrogens have been described in the preceding paper (18).

Source of Pyrogens:

 1. Leucocytic Pyrogen.—Leucocytic exudates were produced by the same technique as previously described (18), except for minor modifications. Physiologic saline, administered intraperitoneally by constant injection over a 3½ hour period, was given in 300 ml. amounts to each of 3 rabbits. The exudates were harvested 7½ hours after injection, pooled, and reduced to one-quarter of their original volume by being dialyzed against 12 per cent dextran solution.[3] The concentrated pooled exudates were incubated for 18 hours at 37°C. and were centrifuged at 625 g for 20 minutes. The supernatant fluid was decanted and stored at 4°C. The centrifugate was discarded. The pyrogen present in the supernatant fluid was shown in preliminary experiments to be inactivated when heated at 90°C. for 30 minutes (3).

 2. Circulating Endogenous Pyrogen (2 Hour Serum).—Sensitized donors, described in a previous study (5), were used to provide 2 hour serum (5, 6) because they are known to clear exogenous pyrogen (typhoid vaccine) more rapidly than do normal (*unsensitized*) donors (6). They are even less likely than normal donors, therefore, to have residual exogenous pyrogen in their blood streams after 2 hours. Each donor rabbit was given a standard inoculation of 1 ml. of undiluted vaccine; 120 minutes later it was bled by intracardiac puncture. The blood was allowed to clot in a flask for 1 hour at 37°C., and the serum was removed and cleared by centrifugation after storage overnight at 4°C. The sera collected from several rabbits were pooled for each test. Before use the pools were shown to be bacteriologically sterile.

 3. Exogenous Pyrogen (5 Minute Serum).—Normal rabbits were inoculated with 1 ml. of undiluted typhoid vaccine and blood was withdrawn 5 minutes later. Five minute serum (5, 6) was used as the source of exogenous pyrogen for two reasons: first, because in such samples the bacterial pyrogen presumably has had time to combine with factors in the serum which influence its actions (19), and secondly because *its pyrogenic activity is approximately equivalent to that of 2 hour serum (5) and is, therefore, suitable for comparison.* The methods of withdrawing blood and separating the serum were the same as for 2 hour serum.

Rationale of Comparing Responses to Intracarotid and Intravenous Injections:

When a pyrogen is injected directly into the carotid circulation, the concentration reaching the thermoregulatory centers of the brain will, during the period of injection, be greater than when the same substance is injected intravenously (see Fig. 1). A pyrogen acting directly on the thermoregulatory centers of the brain would be expected, therefore, to cause a prompter and greater response than when injected intravenously. A pyrogen not acting directly on the brain would, on the other hand, be expected to cause a relatively slow response which would be the same regardless of the route of injection.

Exploratory Experiments to Determine Optimal Position of Intracarotid Catheter:

The point at which the common carotid artery divides to send major branches to (*a*) the ear, (*b*) the tongue and face, and (*c*) the brain (internal carotid) is difficult to visualize in the living rabbit. The approximate position of an intracarotid catheter, however, can be deter-

---

[2] The method of calculating the fever indices was the same as in the preceding study (18) except that a cut-off point of 240 minutes was arbitrarily used because of the consistently longer responses produced by exogenous pyrogen.

[3] Injection plavolex dextran (6 per cent), Wyeth Laboratories, Inc., Philadelphia, concentrated to 12 per cent by boiling.

mined by injection of 1 per cent aqueous methylene blue through the catheter. If the dye enters the internal carotid artery, the eye alone turns blue, whereas injection into the common carotid artery produces a blue color in the ear, eye, and tongue. This method was used to determine the initial position of the tip of the catheter and to check its intracarotid location at the time of testing several weeks after the original operation.

In the earliest experiments a single catheter was placed in the carotid artery. Unilateral catheterization, however, was found to be unsatisfactory in that the differences obtained in

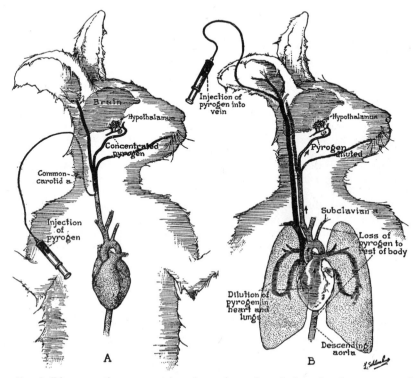

FIG. 1. Diagrammatic representation of experimental method employed to compare the sites of action of endogenous and exogenous pyrogens. (Drawing by Mr. Leon Schlossberg of the Department of Art as Applied to Medicine, Johns Hopkins University.)

the febrile response to intracarotid and intravenous injections were inconsistent. It was reasoned that the variability of results might be due to the fact that in some of the operated animals flow of blood through the catheterized vessel was so reduced as to allow the opposite artery to dominate the circulation of the circle of Willis. Such an imbalance of carotid blood flow might be expected to cause the injected pyrogen to pass directly into the middle cerebral artery and thus enter the venous circulation without reaching the thermoregulatory centers. For this reason bilateral catheterization was attempted.

Both common carotid arteries were first ligated in 2 rabbits to determine the possible ill effects upon the central nervous system. The rabbits appeared healthy postoperatively, but manifested a greatly depressed responsiveness to intravenous injections of both endogenous and exogenous pyrogens. Similarly, catheterization of both *internal* carotid arteries or catheterization of one common carotid and ligation of the other resulted in cerebral ischemia which irreversibly depressed the thermoregulatory centers.

Method Finally Adopted for Preparing Recipients:

A successful method was eventually devised which involved catheterization of both *common* carotid arteries. Although this procedure often caused some degree of cerebral ischemia, its depressive effect upon the thermoregulatory centers proved to be consistently reversible. Within 3 to 4 weeks after operation, the centers were regularly found to be just as sensitive to intravenously injected pyrogen as they had been prior to operation. The operative technique finally adopted was as follows. Each rabbit was anesthetized with intravenous pentobarbital and placed on its back. After the skin of the neck had been shaved with electric clippers, the carotid arteries were exposed through an anterior midline incision. With a small half-circle eye needle, purse string sutures (size 5-0) were placed in each arterial wall in such a way that the lumens of the vessels were not entered. A small hole was then made in each vessel wall, within the purse string suture, by means of a 20 gauge needle. Polyvinyl catheters[4] were inserted cephalad for a distance of 3 cm. and were tightly secured by the purse string sutures. Before being inserted the catheters were filled with heparin and their distal ends were firmly tied. A loop was then made in each catheter between the point of exit from the vessel and an anchoring suture placed in the subcutaneous tissues. From these sutures the catheters were brought subcutaneously around to the back of the neck and out through a small skin incision. The wound in the anterior neck was then closed. A 15 cm. length of each catheter was coiled and deposited in a subcutaneous "bank," a length of only 2 cm. being allowed to emerge from the skin. The incision was loosely sutured in such a manner as to permit additional lengths of the catheters to be withdrawn from the subcutaneous bank for future use. The rabbits were returned to their cages for a period of 3 to 5 weeks before being used for definitive experiments. In order to prevent obstruction of the catheters, it was found necessary to refill them with heparin every 3 or 4 days.

Method of Injection.—

The bilateral intracarotid injections were performed simultaneously through 25 gauge needles inserted into the ends of the carotid catheters. Each intravenous injection was made through a polyethylene catheter placed in the marginal ear vein at the time of injection.

When 10 ml. samples of 2 hour serum were rapidly injected *via* the two routes, no significant difference was observed in the febrile responses. It was postulated that the failure of a difference to occur might be due to the rapidity with which the intracarotid dose had been carried past the thermoregulatory centers. When the injections were slowed, the intracarotid responses without exception exceeded the intravenous. The optimal injection time was eventually determined to be 45 minutes. All injections, both intracarotid and intravenous, were accordingly administered by constant infusion over a period of 45 minutes.

RESULTS

1. Intravenous versus Intracarotid Injection of Leucocytic Extract (Endogenous Pyrogen).—

The febrile responses to intravenous and intracarotid injections of 10 ml. of leucocytic extract were compared in each of 3 rabbits. The routes of injection were alternated on succes-

---

[4] Surprenant Manufacturing Company, Clinton, Massachusetts; Internal diameter 0.015; outside diameter 0.025 inches. Use of this type of catheter was kindly suggested to us by Dr. Clifford Barger of the Department of Physiology of the Harvard Medical School.

sive days in each recipient. Two of the rabbits received the intracarotid injection on the 1st day and the intravenous injection on the 2nd day. In the third rabbit the injection sequence

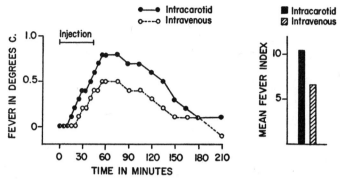

FIG. 2. Comparison of febrile responses to intracarotid and intravenous injections of leucocytic pyrogen. On the left is shown a representative pair of fever curves recorded in a single rabbit; on the right are the mean fever indices from 3 rabbits given the pyrogen by the two routes on successive days.

TABLE I

Comparison of Febrile Reactions to Endogenous and Exogenous Pyrogens Given Slowly by Intracarotid and Intravenous Injections

Source of pyrogen	Rabbit No.	Fever index*		Latent period (min.)‡	
		Intracarotid injection	Intravenous injection	Intracarotid injection	Intravenous injection
Leucocytes (endogenous)	2-00	10.6	6.3	10	20
	2-01	11.6	6.3	10	20
	2-03	8.6	7.2	10	15
2 hour serum (endogenous)	2-00	11.7	7.1	5	25
	2-01	13.0	6.9	10	30
	2-03	10.9	5.4	15	20
5 minute serum (exogenous)	2-00	14.2	15.7	35	35
	2-01	15.0	21.2	35	45
	2-03	18.9	11.6	40	30

* See Methods.

‡ Measured from time of start of injection to beginning of continuing temperature rise (first 0.1°C.).

was reversed. In every instance, regardless of the sequence, the intracarotid response was greater and more prompt than the intravenous.

The results, which are summarized in Table I and are depicted diagrammatically in Fig. 2, clearly indicate that leucocytic pyrogen acts directly upon the thermoregulatory centers of the brain.

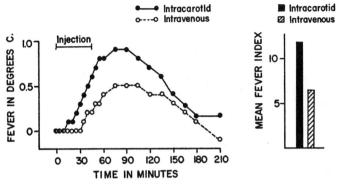

FIG. 3. Comparative pyrogenic effects of 2 hour serum given intravenously and *via* carotid artery. Graphs on left and right as in Fig. 2.

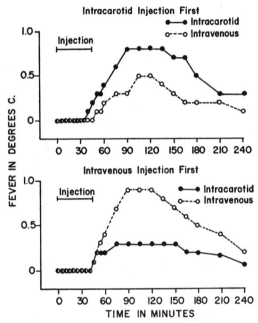

FIG. 4. Comparative fever responses to 5 minute serum injected by intracarotid and intravenous routes on successive days. In experiment shown in upper chart the intracarotid injection was given first; in that shown in the lower chart the intravenous injection was given first. Note depressive effect of tolerance upon response to second injection.

2. Intravenous versus Intracarotid Injection of 2 Hour Serum (Endogenous Pyrogen).—

Identical experiments were performed with 10 ml. samples of 2 hour serum (endogenous pyrogen). The results, which were the same as those obtained with leucocytic pyrogen, are shown in Table I and Fig. 3. It is apparent that 2 hour serum also acts directly upon the central nervous system.

3. Intravenous versus Intracarotid Injection of 5 Minute Serum (Exogenous Pyrogen).—

Analogous experiments with 10 ml. samples of 5 minute serum (exogenous pyrogen) yielded strikingly different results. As illustrated in Fig. 4, when the injections were performed on successive days, the phenomenon of tolerance (10, 11) was regularly encountered on day 2. Its presence caused a definite depression of the 2nd day response regardless of the route of injection (see Table

TABLE II

Effect of Injection Sequence upon Febrile Response to Exogenous and Endogenous Pyrogens Given Slowly by Intracarotid and Intravenous Routes

Source of pyrogen	Injection sequence*	Rabbit No.	Fever index‡		Difference (x) minus (y)
			Intracarotid (x)	Intravenous (y)	
5 min. serum (exogenous)	Vein-artery	2-00	6.3	15.7	−9.4
	Artery-vein	2-01	15.0	6.9	+8.1
	Vein-artery	2-03	6.0	11.6	−5.6
	Artery-vein	2-00	14.2	11.1	+3.1
	Vein-artery	2-01	16.0	21.2	−5.2
	Artery-vein	2-03	18.9	11.8	+7.1
2 hr. serum (endogenous)	Artery-vein	2-00	11.7	7.1	+4.6
	Vein-artery	2-01	13.0	6.9	+6.1
	Artery-vein	2-03	10.9	5.4	+5.5
	Vein-artery	2-00	10.1	5.3	+4.8
	Artery-vein	2-01	13.6	9.3	+4.3
	Vein-artery	2-03	10.1	6.3	+3.8

* Injections given on successive days.
‡ See Methods.

II). In order to avoid the interfering effect of tolerance, the interval between the first and the second injections was lengthened to 3 weeks (10, 11). When tolerance was thus avoided, no consistent difference could be demonstrated between the responses to the intracarotid and intravenous injections (see Fig. 5 and Table I). Furthermore, the latent period was significantly longer than that encountered with either leucocytic pyrogen or 2 hour serum, and the length of the latent period was uninfluenced by the route of injection (see Table I).

These findings indicate that circulating exogenous pyrogen, even when previously exposed to serum *in vivo* (19), acts, in the dosage range used in the present experiments, by a less direct mechanism than does endogenous pyrogen.

4. Repetition of Experiments with 2 Hour Serum (Endogenous Pyrogen) Using Interval of 3 Weeks between Injections.—

Inasmuch as it was necessary to interpose an interval of 3 weeks between the first and second injection of 5 minute serum in order to avoid interference from

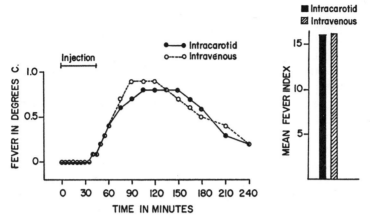

FIG. 5. Comparison of febrile responses to 5 minute serum resulting from successive intracarotid and intravenous injections separated by an interval of 3 weeks to avoid tolerance. Graphs on left and right as in Fig. 2.

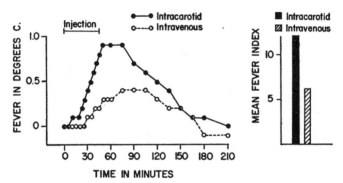

FIG. 6. Comparison of febrile responses to 2 hour serum resulting from successive intracarotid and intravenous injections separated by an interval of 3 weeks in order to duplicate conditions of experiment summarized in Fig. 5. Graphs on left and right as in Fig. 2.

tolerance, the experiments with 2 hour serum were repeated in such a way as to make them strictly comparable. The results obtained with a 3 week interval are shown in Fig. 6. It will be noted that they were exactly the same as those observed following daily injections (Fig. 3).

The conclusion that endogenous pyrogen acts directly upon the central nervous system is thus further substantiated.

DISCUSSION

The method employed in the present study was designed to distinguish pyrogens acting directly upon the central nervous system from those operating through a less direct mechanism. It was based upon the assumption that a directly acting pyrogen will cause a more immediate and greater febrile response when injected *via* the carotid artery than when given intravenously, whereas an indirectly acting one will cause the same response when given by either route. As might have been predicted *a priori*, the method effectively differentiates between the two types of pyrogens only when they are given by relatively slow infusions. The slow injections are required because the increased intracerebral concentration of pyrogen, resulting from intracarotid injection, must be maintained long enough for the thermoregulatory centers to become activated, if an augmented response is to occur. Thus it is not surprising that in preliminary experiments rapid intracarotid and intravenous injections failed to differentiate between the two types of pyrogens, which later were shown to be clearly distinguishable by slower injections.

The method finally adopted revealed that endogenous pyrogen, as it exists both in leucocytic exudates (3, 18) and in 2 hour serum (5, 6) causes a more rapid and a more pronounced fever when introduced *via* the carotid artery than when injected intravenously. The exogenous pyrogen in 5 minute serum (5, 6), on the other hand, caused a relatively slow response which was uninfluenced by the route of injection. From these findings it is concluded that endogenous pyrogen acts directly upon the central nervous system and that exogenous pyrogen, in the quantities present in 5 minute serum (5), exerts a different action, which involves a less direct mechanism.

The question may be raised as to whether these same results might not also be explained by a difference in the rate at which the two forms of pyrogen penetrate the thermoregulatory centers. It might be reasoned that the failure of exogenous pyrogen to cause a greater and more prompt response when injected into the carotid artery than when given intravenously is due to the fact that it does not activate the centers until after the injections have been completed. As will be noted from Figs. 4 and 5 and Table I, however, the febrile response to exogenous pyrogen has been repeatedly shown to begin before the injection is discontinued. Since the concentration of pyrogen in the cerebral circulation is greater throughout the period of intracarotid injection than it is during the intravenous injection, a difference in the responses, similar to that observed with endogenous pyrogen, should occur if the exogenous pyrogen acted directly upon the thermoregulatory centers. Inasmuch as no such difference could be demonstrated, it must be concluded that circulating exogenous pyrogen acts by a different and less direct mechanism.

Under special circumstances, on the other hand, exogenous pyrogens do act directly upon the thermoregulatory centers to cause fever.

By means of an ingenious technique Keene (20) has recently shown, for example that bacterial endotoxins, when injected into the spinal fluid in relatively minute quantities, produce prompt and striking febrile responses. This effect may be explained by three facts. First, bacterial endotoxins are notoriously cytotoxic substances (10, 15, 16); secondly, the thermoregulatory centers at the base of the brain are in close apposition to the subarachnoid space (21); and thirdly, these centers are highly sensitive to noxious stimuli (22).

It has also been recently demonstrated by Bennett et al. (23) that bacterial endotoxins regularly "spill over" into the spinal fluid when given in relatively large intravenous doses to animals previously made leucopenic with nitrogen mustard. Under these circumstances, the usual febrile responses to the endotoxin occur despite the fact that endogenous pyrogens may not be easily demonstrated in the circulation. This observation has led to the suggestion that the endotoxin in the spinal fluid, rather than the endogenous pyrogen in the blood stream, acts upon the thermoregulatory centers to cause the fever (23). A similar apparent dissociation of fever from endogenous pyrogen has been described in animals previously made tolerant to endotoxin (9). Although the 2 hour sera of such animals contain relatively little endogenous pyrogen, the fever which they exhibit may be comparable to those of normal animals, provided the doses of endotoxin injected are sufficiently large.

Experiments of this type must be interpreted with caution. First of all, they bear primarily upon the intricacies of the endotoxin model and are probably not relevant to the pathogenesis of other common forms of fever, particularly those resulting from natural causes. Secondly, they pertain only to special situations in which the normal production of endogenous pyrogen has been interfered with, and in which the dose of endotoxin is relatively large. Thirdly, they involve a very insensitive method of detecting endogenous pyrogen, namely the method of passive transfer (6) (see study that follows (24)). Even though the evidence is conclusive that endotoxin may, under certain circumstances, exert a direct pyrogenic action upon the central nervous system (20, 23), the possibility remains that the dissociation emphasized by Bennett et al. (8, 9, 23) is more apparent than real. Indeed, in normal rabbits made febrile with typhoid vaccine the degree of fever present throughout the reaction has already been shown to be closely related to the amount of endogenous pyrogen in the blood stream (6).

The fact that the leucocytic and the 2 hour serum pyrogens act by the same direct mechanism adds further support to the concept that they are one and the same substance (6). The conclusion that the endogenous pyrogen is of leucocytic origin is also confirmed by the recent observation of Bennett et al. (8, 23) that induction of leucopenia with nitrogen mustard decreases the amount of endogenous pyrogen produced by the injection of bacterial endotoxin. Both of these findings strongly support the hypothesis previously advanced (6) that endogenous pyrogen derived from injured leucocytes plays a central role in the pathogenesis of endotoxin fever.

The endotoxin fever model, however, is extremely complex. In many ways it is unsatisfactory for studies dealing with the pathogenesis of fever in general. Not only may a double pyrogenic mechanism be involved under certain circumstances, as postulated by Bennett and his colleagues (8, 9, 20, 23), but the whole phenomenon of tolerance, which depends in part at least upon varying rates of clearance of the injected endotoxin (10, 11), introduces variables which appear to have no counterpart in naturally occurring fevers.

The primary objective of the present investigations (5, 6, 18) has been to study the pathogenesis of fever, particularly as related to the febrile states most commonly encountered in clinical medicine. Because of the complexities of the endotoxin model, which appear to set it apart from other forms of fever, the decision was made to turn directly to the study of fever caused by experimental bacterial infections. The results reported in the paper which follows (24) indicate that in this form of fever also endogenous pyrogen plays a central role.

SUMMARY

By means of a method designed to compare the febrile responses produced by intracarotid and intravenous injections, the endogenous pyrogen, which is contained in leucocytic exudates and is present in the serum of rabbits 2 hours after intravenous injections of typhoid vaccine, has been shown to act directly upon the thermoregulatory centers of the brain. In contrast, the exogenous bacterial pyrogen present in serum obtained 5 minutes after vaccine injections was found to act by a different and less direct mechanism. These observations add strong support to the original hypothesis that endogenous pyrogen, presumably derived from polymorphonuclear leucocytes, is an essential factor in the pathogenesis of endotoxin fever.

BIBLIOGRAPHY

1. Beeson, P. B., Temperature-elevating effect of a substance obtained from polymorphonuclear leucocytes, *J. Clin. Inv.*, 1948, **27,** 524 (abstract).
2. Bennett, I. L., Jr., and Beeson, P. B., Studies on the pathogenesis of fever. I. The effect of injection of extracts and suspensions of uninfected rabbit tissues upon the body temperature of normal rabbits, *J. Exp. Med.*, 1953, **98,** 477.
3. Bennett, I. L., Jr., and Beeson, P. B., Studies on the pathogenesis of fever. II. Characterization of fever-producing substances from polymorphonuclear leukocytes and from the fluid of sterile exudates, *J. Exp. Med.*, 1953, **98,** 493.
4. Petersdorf, R. G., and Bennett, I. L., Jr., Studies on the pathogenesis of fever. VII. Comparative observations on the production of fever by inflammatory exudates in rabbits and dogs, *Bull. Johns Hopkins Hosp.*, 1957, **100,** 277.
5. Atkins, E., and Wood, W. B., Jr., Studies on the pathogenesis of fever. I. The presence of transferable pyrogen in the blood stream following the injection of typhoid vaccine, *J. Exp. Med.*, 1955, **101,** 519.

6. Atkins, E., and Wood, W. B., Jr., Studies on the pathogenesis of fever. II. Identification of an endogenous pyrogen in the blood stream following the injection of typhoid vaccine, *J. Exp. Med.*, 1955, **102,** 499.
7. Petersdorf, R. G., and Bennett, I. L., Jr., Studies on the pathogenesis of fever. VI. The effect of heat on endogenous and exogenous pyrogen in the serum of dogs, *Bull. Johns Hopkins Hosp.*, 1957, **100,** 197.
8. Petersdorf, R. G., and Bennett, I. L., Jr., Studies on the pathogenesis of fever. VIII. Fever-producing substances in the serum of dogs, *J. Exp. Med.*, 1957, **106,** 293.
9. Petersdorf, R. G., Keene, W. R., and Bennett, I. L., Jr., Studies on the pathogenesis of fever. IX. Characteristics of endogenous serum pyrogen and mechanisms governing its release, *J. Exp. Med.*, 1957, **106,** 787.
10. Bennett, I. L., Jr., and Beeson, P. B., The properties and biological effects of bacterial pyrogens, *Medicine*, 1950, **29,** 365.
11. Beeson, P. B., Tolerance to bacterial pyrogens. I. Factors influencing its development, *J. Exp. Med.*, 1947, **86,** 29.
12. Bennett, I. L., Jr., Comparison of leukocyte changes produced by pyrogens and by anaphylaxis in the guinea pig, *Proc. Soc. Exp. Biol. and Med.*, 1951, **77,** 772.
13. Bennett, I. L., Jr., and Beeson, P. B., The effect of cortisone upon reactions of rabbits to bacterial endotoxins with particular reference to acquired resistance, *Bull. Johns Hopkins Hosp.*, 1953, **93,** 290.
14. Atkins, E., Allison, F., Jr., Smith, M. R., and Wood, W. B., Jr., Studies on the antipyretic action of cortisone in pyrogen-induced fever, *J. Exp. Med.*, 1955, **101,** 353.
15. Berthrong, M., and Cluff, L. E., Studies of the effect of bacterial endotoxins on rabbit leucocytes. I. Effect of intravenous injection of the substances with and without induction of the local Shwartzman reaction, *J. Exp. Med.*, 1953 **98,** 331.
16. Kirby, G. P., and Barrett, J. A., Jr., The effect of hydrocortisone and of Piromen *in vitro* on leukocytes of patients receiving A.C.T.H. and cortisone therapy, *J. Clin. Inv.*, 1954, **33,** 725.
17. Braude, A. I., Carey, F. J., and Zalesky, M., Studies with radioactive endotoxin. II. Correlation of physiological effects with distribution of radioactivity in rabbits injected with lethal doses of *E. coli* endotoxin labelled with radioactive sodium chromate, *J. Clin. Inv.*, 1955, **34,** 858.
18. King, M. K., and Wood, W. B., Jr., Studies on the pathogenesis of fever. III. The leucocytic origin of endogenous pyrogen in acute inflammatory exudates, *J. Exp. Med.*, 1958, **107,** 279.
19. Grant, R., and Whalen, W. J., Latency of pyrogen fever. Appearance of a fast-acting pyrogen in the blood of febrile animals and in plasma incubated with bacterial pyrogen, *Am. J. Physiol.*, 1953, **173,** 47.
20. Keene, W. R., The pathogenesis of fever. Fevers produced by intrathecal injection of endotoxin, *Bull. Johns Hopkins Hosp.*, 1956, **99,** 103 (abstract).
21. Ranson, S. W., and Clark, S. L., The Anatomy of the Nervous System, Philadelphia, W. B. Saunders Company, 1953, 9th edition.

22. Selle, W. A., Body Temperature, Springfield, Illinois, Charles C. Thomas, 1952.
23. Bennett, I. L., Jr., Petersdorf, R. G., and Keene, W. R., Pathogenesis of fever: evidence for direct cerebral action of bacterial endotoxins, *Tr. Assn. Am. Physn.*, 1957, **70,** 64.
24. King, M. K., and Wood, W. B., Jr., Studies on the pathogenesis of fever. V. The relation of circulating endogenous pyrogen to the fever of acute bacterial infections, *J. Exp. Med.*, 1958, **107,** 305.

FEVER IN THE LIZARD *DIPSOSAURUS DORSALIS*

L. K. Vaughn, H. A. Bernheim, and M. J. Kluger

*Department of Physiology,
University of Michigan Medical School*

FEVER is considered to be a universal response of warm-blooded animals to endotoxins[1]. Although during a fever a mammal uses behavioural as well as physiological means to increase its body temperature[2], it is not known whether fever develops in an animal such as a lizard which regulates its body temperature largely by behaviour. For example, the desert iguana (*Dipsosaurus dorsalis*) regulates its body temperature close to 38.5° C if placed in a chamber with a temperature gradient[3]. If this lizard is placed in a temperature chamber in which one end is heated to above the animal's lethal body temperature (50° C) and the other end is maintained at room temperature, the lizard regulates its temperature by moving back and forth between the two sides[4]. Under these conditions, one can determine its high and low set-points (Fig. 1). The central nervous control of temperature in an ectotherm, such as a lizard, and an endotherm, such as the rabbit, appears to be quite similar. For example, both possess a hypothalamus which is thermally sensitive[5,6], and lesions in the posterior hypothalamus in both lizards[4] and mammals[7] lead to an inability to maintain a high body temperature.

Because of the similarities in the central nervous control of thermoregulation in reptiles and mammals, and because fever in mammals is accompanied by major behavioural adjustments, we suspected that fever could be produced in a reptile. We now report that bacteria that produce fever in a rabbit will produce a similar fever in the lizard *Dipsosaurus dorsalis*.

Lizards weighing 25–60 g (Hermosa Reptile Farm, Hermosa, California) were housed in circular cages and fed meal worms, lettuce and water *ad libitum*. The cages and experimental chambers were kept on a 12-h light and 12-h dark photoperiod. The chamber was heated by a 250 W heat lamp which was also on a 12:12 cycle.

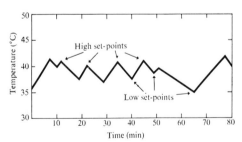

Fig. 1 Record of the cloacal temperature of *Dipsosaurus dorsalis* regulating its temperature in the wooden box in which the substrate at one end was maintained at 30° C and the other end at 50° C. At the high set-point, the lizard moved from the warm side of the chamber to the cool side. At the low set-point the lizard moved back to the warm side of the chamber. Data were recorded every 30 s. The high and low set-points, for any time period, represent the average of these points over that time period.

Experiments were carried out in a temperature-controlled room. Each lizard was placed in a wooden box (30 cm × 140 cm × 30 cm) of which one end was at the room temperature of 30° C and the other end at 50° C. This high temperature was provided by heating coils taped to the undersurface of the floor of one end of the box. The two sides of the box were separated by a small wire mesh bridge to provide a clear boundary between the two temperature extremes. A copper–constantan thermocouple covered with polyethylene tubing (PE 100) was placed about 3 cm into each lizard's cloaca and taped to its tail; this did not noticeably impair the movement of the lizards. Thermocouples were connected to a Honeywell Electronik 112 multipoint recorder which recorded the temperature of each lizard ($\pm 0.1°$ C) every 30 s.

Aeromonas hydrophila, Gram-negative bacteria pathogenic to reptiles and amphibians[8], were grown on blood agar and killed by washing in 70% ethyl alcohol. They were then centrifuged and resuspended in physiological saline. The concentration of bacteria was determined by a turbidity test.

Lizards were allowed 1 d to adapt to the wooden box, and on the second day control data were recorded. On the third day either 0.2 ml of a solution containing 2×10^{10} bacteria per ml of saline or 0.2 ml of sterile physiological saline alone was injected into the heart using a 1.5-inch 26-gauge needle. Blood was redrawn into the syringe to ensure that the needle was in the heart. All injections were done at the same time of day to avoid possible effects of diurnal variations.

To determine whether increases in body temperature were due entirely to behavioural modifications, or whether an increased internal production of heat was part of the response, three lizards were given injections of bacteria identical to the above. They were then placed in a wooden chamber (14 cm × 30 cm × 30 cm) which was held at 30° C. Cloacal temperatures were measured as before.

Four New Zealand white rabbits (*Oryctolagus cuniculus*) were used to determine whether *A. hydrophila* could also produce fever in mammals. The rabbits' tails were shaved and a thermocouple was inserted 10 cm into the rectum and taped to the tail. The rabbits were placed in a restrainer and allowed 1 h to acclimatise to room temperature (22° C). Then 0.2 ml of saline was injected into the marginal ear vein. After 1 h (control period), 1 ml of 3×10^9 bacteria per ml saline was injected into the marginal vein of the other ear.

In 10 lizards, given free choice of either the 50° C or 30° C environment, injection of 4×10^9 bacteria produced little increase in temperature during the first few hours. Body temperature increased approximately 2° C between the fourth and sixth hours (Fig. 2). The increases of the average high (41.1° C to 42.7° C) and low (37.4° C to 39.7° C) set-points were highly statistically significant ($P < 0.001$ using paired sample analysis).

When the same concentration of bacteria was injected into three lizards maintained in a constant temperature chamber at 30° C, their temperatures did not change throughout the

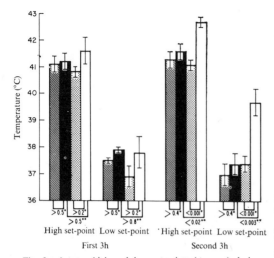

Fig. 2 Average high and low set-points (±s.e.m.) during control periods (day 2) and after intracardiac injection of 0.2 ml isotonic saline or 0.2 ml of 2 × 10¹⁰ *Aeromonas hydrophila* ml⁻¹ (day 3) into *Dipsosaurus dorsalis*. *, Level of significance using paired sample analysis; †, level of significance using Student's *t* test. Hatched columns, control period before saline injections; solid columns, period after saline injections; stippled columns, control period before *Aeromonas* injections; open columns, period after *Aeromonas* injections.

control and experimental periods. Injection of saline into nine lizards, given free choice of the warm or cool environment, produced no changes in high or low set-points (Fig. 2). Comparison of lizards injected with saline to those with *A. hydrophila* revealed a statistically significant increase in the high ($P < 0.02$) and low ($P < 0.003$) set-points during the second 3 h (Student's *t* test). Injection of saline into the four rabbits produced no increase in rectal temperature. Injection of the bacteria produced a fever with a latency of about 20 min and a mean maximum rise of 2.2° C within 3 h.

These data indicate that a bacterium that causes fever in a rabbit has a similar effect in a lizard. Since the lizards could not increase their temperature in response to the bacteria when the ambient temperature was held constant, fever in these lizards cannot be developed by increased internal production of heat. The relatively long latency before the onset of fever in the lizards might be due to the lower metabolic rate of an ectotherm in comparison with that of an endotherm.

These results demonstrate that fever can be sustained by behavioural regulation in an ectotherm. Also, since the hypothalamic control over thermal responses in reptiles and mammals are similar[4-8] and as we have shown, a similar concentration of bacteria will produce a similar increase in temperature in both a reptile and a mammal, we suspect a common origin for reptilian and mammalian fever. If this is true, then the ability to develop a fever existed before the time when evolutionary lines for mammals and reptiles diverged. This suggests that at least the behavioural component of fever had evolved by the late Palaeozoic or early Mesozoic, or perhaps even earlier. We cannot rule out the possibility, however, that fever might have independently and perhaps recently evolved in reptiles and mammals.

These findings also open new possibilities for ascertaining the adaptive value of fever. The question of whether fever is beneficial or harmful to the host has been difficult to resolve in mammals. Since the increase in body temperature in response to bacterial infection might have evolved in primitive reptiles, or even in amphibians, the adaptive value of fever might be revealed by a careful study of the role of the febrile response in these classes of vertebrates.

We thank Drs B. Cohen and D. Ringler and the Unit for Laboratory Animal Medicine for bacteria. We thank L. D'Alecy, D. Mouw and A. Vander for critically reviewing our manuscript.

Received September 16; revised October 17, 1974.

[1] Atkins, E., and Bodel, P., *New Engl. J. Med.*, **286**, 27–34 (1972).
[2] Cabanac, M., in *Essays on temperature regulation* (edit. by Bligh, J., and Moore, R.), 19–36 (North-Holland, Amsterdam, 1972).
[3] DeWitt, C. B., *Physiol. Zool.*, **40**, 49–66 (1967).
[4] Kluger, M., Tarr, R. S., and Heath, J. E., *Physiol. Zool.*, **46**, 79–84 (1973).
[5] Cabanac, M., Hammel, T., and Hardy, J. D., *Science*, **158**, 1050–1051 (1967).
[6] Hammel, H. T., Caldwell, F. T., jun., and Abrams, R. M., *Science*, **156**, 1260 (1967).
[7] Ranson, S. W., and Magoun, H. W., *Ergeb. Physiol. Biol. Chem. exp. Pharmakol.*, **41**, 56–163 (1939).
[8] Reichenbach-Klinke, H., and Elkan, E., in *The principal diseases of lower vertebrates*, 396 (Academic Press, London, 1965).

BEHAVIORAL FEVER IN NEWBORN RABBITS

E. Satinoff, G. N. McEwen, Jr., and B. A. Williams

*Environmental Control Research Branch,
Ames Research Center, NASA*

Unlike adults, newborn mammals do not develop a fever upon initial exposure to a bacterial pyrogen (1). Neonates do not respond well to thermal stresses in general, and can maintain normal body temperatures only within a narrow range of environmental temperatures. In this respect, infant mammals are very much like reptiles, which have inadequate or nonexistent physiological thermoregulatory mechanisms. Reptiles, however, will maintain their body temperatures within narrow limits when given the opportunity to do so behaviorally (2). Many infant mammals also show thermal preferences from the day of birth and will maintain normal body temperatures by moving to the appropriate place in a thermal gradient (3). In response to a bacterial infection, iguanas will develop a fever by spending more time at higher environmental temperatures (4). Fish also show a behavioral fever when injected with a pyrogen (5). We report here that, in response to an initial challenge with a pyrogen, newborn rabbits, unable to develop a fever physiologically, will do so behaviorally by selecting higher environmental temperatures than nonchallenged controls do.

Eleven New Zealand white rabbit pups from two litters were used in these experiments. They were 12 to 72 hours old and weighed between 50 and 89 g. The pups were removed from their mothers and divided into three groups matched as closely as possible for body weight. One group was given an intraperitoneal injection of *Pseudomonas* polysaccharide (Piromen), 500 μg per kilogram of body weight, dissolved in sterile saline (250 μg/ml). The second group received the saline vehicle alone (2 ml/kg), and the third group was given no treatment. After injection, the pups were placed in individual containers in an incubator kept at 32°C. Two hours after injection, rectal temperatures were measured with a 36-gauge, copper-constantan thermocouple inserted 2.0 cm into the rectum. The pups were then placed, two at a time, in a temperature gradient apparatus similar to that used by Ogilvie and Stinson (3). This consisted of an alleyway whose bottom was a copper bar (183 by 15 by 0.64 cm) with aluminum sides (122 cm long, 15 cm high) and a hinged Plexiglas top. At each end of the alleyway, the bar extended 31 cm. Heating tape was wrapped around one end, and temperature was controlled with a Variac voltage transformer. The temperature gradient along the alleyway ranged from 22° to 55°C. Thermocouples were placed every 2.5 cm along the gradient, and 49 divisions were marked along the sides of the enclosure corresponding to the thermocouple placements. The pups were placed in the gradient with their noses touching a point corresponding to 30°C; half were positioned facing the hot end and half facing the cool end. The position of each animal in the gradient was recorded every minute, as were the temperatures under the thermocouples. An experiment, which generally took no more than 25 minutes, was terminated when a pup remained at the same place in the gradient for at least 5 minutes. When a pup was removed from the gradient, its rectal temperature was again recorded, and it was returned to its mother. No rabbit in any group was given more than a single injection of Piromen. The rabbits getting no injection on day 1 received Piromen on day 2 and saline on day 3; those receiving saline on day 1 got no injection on day 2 and Piromen on day 3.

To ensure that the dose of Piromen used was sufficient to give older rabbits a fever, the pups were divided into two groups, matched for body weight, when they were 14 days old. Five were injected with Piromen (500 μg/kg) and six with a similar volume of saline alone. All the pups were maintained individually in the incubator at 27°C, and rectal temperatures were taken before and 2 hours after injection.

All data are reported as the mean ($\pm$ standard error of the mean) unless otherwise indicated. Student's t-tests were performed on the data, and the null hypothesis was rejected when $P \leq .05$.

The main results of the experiment are shown in Fig. 1a. Rabbit pups injected with saline selected a gradient temperature of 36.5° $\pm$ 0.47°C. Pups injected with Piromen selected a significantly higher gradient temperature, 40.4° $\pm$ 0.76°C ($P < .001$). No pup injected with saline selected a gradient temperature higher than 38.0°C, whereas nine of the 11 pups injected with Piromen selected temperatures higher than 39.9°C. Two of the 11 Piromen-injected animals did not respond to the pyrogen. The reasons for this are not obvious,

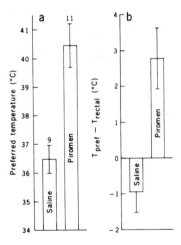

Fig. 1. (a) Preferred gradient temperature (°C) as a function of saline or pyrogen injection. Numbers over bars are number of pups in each group; bars are $\pm$ 1 S.E.M. (b) Preferred gradient temperature (T) minus pretest core temperature as a function of saline or pyrogen injection.

since the body weights, initial rectal temperatures, and general mobility of all the pups were the same. Uninjected pups selected a slightly lower mean gradient temperature (35.4° ± 0.46°C) than did saline controls, but the difference was not significant.

The mean rectal temperatures of all three groups were within 0.5°C of each other (37.2° to 37.7°C) just before they were placed in the thermal gradient. Since this was 2 hours after treatment, it indicates that the group receiving Piromen was not able to develop a fever physiologically at 32°C within this time period. When the pups were removed from the alleyway, after 5 minutes at their selected positions, the mean rectal temperature of the Piromen-treated group was significantly higher than it had been prior to testing (38.6° ± 0.29°C compared to 37.7° ± 0.25°C; $P < .005$). There was no significant change in rectal temperature in either of the control groups.

The increase in body temperature was achieved solely by selecting higher gradient temperatures. Figure 1b shows the gradient temperature selected in relation to the pups' core temperatures just prior to testing. The saline-treated group selected a gradient temperature about 1°C below its starting core temperature. The Piromen-treated group selected a gradient temperature 2.8° ± 0.9°C above its body temperature before the test. This difference was significant ($P < .001$).

The pups were extremely mobile, even 12 hours after birth, and their behavior was highly consistent. When first placed in the alleyway at the 30°C position, they immediately moved toward the hotter end, regardless of whether they had been placed facing toward or away from that end. They often moved past the point corresponding to their final preferred position. Then they either turned immediately and began moving toward the cooler end, or rested at the high temperatures for up to 2 minutes before moving down to a position cooler than their final choice. This oscillatory behavior lasted for 10 to 20 minutes, after which the animals remained in their final selected positions for at least 5 minutes. There was no significant difference between the two groups either in the time taken to reach the final positions or in the time spent at temperatures warmer or cooler than the final selections. When the pups rested at nonpreferred temperatures, they made postural adjustments similar to those of adults. At high gradient temperatures they assumed a sprawled posture; at cooler temperatures they curled up into a ball-like posture identical to that reported for adult rabbits (6). At the final selected temperatures the pups usually lay on their sides, often in a semicurled position.

When they were 14 days old, five rabbits were injected with Piromen and six with the saline vehicle. All were kept in the incubator at 27°C. At the end of 2 hours, the rectal temperatures of the animals injected with saline rose from a mean of 38.1° ± 0.19°C to 39.1° ± 0.09°C, a rise of 1° ± 0.23°C. The rectal temperatures of the Piromen-injected animals rose from 37.9° ± 0.11°C to 39.9° ± 0.12°C, an increase of 2.0° ± 0.19°C. The difference between the initial and final temperatures of the animals injected with saline in contrast to that for animals injected with Piromen was significant ($P < .005$). Thus, the drug, dose, and route of administration used for neonates were adequate to generate a fever physiologically when the pyrogen was injected a second time when the rabbits were older.

This experiment demonstrates that exogenous pyrogen will produce a fever in naive newborn rabbits if the opportunity for thermoregulatory behavior is present. No fever develops if the pups must rely solely on internal thermoregulatory mechanisms. According to current concepts, fever develops when a bacterial pathogen, interacting with white blood cells, causes the release of leukocytic pyrogen. The leukocytic pyrogen travels through the bloodstream to the hypothalamus where it changes the activity of temperature-sensitive neurons in such a way that heat production is increased and heat loss is suppressed. This, in turn, raises body temperature (7). Until the present experiment, the absence of fever in newborn mammals could have been attributed to (i) the leukocytes' inability to manufacture endogenous pyrogen, (ii) the immaturity of the hypothalamic thermal control system, or (iii) the inadequacy of the effectors to increase heat production reflexively and so offset the heat lost because of lack of insulation (newborn rabbits have little fur) and an unfavorable surface-to-volume ratio. The present experiments suggest that the third possibility is most probable; the drive toward fever is present, but heat-producing and heat-conserving mechanisms are inadequate. Alternatively, early in development there may be separate neural mechanisms for behavioral and physiological thermoregulation. The behavioral system for producing a fever is clearly mature at birth, but an adequate system of internal reflexes does not appear to develop for some days.

References and Notes

1. K. E. Cooper, Q. J. Pittman, W. L. Veale, in *Temperature Regulation and Drug Action*, P. Lomax, E. Schönbaum, J. Jacob, Eds. (Karger, Basel, 1975), p. 43; C. M. Blatteis, *J. Appl. Physiol.* **39**, 251 (1975).
2. C. B. DeWitt, *Physiol. Zool.* **40**, 49 (1967).
3. For pigs, see L. E. Mount, *Nature (London)* **199**, 1212 (1963); for mice, D. M. Ogilvie and R. H. Stinson, *Can. J. Zool.* **44**, 511 (1966); and for rabbits, E. Jeddi, *Int. J. Biometeorol.* **15**, 337 (1971).
4. L. K. Vaughn, H. A. Bernheim, M. J. Kluger, *Nature (London)* **252**, 473 (1974).
5. W. W. Reynolds, M. E. Casterlin, J. B. Covert, *ibid.* **259**, 41 (1976).
6. G. N. McEwen, Jr., and J. E. Heath, *J. Appl. Physiol.* **35**, 884 (1973).
7. R. J. Hellon, *Pharmacol. Rev.* **26**, 289 (1975).
8. This work was completed at Ames Research Center under an NRC senior postdoctoral fellowship to E.S. and an NRC postdoctoral fellowship to G.N.M., Jr. It was supported in part by PHS grant NS-12033 to E.S. We thank Travenol Laboratories for generously supplying us with the Piromen.
* Present address: Department of Psychology, University of Illinois, Champaign 61820.

Part V

THERMOREGULATORY BEHAVIOR

Editor's Comments on Papers 19 Through 24

19 **NEWPORT**
Excerpts from *On the Temperature of Insects, and Its Connexion with the Functions of Respiration and Circulation in This Class of Invertebrated Animals*

20 **ROZIN and MAYER**
Thermal Reinforcement and Thermoregulatory Behavior in the Goldfish, Carassius auratus

21 **REGAL**
Voluntary Hypothermia in Reptiles

22 **KINDER**
Excerpts from *A Study of the Nest-building Activity of the Albino Rat*

23 **WEISS**
Thermal Behavior of the Subnourished and Pantothenic-acid-deprived Rat

24 **STRICKER and HAINSWORTH**
Evaporative Cooling in the Rat: Interaction with Heat Loss from the Tail

The papers in this section make a number of points. First, they demonstrate the ubiquity of behavioral thermoregulatory responses in the animal kingdom, from insects to mammals. George Newport's article, from which Paper 19 is excerpted, shows that several different species of colonial insects not only regulate the temperature of their nests but also their own individual body temperatures. Given the thermometers available at the time, his measurements are remarkable. Second, they demonstrate the precision of the regulation. Paper 20 by Paul Rozin and Jean Mayer describes how goldfish maintained the temperature of their aquarium within a range of 3°C (out of a possible range of 16.5°C). This experiment unambiguously demonstrates that fish are capable of operant thermoregulation, that is, learned, motivated responses. Paper 21, by Philip J. Regal, illustrates that the preferred temperature in lizards varies with the time of day. Such circadian rhythmicity of body temperature has long been known in endotherms.

The three remaining papers in this section demonstrate how behavior is precisely attuned to the regulation of body temperature in rats—a well-studied species of mammal. This is as true in the cold, whether the rats are building nests (Paper 22 by Elaine Kinder, who was working in Curt Richter's laboratory) or pressing a bar to turn a heat lamp on (Paper 23 by Bernard Weiss) as it is in the heat, where the major method of heat loss is evaporation of saliva, which the animals groom onto their fur (Paper 25 by Edward Stricker and F. Reed Hainsworth).

ON THE TEMPERATURE OF INSECTS, AND ITS CONNEXION WITH THE FUNCTIONS OF RESPIRATION AND CIRCULATION IN THIS CLASS OF INVERTEBRATED ANIMALS

George Newport

EVERY naturalist is aware that many species of insects, particularly of hymenopterous insects, which live in society, maintain a degree of heat in their dwellings considerably above that of the external atmosphere, but no one, I believe, has hitherto demonstrated the interesting facts that every individual insect when in a state of activity maintains a separate temperature of body considerably above that of the surrounding atmosphere, or medium in which it is living, and that the amount of temperature varies in different species of insects, and in the different states of those species. Previously, therefore, to considering the connection which subsists between the evolution of animal heat and the functions of respiration and circulation in insects, I shall endeavour to prove that every species maintains a distinct temperature of body, the amount of which differs in the different states of the insect.

I was first led to the particular consideration of the subject of temperature in insects by some observations on the temperature of wild bees in their natural haunts, which were made by myself at Richborough, near Sandwich in Kent, in the autumn of 1832, at the suggestion of Dr. MARSHALL HALL, for the purpose,—similar to that of my observations on respiration, as noticed on a former occasion[*],—of ascertaining what relation, if any, subsists between the natural heat of these insects in their hybernating condition and the irritability of their muscular fibre. The results of these observations on the temperature of Bees are shown on Table III., Nos. 1 to 14, and together with many other facts connected with the physiology of insects were communicated to Dr. HALL a short time afterwards[†]. These observations were

---

[*] Philosophical Transactions, Part II. 1836, p. 551.

[†] In submitting these observations on the Temperature of Insects to the consideration of the Royal Society, I have felt myself imperatively called upon to make the above remark, in explanation of the nature of my supposed obligations to Dr. MARSHALL HALL, with regard to this and other subjects connected with the Physiology of Insects, in consequence of certain misrepresentations which were made on a recent occasion respecting my communications with that gentleman; and I beg further to state, that many of the views here advanced respecting the temperature of insects, and also most of the subjoined Tables, particularly those on the temperature of the Hive Bee, from the commencement of my observations to the month of May 1836, were communicated by myself to Dr. MARSHALL HALL, at his own particular request, in the beginning of July 1836, in the presence of my intelligent friend, and late pupil, Mr. JOHN OSBORN, who assisted me in making the observations, and unto whom I am indebted for much valuable assistance during my investigations.

made in the usual manner, by placing a considerable number of insects of the same species together, and then introducing the thermometer among them. But it was a few days previously to making these observations that I first noticed the interesting fact, that each individual insect maintains its own temperature, which is perceptible externally by the thermometer, and that the amount of this varies in the different conditions of the same insect. The observation was first made on the larva of *Sphinx Atropos*, LINN., and on that of *Pygæra bucephala*, STEPH., as will presently be shown.

During the time I have been engaged in preparing the present communication I have become acquainted, through the kindness of Dr. FORBES of Chichester, with the recently published views of Dr. BERTHOLD, of Gottingen, who has made a series of observations on the temperature of cold-blooded animals[*], and among them several on insects, somewhat similar to those which I now have the honour of submitting to the Society. But excellent as are the views of that gentleman, he does not appear to have paid sufficient attention to the conditions of activity or rest in the insects at the time of making his experiments, and consequently has omitted to observe the important fact of the existence of a distinct temperature of body in individual insects[†], and also those circumstances which augment or lessen its amount, and has estimated the temperature by placing many individuals together, which, as will presently be seen, is open to several objections. Dr. BERTHOLD has, however, anticipated me in the expression of one opinion, unto which we have mutually been led by our observations, viz. that at all events the higher classes of invertebrated animals ought not to be considered as *cold-blooded*, since it is found that under certain conditions they have a temperature of body higher than that of the surrounding medium. HAUSMANN[‡] made an observation as long ago as the year 1803, which ought to have led to a proper understanding of the nature of the temperature of insects. He placed a perfect specimen of *Sphinx Convulvuli*, LINN. in a small glass phial when the temperature of the atmosphere was 17° REAUM. (70°·25 FAHR.), together with a small thermometer, and at the expiration of half an hour the temperature of the phial was 19° REAUM. (74°·75 FAHR.), but soon afterwards he found that the temperature of the phial had sunk again to the previous standard 17° REAUM. He then repeated the observation with six specimens of *Carabus hortensis*, LINN. with similar results. From what will subsequently be shown respecting the temperature of *Carabi*, which do not develop so large a quantity of heat, it is very probable, as suggested by Dr. BERTHOLD[§], that the results obtained by HAUSMANN arose from the bottle which contained the insects being touched by the hand of the operator. Dr. BERTHOLD has observed this in his experiments, and I have constantly remarked the same thing myself when proper care was not

---

[*] New Experiments on the Temperature of Cold-Blooded Animals, by A. H. BERTHOLD, M.D., Gottingen, 1835.

[†] Ibid. p. 36. Experiment 59.

[‡] De Animalium exsanguinum Respiratione. Götting. 1803. p. 68.

[§] Neue Versuche, &c., p. 11.

taken to guard against its occurrence. RENGGER* observed a distinct temperature in *Melolonthæ* when many of them were collected together in an earthen vessel, but could not detect a distinct temperature in water-insects, or in Caterpillars. JUCH† likewise made observations on the temperature of the bee-hive, the ant-hill, and on the common Blister-flies. In a vessel containing a large quantity of the latter insects, the *Lyttæ*, he found the thermometer rise several degrees above the temperature of the atmosphere. Dr. DAVY, according to BERTHOLD‡, in making observations on several species of insects, *Scarabæus pilularis, Lampyris, Blatta, Gryllus,* and *Apis,* found only a slight difference, except in the *Gryllus,* in which the difference amounted to five or six degrees, while in the Scorpion and Centipede he found a temperature lower than that of the atmosphere. Dr. BURMEISTER, in his Manual, recently translated by Mr. SHUCKARD, has spoken of the temperature of insects, but only of insects in society, and has referred to the observations of JUCH, REAUMUR, &c., and although he believes in the existence of individual temperature in insects, has given no observation of his own to prove the fact, while Dr. BERTHOLD, in the work just noticed, (experiment 59,) made on a single insect, could not detect it, nor could he do so in every species when the observation was made on a number of individuals collected together. It is evident, therefore, that although the existence of individual temperature is inferred from experiments on insects collected together, it yet remains to be proved that every individual insect in a state of activity invariably maintains a certain amount of temperature, which is readily appreciable by the instruments we are enabled to employ.

Before detailing the results of my observations it is necessary to explain the manner in which the observations themselves have been made, and to point out those circumstances which seem to have been overlooked by other inquirers in their experiments on the temperature of insects. It is only by a careful attention to those circumstances that we are enabled to detect the existence of temperature in single insects, and to understand the causes of its variations at different periods.

The thermometers employed by me on every occasion are of the smallest possible calibre, with cylindrical bulbs about half an inch in length, and scarcely larger than crow-quills, and are similar to those employed by Professor DANIEL for the purpose of ascertaining the dew point. They were made by Mr. NEWMAN of Regent Street, and are graduated from zero, or from a few degrees below freezing to about 110° or 120°. Whenever great delicacy of observation is required, in order to observe the varying temperature of an insect during a state of partial rest, it is necessary to use the same instrument for ascertaining the temperature of the atmosphere as for that of the insect, otherwise a great difficulty will arise, from the well known circumstance that two thermometers, be they ever so delicately constructed, and carefully compared

---

* Physiologische Untersuchungen über die thierische Haushaltung der Insecten. Tübingen, 1817, p. 39.
† Ideen zu einer Zoochemie, Bd. 1. 1800, p. 92.
‡ Neue Versuche, &c. p. 12, 13.

with each other, will seldom if ever both indicate precisely the same amount of temperature in exactly the same space of time. The mode of taking the temperature is either by allowing the insect to remain with the soft ventral surface of its abdomen pressing against the bulb of the thermometer when in a state of rest, or by pressing the thermometer firmly against its body when in a state of excitement, the insect being held during the time between a pair of forceps covered with woollen, in order that the contact of the fingers of the operator may not interfere with the correctness of the observation by unnaturally increasing the temperature of the insect. It is also further necessary to guard the hand with a glove, or non-conducting substance, to prevent the thermometer itself from becoming affected by it during the experiment. Much caution also is necessary when the same thermometer is employed to ascertain the temperature both of the atmosphere and of the excited insect, to guard against one very material source of error. It is necessary *first* to ascertain the temperature of the atmosphere, and *then* that of the insect, because if this be not attended to, and the experiment be made by taking the temperature of the insect *before* observing that of the atmosphere, the moisture on the bulb of the instrument occasioned by the condensation of the *cutaneous perspiration* from the body of the animal will occasion during its drying or evaporation, while taking the temperature of the atmosphere, an indication of a lower amount of atmospheric temperature than what really exists, and consequently the apparent difference between the temperature of the insect, previously taken, and that of the atmosphere, will be much too great, and thereby appear to indicate a higher temperature than what the body of the insect really possesses. When the temperature is taken during a state of rest, the thermometer is placed beneath, and as completely covered by the abdomen of the insect as possible, while a second thermometer, which has been very carefully compared with the first, is placed on the same level with and at a short distance from it to indicate the temperature of the atmosphere. When the temperature of active volant insects is to be taken, it is preferable to inclose them singly in a small phial, introducing them with the forceps as before, and being particularly careful not to touch the phial with the fingers. The degree of activity or quiescence of the insect must always be particularly noticed, and also the number of inspirations. By attending to these facts we acquire a knowledge of the amount of respiration compared with the quantity of heat evolved, as indicated by the thermometer. The temperature of the insect taken on the exterior of the body is always a little lower than that of the interior; but the difference is not so great as might at first be imagined, so that I have generally preferred taking the exterior temperature, because the observations are then less complicated by unnatural causes. The interior temperature is seldom if ever more than a degree and a half, or at most two degrees above the exterior, and often not even half a degree, when the insect is in a state of perfect rest. Perhaps it may be urged as an objection, that when the bulb of the thermometer is applied to the exterior of the body, it can seldom be so completely covered as to indicate the whole amount of heat developed. But this objection, although at first plausible, must be considered valid only when the observations are made very quickly. But even were the objection substantiated it would be of but little consequence, because it is only the relative amount of heat developed by one insect as compared with that of another, when the observations on both are conducted in a similar manner, which is ultimately sought for, it being almost impossible to ascertain the exact amount evolved by any single insect. It may also be urged as an objection to this mode of taking the temperature of insects in a state of excitement, that when an insect is respiring very rapidly, the friction of the segments of its body against the bulb of the thermometer may evolve a certain amount of heat independent of the natural heat of the insect, and thereby indicate a higher temperature in the insect than that which really exists. In order to meet this objection, I made a number of trials with my thermometers, by using, as nearly as could be ascertained, about the same amount of attrition against the bulb of the instruments as that which is exerted by the segments of the excited insect during its laboured respiration and efforts to escape, and found that so small a quantity of heat is evolved that it is not in the slightest degree indicated on the scale of the thermometer. Hence I have not in general found it necessary to take the temperature of the interior of the body, although I have done so in a few instances, because there are also other circumstances which interfere with the correctness of the observation. The first of these is the large size of the instrument employed compared with that of the body of the insect into which it is inserted, and the consequent necessary loss of a certain amount of caloric, which becomes latent in the thermometer, before there is any indication of increased temperature on the scale, and because also of the unavoidable escape of a large amount of caloric into the surrounding atmosphere, and because still further it is only at the very instant after the introduction of the thermometer into the body of the insect that the real perceptible amount of temperature is indicated, while the insect under observation is every moment losing the power of generating and of maintaining its temperature, owing to the injury that has been inflicted upon it. These objections do not occur when the observations are made on the exterior of the insect, which from its being uninjured, continues to possess its power of generating heat unaffected by those circumstances which tend very materially to interfere with or destroy it, while a sufficient length of time is afforded for the production of its full amount of heat after a certain quantity has become latent in the thermometer, before the observation of the amount is taken.

These are the principal circumstances to be attended to in ascertaining the temperature of insects, and which have directed me in my observations.

[*Editor's Note:* Material has been omitted at this point.]

IV. *Temperature of Insects which live in Society.*

We pass now to those insects which live in society, all of which belong to that great division the Hymenoptera, which have been shown to possess the highest temperature and greatest amount of respiration. Naturalists hitherto have examined only two genera of this great division with reference to the subject of temperature of these insects in their dwellings. These are the *Apis Mellifica,* or common Honey Bee, and the society of Ants; and the existence of a higher temperature than that of the atmosphere in the other families has only been inferred. Those species unto which I have devoted particular attention are the *Bombus terrestris* and *Apis Mellifica.*

Bombus terrestris.—1. *Temperature of Nests under observation.*

During the summer of 1830, having obtained a colony of this species, with the original parent bee, from the neighbourhood of Richborough, near Sandwich, (which locality had before that time afforded me opportunity of observing the habits of other species of this interesting family of insects,) I removed it from its locality in the earth to my own residence, the distance of a mile, and placed it in a small insect breeding cage for the purpose of more closely watching the economy of this species. The bees at first were somewhat irritable, and of course were kept in close confinement, and were fed with moistened sugar; but within a day or two they became quite accustomed to their new residence, and I had ample opportunity of watching the economy of the nest. On the third day they were placed on a table in my sitting-room near the window, which remained open, and also the door of the cage, that the bees might go abroad and return at pleasure, which they did with as much regularity after the first day or two as if the nest had been placed in its proper locality in the earth. I had thus most ample opportunity of watching their habits. The nest consisted of from forty to fifty individuals, and it gave me great pleasure in being able to confirm many of the statements made respecting these insects by HUBER. During the time the bees were in my possession, a period of nearly three weeks, I observed upon introducing a thermometer among them, that the temperature of the nest varied at different times, and was considerably higher when they were in a state of excitement; but the circumstance did not then attract my particular attention. In the summer of 1834, while engaged with the observations before detailed, I determined to repeat the observation which I then remembered having made in 1830; and accordingly on the 10th of July 1834, having taken a nest of *Bombus terrestris* with brood comb, it was placed on a table near the window of my apartment, in a small box about eight inches square, and four deep, covered with green gauze, and after the first day's confinement the bees were allowed to go and return as on the former occasion. Soon after commencing my observation, I was interested in observing that the bees were at first greatly affected and agitated by the slightest noise, such as the removal of a chair, or one's footsteps about the room, or the passing of carriages along the road, which was at least thirty feet distant from the window of the apartment; but they were not in the slightest degree affected by persons talking loudly in the room, while a gentle tap with one's finger on the table put them immediately into a state of the greatest agitation. Hence during the observations it was necessary to be cautious, and not disturb the bees when wishing to take the temperature of the nest. The bees, however, in the course of a day or two became accustomed to their situation, and were not disturbed by slight noises or vibrations; and I was then enabled to take their temperature under all circumstances. The observations were commenced at 12 A.M., July 10, about two hours after the bees were placed in the box. The temperature of the atmosphere was then 70°·5 FAHR., that of the box and nest 73°; but when they became excited it soon rose to 77° but gradually subsided again to 73° as the bees became quiet. The thermometer was introduced very carefully under the gauze covering, and was not allowed to touch the bodies of the bees in this and the subsequent observations. At 1½ P.M., the insects having remained at rest for more than a quarter of an hour, atmosphere 70°, the thermometer, introduced as before, rose to 75°, and in a few minutes afterwards, when the bees had become much excited, to 80°·2, a difference of 10°·2 between the temperature of the atmosphere and that of the box; and when the body of a bee touched the bulb of the thermometer, even but for an instant, the mercury immediately rose at least a degree on the scale. At 2½, atmosphere 70°·5, bees quiet, atmosphere of the box 76°; but when they became much excited it rose in four minutes to 80°·4. At 12½ midnight, atmosphere 68°·5, interior of the box was 73°, the bees having been quiet during the previous nine hours; but when they became greatly excited it rose to 80°·3, a difference of 11°·8. At 6 o'clock on the following morning, July 11, atmosphere 67°, the interior of the box was 71°, but when the bees became much excited it rose to 77°·3. At 12½ midnight, atmosphere 67°·5, box with bees at rest 73°, when agitated 78°. At 7 A.M., July 13, the box in which the bees were confined had remained closed during the night, which had been perfectly calm and still, and at the time of making the present observation there was not a breath of wind stirring; indeed the air was suffocatingly calm, and its temperature 68°·7; when the thermometer was carefully introduced under the lid of the box the mercury rose to 72°, which was the temperature of the interior of the box around the nest, but when the thermometer was placed in the nest itself the temperature stood at 76°·5.

2. *Nest of Bombus in its natural haunts.*

Having proceeded thus far with my observations on the temperature of the nest, removed from its proper locality in the earth for the purpose of experiment, it became a matter of interest to endeavour to ascertain its temperature while undisturbed in its natural haunts. Having at length discovered the nest of a species of Bombus nearly allied to *Bombus terrestris* situated in a shaded chalk bank near the ground, and about eight inches from the surface, at 10 A.M.,—the temperature of the atmosphere in the shade four feet from the ground being 68°·7, while that of the exterior of the chalk bank in which the nest was situated, and near the entrance to it was 66°,—I very carefully introduced a small thermometer without disturbing the inmates, and found that the temperature of the interior of the nest was 83°, but in a few minutes it rose to 85°; it was thus evident that the temperature of the nest upon which I had made the preceding observations was at about its average temperature in its natural haunts.

3. *Nurse Bees.—Voluntary Power of generating Heat.*

The above experiments on the nest of *Bombus terrestris* thus confirmed the results of my observations made a short time before on individual insects with regard to the rapid transmission of heat from the body of the animal when in a state of excitement, and also in a less degree when in a state of rest; but during the time I was engaged upon them they also afforded me a new and totally unexpected phenomenon, and one which is not a little interesting and important as regards its connection with the origin of animal heat;—it was the capability which these insects possess during the act of incubation on the cells which contain nymphs, of increasing their own temperature many degrees above that of the surrounding medium, of in fact a voluntary power of generating heat through means of respiration. HUBER has stated that there are certain individuals in the nests of the Humble Bees, and among the bees in a hive, which at a particular season of the year are employed to impart warmth from their bodies to the young bees in the combs by brooding over them, and these he called Nurse Bees. It gives me great pleasure in being able to bear testimony to the correctness of his statement, particularly with regard to those in the nest of the Humble Bee, which I had ample opportunity of observing. These individuals are chiefly the young female bees, and at the period of the hatching of nymphs they seem to be occupied almost solely in increasing the heat of the nest and communicating warmth to the nymphs in the cells by crowding upon them and clinging to them very closely, during which time they respire very rapidly, and evidently are much excited. These bees begin to crowd upon the cells of the nymphs about ten or twelve hours before the nymph makes its appearance as a perfect bee. The incubation during this period is very assiduously persevered in by the Nurse Bee, who scarcely leaves the cell for a single minute; when one bee has left another in general takes its place: previously to this period the incubation on the cell is performed only occasionally, but becomes more constantly attended to the nearer the hour of development. The manner in which the bee performs its office is by fixing itself upon the cell of the nymph, and beginning at first to respire very gradually; in a short time its respiration becomes more and more frequent, until it sometimes respires at the rate of 120 or 130 in a minute. I have seen a bee upon the combs perseveringly continue to respire at this rate for eight or ten hours, at the expiration of which time its body has become of a very high temperature, and on attentive observation the insect is often found in a state of great perspiration; when this is the case the bee generally discontinues her office for a time, and another individual will sometimes take her place. Very frequently the Nurse Bee respires with much less rapidity, and remains many hours on the cells. The very high temperature unto which the insects are able to raise their own bodies, and the cells upon which they are incubating at this period, will be best shown by detailing the continued observations on the nest.

At 8 A.M., July 13, when the temperature of the atmosphere was $71°·8$, and the temperature of the interior of the box around the nest $72°·5$, I inserted the bulb of a fine thermometer very carefully between the abdomen of several bees and the cells upon which they were incubating, and which contained nymphs, and found the body of a single nursing bee was $84°·1$, while the exterior of some cells that contained nymphs, but which were not covered, was $76·5$. At $8\frac{1}{2}$ the temperature of the outside of the waxen cover, or top of the nest, was $77°·7$, and that of the atmosphere $72°·5$, while the interior of the nest, where the bulb of the thermometer was introduced among four bees which were nursing upon the cells, was $89°·2$. At $8\frac{3}{4}$, atmosphere as before, when the thermometer was introduced among seven nursing bees at the same spot, three of which were large females, and the others males, which also assist in the process of incubating, the mercury of the thermometer rose to $90°·2$ FAHR. At 9 A.M., atmosphere still $72°·5$, the temperature of the same bees still incubating was $92°·3$, and of others incubating in another part of the same nest $91°·5$; at $9\frac{1}{2}$, atmosphere $72°·7$, that of the bees still nursing was $91°$. At 12 A.M. the observations were resumed: in the interval between the last observation and the present time there had been a gentle shower with light wind, and the atmosphere had sunk to $70°·2$; the temperature of the Nurse Bees on the cell was now $92°·5$. The thermometer was raised to this height within about ten minutes, and was maintained at that standard as long as the bulb of the instrument was allowed to remain in contact with the bodies of the insects, while the temperature of some of the adjoining cells beneath the same cover, but which were not covered by the bees, was maintained at only $80°·2$. Within a quarter of an hour after these observations were made three large female bees were hatched from the cells upon which the seven bees had been incubating; the temperature of the atmosphere was then $72°·2$, while that of the Nurse Bees, which had now desisted from incubating, and consequently were respiring less rapidly, had sunk to $85°$. It was thus evident that the greatest amount of heat is generated by the Nurse Bees just before the young bees are liberated from the combs, at which period they require the greatest amount of invigorating heat. It is at this period also, as before noticed, that the young bee is most susceptible of diminished temperature; it is then exceedingly sleek, soft, and covered with moisture; perspires profusely, and is highly sensitive of the slightest current of air. It crowds eagerly among the combs and among the other bees, and everywhere where there is the greatest warmth. In the course of a few hours it becomes a little stronger, and is less sensitive, and better able to bear a diminished temperature. It then moves about with less circumspection, and its wings, which at first are soft and weak, and bent upon its trunk, become plain and straight. When the young bee first leaves its cell it is entirely of a whitish or pale grey colour, but within half an hour the black markings on the thorax become very distinct, although they retain a tinge of grey colour for a much longer period; the yellow bands on the body and thorax are at first quite white, and it is not until an hour or two has elapsed that the principal yellow band on the thorax begins at length to gain colour, while it is several hours before the yellow bands acquire their full shade or degree of colour. During all this time the bee continues in an enfeebled state and takes no part in the business of the nest, but seeks for itself the warmest place among the combs, and it is not until sometime after it has acquired its proper degree of colour that it becomes active like the other bees, and is able to maintain its own proper temperature. It is thus evident, that the same principle which has been shown by Dr. Edwards to prevail with regard to the young of some of the mammiferous animals, that they are unable for a certain period after birth to generate and maintain within themselves a proper amount of temperature, but require to be cherished by external warmth, regulates also the development of the individuals of this family of Hymenopterous insects, from their pupa or nymph to their perfect state, and further tends to prove to us how universal and simple are the great laws which regulate the continuance of animal life. It is a curious fact that these bees do not incubate on the cells which contain only larvæ, the temperature of the atmosphere of the nest being sufficiently high for them in that condition; consequently the larvæ at an advanced period do not require so high a temperature before changing into nymphs as that which has just been shown to be required by the nymphs before coming forth as perfect insects. This will be shown in some observations made on larvæ in the nest now under examination, at the same time with those just described, and also with others which were made on nymphs. The temperature of the atmosphere being 76°, some of the cells which were open and contained larvæ were exposed in the nest, and the Nurse Bees therefore covered them lightly with dried grass, of which the nest of this species of Bombus is usually composed; but when the temperature of the atmosphere a few hours afterwards had risen to $73°·5$, most of the dried grass with which these cells had been covered was removed, and the larvæ were more exposed; the temperature of these cells and the larvæ being $77°·4$, while that of the cell of a nymph, with the Nurse Bee upon it, in another part of the nest was 92°, and subsequently when four large females were nursing around it was $94°·1$, the temperature of the atmosphere being still $72°·5$.

When there are no longer any nymphs which are soon to be developed into perfect insects the necessity for generating a larger amount of heat is diminished, and the Nurse Bees remain in a state of quietude; the temperature of the nest is then much lower than when young bees are about to be produced. This was the case on the 14th of July; the atmosphere was then 69°, while that of the nest was in no part higher than $72°·5$; and even when the bulb of the thermometer was in contact with the bodies of several of the bees, the mercury scarcely rose to $73°·5$, while at 12 o'clock on the preceding night, when the atmosphere was 68°, and several young bees were soon to come forth, the temperature of the box was $70°·5$, and that of some bees very moderately excited in the act of nursing 83°·2. It is not only at the moment when the young bee is about to come forth that the Nurse Bees produce a larger amount of heat; they keep up the heat to a considerable amount for some time after the young bee is developed. At 1½ P.M., July 14, the bees were again incubating, the atmosphere 69°·5; the cells immediately beneath the cover of the nest were 89°·4. At 2 P.M., atmosphere 69°·5, the same cells were 92°·2, at which time most of the bees were crowding around this part of the comb, from which at 6 P.M. several young ones came forth. At 3 P.M., atmosphere 69°·5, the temperature of a single bee nursing on these cells was 91°. At 5 P.M., atmosphere 73°·4, atmosphere of the box was 75°·3, and that of four bees nursing 94°·2; while at 11 P.M., five hours after the young bees had been developed from this part of the comb and when no bees were present, the temperature at the very same spot was only 68°, exactly that of the open atmosphere; but in another part of the nest where the bees were again nursing it stood at 83°. It was in this way that the nurse bees constantly raised their own temperature and that of the cells upon which they were incubating whenever new bees were to be produced. In order to prove that this great amount of heat resulted directly from the temperature of the nursing bee, I placed the bulb of a thermometer on the back of a single individual that was nursing on the upper surface of a comb that was exposed to the temperature of the atmosphere, 71°·6, when it rose to and was maintained exposed as it was at 85°, while the temperature of the cell immediately after the bee had quitted it was 75°·3, and it was maintained at that temperature several minutes. In other observations I found that on one occasion, when the atmosphere was 72°·5, a single female bee while nursing upon a single cell, from which a perfect insect was developed about eight hours afterwards, had a temperature of 92°·3: the bulb of the thermometer in this instance was placed upon the cell immediately beneath the abdomen of the bee, which was respiring at the rate of 120 per minute. In another observation, when the temperature of the atmosphere was still the same, 72°·5, a single bee while nursing had a temperature of 94°·5, but a little while afterwards when the atmosphere was 72°·7 it had subsided to 91°.

These facts distinctly prove that bees have a voluntary power of evolving heat, while it seems only fair to conclude, on comparing the facts, that the quantity of heat produced in a given time and space, has relation to the number of respirations performed by the individual; and from the quantity of atmospheric air consumed, and of carbonic acid gas evolved, that animal heat is greatly and perhaps almost entirely dependent upon the chemical changes which take place in the air respired.

[*Editor's Note:* Material has been omitted at this point.]

Conclusion.

The very great length unto which this paper has already been extended, necessarily prevents me from entering so fully into all the circumstances connected with the evolution of heat in insects as the great importance attached to this interesting subject demands; I shall, therefore, review the contents of this paper, and other circumstances connected with the production of animal heat, with as much conciseness as possible.

On comparing the whole of the facts we have just examined, we cannot fail to observe the very close relation which subsists between the amount of heat developed, and the quantity of respiration. We have seen in the larva, the pupa, and the perfect insect, that when the respiration is accelerated the temperature is also increased, and that when respiration is diminished the temperature subsides. When the insect is sleeping, its respiration gradually becomes slower, and its temperature continues to lessen until the insect is aroused, when immediately after the first respirations it is again increased. When the insect falls into a state of hybernation, and its respiration is suspended, its evolution of heat becomes so likewise. When the insect is most active, and respiring most voluminously, its amount of temperature is at its maximum, and is very great, and corresponds with the quantity of respiration, and, as in the Bee, an immense quantity of heat passes off into the surrounding medium. When the insect wishes to impart heat to its young it can do so at pleasure, and can voluntarily increase its own temperature. It does this by accelerating its respiration. At those times, as shown in the comparative observations, the insect evolves in one hour, in this state of activity and excitement, at least twenty times the amount of heat, and consumes nearly twenty times the quantity of air, which it consumes at the same temperature when in a state of repose. In insects which live in society the temperature of their dwellings is increased in proportion to the activity of the inmates, and consequent amount of their respiration. In the hive it is steadily increased until the time of swarming. In the winter when the bees are quiet, and their respiration is exceedingly low, and when not a bee is observed ventilating at the entrance, the temperature of the hive may be raised in a few minutes, very many degrees, by disturbing the inmates, and thereby increasing their respiration, until such an amount of heat is evolved, and so much air is deteriorated, as to become oppressive and noxious to the bees, many of whom, although the open atmosphere be too cold for them to venture abroad, will come to the entrance of the hive and begin as laboriously to ventilate the interior, by vibrating their wings, as in the midst of summer. The quantity of free heat is always greatest in the hive when the bees are most active, and least when they are most quiet. With regard to the habits and anatomical structure of insects, the amount of heat is by far greatest in volant insects; these always have the largest respiratory organs, and breathe the greatest quantity of air. In the terrestrial insects the amount of heat is greatest in those which have the largest respiratory organs, and breathe the greatest quantity of air, whatever be the condition of their nervous system. In the larva state the respiratory organs are smaller than in the perfect insect, compared with the size of the body, and the larva, we have seen, has the lowest temperature. But in these comparisons we must observe that the activity of respiration is equal in the individuals which are compared. Thus although the respiratory organs are larger in the pupa than in the larva, the physiological condition of the insect is lower, its respiration is inactive. These facts, it will be seen, are all in strict accordance with each other, and point to the chemical changes in the air during respiration as the immediate source of animal heat. But it may be matter of inquiry how it is that the heat evolved within the body of the insect, during respiration, becomes evident so rapidly. This, it may be urged, tends to show that it results from the influence of the nervous system. But when we remember that in insects the circulatory vessels are in close and most extensive communication with the respiratory organs over the whole body of the individual, and that, unlike the vessels in those vertebrated warm-blooded animals which have extensive respiration, they are neither strictly venous nor arterial, but probably intermediate between the two, may it not arise from only a very small amount of heat evolved at each respiration becoming latent, while nearly the whole becomes free, and is liberated as quickly as produced, and that this is the occasion of the temperature of the insect being so quickly raised during its respiration, and so rapidly diminished as the acts of respiration become less frequent? That, in other words, in insects the capacity of the blood for caloric is but very little increased during respiration? With these facts in consideration, and looking at the analogical condition of insects, and with Professor GRANT[*] and Mr. OWEN[†], comparing the vast extent of their respiratory organs, distributed over the whole body, with a like extensive respiration in birds, and finding that, like birds, insects have also a greater activity of respiration, and a higher temperature of body than any other class in the division of animals unto which they respectively belong, we can hardly withhold our assent to the opinions which have long been advocated by many of our best physiologists, that animal heat is the direct result of the chemical changes which take place in the air respired. But it may be urged that activity of respiration is coincident with increased rapidity of circulation, and hence that the latter may, perhaps, precede the former, and be in reality the source of heat. Unto this it may be replied that the larva in its earlier state has a more rapid circulation, but develops less heat than in its latter. In many of the observations on the Tables it is shown that the pulse may be rapid with a low amount of heat. It is shown in the larva, when arousing, that the pulse is not increased until

---

[*] Lectures on Comparative Anatomy.—*Lancet*, 1833–34.
[†] Cyclopædia of Anatomy and Physiology, vol. i. p. 341.

after the first respirations*, when the heat is becoming apparent. With regard to the digestive process, we have seen that when the animal is taking food it has the greatest amount of gaseous expenditure from its body, and that the greatest amount of heat, when in a state of quietude, is then generated. But a greater quantity of air is then consumed, in assimilating the new matter which has been taken into the system, and the quantity of heat is still further increased if the animal becomes active, and this is regulated by the increased expenditure from the surface of the body. Lastly, we have seen that in the more perfect volant insects, the Bees, Sphinges, &c., there is the largest amount of heat produced, and the greatest quantity of air consumed, but the nervous system is also largely developed, and hence it may fairly be supposed to have much influence in the development of heat. But on the other hand we find many insects, as the *Melöe* and its congeners, which produce a large amount of heat, in which the nervous system is comparatively small, while these insects have large respiratory organs, and a large amount of respiration. In the *Staphylinus* the nervous system is exceedingly large, compared with the size of the body, but the respiratory organs are by no means small, while the amount of heat is very moderate. In the *Carabus* the nervous system is also large, as are likewise the organs of respiration, but the amount of heat and activity of respiration are low, and the same is the case in the *Blaps*, in which the nervous system is rather small. If the development of heat depends upon the nervous system, or the number of ganglia, the *Leech*, which has twenty-two ganglia, ought to generate more heat than the larvæ of lepidopterous insects, which have but ten or twelve, and the larva ought to generate as much as the perfect insect. In the larvæ of the Bee, the Hornet, Ichneumon, and Tenthredo, which generate so large an amount of heat, the nervous system is exceedingly small; and if, as some suppose, heat is the result of muscular contraction, surely it ought to be most developed where there is the greatest amount of muscular contractility; it ought to be generated more in the *Leech* than in other articulated animals, and in those *Vertebrata* which are peculiarly noted as cold-blooded. These facts con-

---

* This is in perfect accordance with the condition of the circulation in the human body during sleep, and at the moment of waking, as noticed by BLUMENBACH, and as I myself once had an opportunity of observing in a female patient who was suffering from severe fracture of the skull, for which she had been trephined; subsequently to which, a large portion of the bone (the right parietal) became affected with necrosis and was removed by operation, and the patient afterwards gradually recovered. At least one-third of the whole parietal bone had been removed, and a large surface of the dura mater being thus exposed, the activity of the circulation in the brain was readily observed. I thought this a fair opportunity, as the patient was recovering, for observing the state of the circulation during sleep, and at the moment of waking. The patient was sleeping soundly at the time of the observation, and while she remained entirely undisturbed, the pulsations in the arteries of the dura mater were at the rate of ninety-four beats per minute, and were perfectly synchronous with the pulsations at the wrists; but immediately she began to inspire deeply at the moment of waking, the pulsations became much accelerated. At the instant of waking, the patient fetched a full and deep inspiration, and in less than a minute and a half after this, the patient being perfectly awake, the pulsations amounted to 104 beats per minute, thus making a difference of about 600 beats per hour in the rate of pulsation when sleeping and immediately after waking.

sidered, and connected with that very remarkable one, the voluntary power of producing heat possessed by the Bee, must lead us to conclude, that although, doubtless, the whole of the functions of the body are more or less remotely concerned in the production of heat, yet that the immediate source of its evolution seems to be chemical changes effected during respiration, and that the nervous system is only secondarily concerned.

Appendix.

Since the preceding paper on the Temperature of Insects was submitted to the Royal Society, circumstances have enabled me to ascertain a few additional facts respecting the temperature of some other species which I had not heretofore any opportunity of examining, and these the Council have kindly permitted me to subjoin to my paper.

I am not aware that the temperature of the nest of the common wasp has ever before been examined, and it is therefore pleasing to find that all the circumstances connected with the evolution of heat in the nest of this species are in perfect accordance with the observations made on the neighbouring families of hive and humble bees.

On the 11th of August, during the past summer (1837), I dug away the soil from the top of a nest of *Vespa vulgaris* which was situated in a bank of earth at the depth of about seven inches from the surface. The nest was nine inches in diameter, so that the colony was by no means a small one. The temperature of the atmosphere, when the covering of the nest was removed, at $4\frac{1}{2}$ P.M. was 70° Fahr. When the thermometer was passed through the top of the nest the mercury rose immediately to 80°. In about ten or fifteen minutes afterwards, when the colony had become disturbed, and the thermometer was passed a little deeper into the nest, the mercury rose to 95°. This distinctly proves that the evolution of heat in the wasps' nest is greatly increased, as in the beehive, when the insects have become excited. At $6\frac{1}{2}$ P.M. the temperature of the atmosphere was 65° Fahr., and the wasps having now become more quiet, the temperature of the nest, which had remained with its upper surface exposed since the last observation, was only 90° Fahr.; but an hour afterwards, when the temperature of the atmosphere had sunk to 63°, that of the nest had risen to 91°, the thermometer having remained undisturbed in the nest since the last observation. This increase of temperature was readily explained by a great number of the excited insects, which had been flying around the spot, having now returned to the nest. Thus the circumstances connected with the evolution of heat in the nests of the predaceous and in the melliferous Hymenoptera are precisely similar; and they are similar also in another interesting family of this order—the ants. It is elsewhere noticed* that Juch found the temperature of an ant-hill about 15° Fahr. above that of the atmosphere. My own observations are in accordance with this statement. On the 27th of July 1837 I examined the temperature of the nest of *Formica herculanea*, Linn. The temperature of the atmosphere in the shade, at 11 A.M., was 76° Fahr., but when the thermometer was exposed on the ground to the full rays of the sun the mercury rose to 95° Fahr. The nest was rather a small one, and at the time of commencing the observations was completely undisturbed. When the thermometer was first passed into it, to the depth of five inches, the temperature was maintained steadily at 84° Fahr.; but within six or eight minutes afterwards, when the insects had become excited by the presence of the thermometer, and were running about in every direction in a state of the greatest agitation, the temperature of the nest rose to 93° Fahr., and in a few minutes after this, when the insects were still more excited, to 95°·5, and a little nearer the surface, where the commotion was greatest, to 98°·6 Fahr. During these observations the ant-hill was carefully shaded from the rays of the sun, in order to avoid all source of error. When the ant-hill was again exposed to the sun, and the thermometer placed upon its surface, the mercury rose to 108° Fahr. This was a temperature much too great

* Page 283.

for the insects to bear, since nearly the whole of them immediately retired beneath the covering of the nest, and there was scarcely a single ant to be seen. On the 2nd of September I repeated my observations on the same ant-hill. On this occasion the day was very gloomy, with steady light rain, and the temperature of the atmosphere at 11 o'clock A.M. was only 54°. The temperature of the ant-hill varied but little in its different parts but it was now greatest near the surface. At a depth of one inch it was 65°, at two inches 66°, below which it gradually diminished. At this time I also examined another nest of the same species, but which was about twice the size of the first. The atmosphere being, as before stated, 54°, the mean temperature of this nest, when the insects were a little excited, was 74°.

During the summer and autumn of the present year I have repeated my observations on the temperature of the bee-hive, and have found but little variation in its average amount at similar periods in the two years. I have also examined the nests of *Bombus lapidarius*, and *Bombus sylvarum*, and in both have found that the ordinary temperature, which is about 10° or 15° above that of the atmosphere, is considerably increased during the period of incubation, exactly the same as in the nest of *Bombus terrestris*.

On the following day after examining the nest of the wasp, I examined the temperature and pulsation of the larva of the same species. The specimens examined had been removed from the nest on the previous evening, but had not been removed from their cells. The results are given on the accompanying table. I examined also the larva of the hornet, *Vespa Crabro*, Linn., which was still contained in its cell, but had been some days removed from the nest. In this instance the temperature of the larva was found to be about 2°·5 Fahr. above that of the atmosphere, but its rate of pulsation was only thirty-two beats per minute. I should have attributed this low rate of pulsation to the specimen having been so long removed from the nest, had not the rate of pulsation in this larva been examined by my friend Mr. Orsborn a few days before, and almost immediately after the specimen was obtained from its nest, and found at that time not to exceed thirty-three or thirty-four beats per minute. These facts therefore are in accordance with the observations on the larva of *Anthophora retusa* and *Bombus terrestris*, and also accord with other observations on the larvæ of that very destructive tenthredo or saw fly *Athalia centifolia*, Klug; which has been so obnoxious to the agriculturist by destroying his crops of turnips during the last three summers.

London, November 7th, 1837.

TABLE.—TEMPERATURE OF LARVÆ.

No. of Exp.	Name of Species.	Period of observation.	No. of Specimens.	Atmosphere.	Soil.	Insect.	Difference.	Pulsation.	Remarks.
1	Vespa Crabro (larva)	July	1	70·	....	72·5	2·5	32	Full grown; has fasted three or four days.
2	Vespa vulgaris (larva)	Aug. 12	1	72·7	....	75·8	3·1	56	Nearly full grown; very active.
3	Vespa vulgaris (larva)	12	1	72·7	....	74·	1·3	52	Full grown.
4	Vespa vulgaris (larva)	12	1	72·7	....	75·2	2·5	52	Full grown.
5	Athalia (larvæ)	Sept. 6	50	64·5	....	66·5	2·0	....	Larva nearly full grown; very active.
6	Athalia (larvæ)	6	....	64·7	....	66·	1·3	....	
7	Athalia (larvæ)	6	6	66·3	....	66·8	0·5	....	Larva inactive.
8	Athalia (larvæ)	6	200	65·3	....	67·3	2·	....	Full grown; active.
9	Athalia (larvæ)	6	50	65·3	....	67·3	2·	....	
10	Melolontha vulgaris (larva)	Oct. 7	1	61·5	59·7	60·2	0·5	....	No. 1.
11	Melolontha vulgaris (larva)	7	1	61·5	59·7	60·3	0·6	....	No. 2. Full grown larvæ; the temperature
12	Melolontha vulgaris (larva)	8	1	64·7	64·6	64·7	0·1	....	No. 1. taken while lying in their cells
13	Melolontha vulgaris (larva)	8	1	64·7	64·6	64·8	0·2	....	No. 2. and compared with the temperature of the soil.
14	Melolontha vulgaris (larva)	12	1	63·5	63·5	63·7	0·2	....	No. 1.
15	Melolontha vulgaris (larva)	12	1	63·5	63·5	63·7	0·2	....	No. 2.

THERMAL REINFORCEMENT AND THERMOREGULATORY BEHAVIOR IN THE GOLDFISH, *CARASSIUS AURATUS*

Paul N. Rozin and Jean Mayer

Psychological Laboratories, Harvard University
School of Public Health, Harvard University

The rate of activity and metabolism of poikilotherms is largely determined by the temperature of their environment. Yet thermal adaptation in these animals tends to reduce the effects of temperature and poikilotherms can also change their body temperature by moving from one environment to another.

The process of temperature selection has been investigated in a number of poikilotherms, including the goldfish. Fry (1) has found that goldfish, when placed in water containing a temperature gradient, spend most of their time in water within a certain temperature range. This finding suggests that temperature might be used to reinforce learning in these fish. If a goldfish is placed at a temperature that is considerably different from its preferred temperature, will it perform some arbitrary response in order to bring the temperature of its environment closer to its preferred temperature? Furthermore, if temperature change can be used as a reinforcement, will the fish regulate its body temperature by regulating the temperature of its environment? Weiss and Laties (2) have shown that the albino rat, when placed in a cold environment, will press a lever for heat reinforcement. No similar experiment has been performed with a poikilotherm. In the experiment presented here, it is demonstrated that goldfish will work to produce certain temperature changes in their environment, and that, when given the opportunity to control their body temperature, they will do so to a certain extent.

The experimental apparatus is shown in Fig. 1. A small goldfish (3 to 8 g) was placed in a 1-pint container of water. This container rested in a constant-temperature water bath. During training the bath was initially at a temperature of 24.5° ± 0.5°C. The home container of the fish was kept at 23° ± 1°C. The

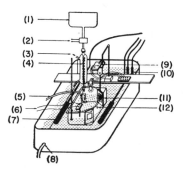

Fig. 1. Device for the study of regulatory behavior in the goldfish. 1, Water supply; 2, electric valve; 3, cold water; 4, distilling tube; 5, air line; 6, wires from thermistor; 7, heater; 8, "constant level" hole; 9, lever assembly; 10, thermostats; 11, lever guard; 12, "constant level" hole.

fish was given 10 minutes to adapt to the experimental container, and then the temperature of the water bath was gradually raised to 41°C over a period of about ½ hour. The lethal temperature for these goldfish is approximately 41°C (3). When the temperature in the experimental container reached between 30° and 35°C, training was begun. Measured amounts of cold water were introduced into the container at irregular intervals. Each cold reinforcement consisted of a 1-sec flow of cold water (2 to 3 ml) from the distilling tube mounted above the container (see Fig. 1) and produced a transient drop in temperature of approximately 0.3°C. A small light bulb mounted above the container was lighted during the 1-sec reinforcement period. Each fish received approximately 50 reinforcements in each of two training sessions.

In the third session, the lever was placed in its appropriate position, and the lever target was located behind the hole in a Plexiglas lever guard (Fig. 1). In order to actuate the lever, the fish had to insert its head through the hole and push at the target. The lever guard prevented chance operation of the lever by the swimming movements of the fish. When the temperature rose to above 30°C, training for lever pressing was begun. The method of "successive approximations" was employed (4). In this method, the reinforcement is first given whenever the animal is near the lever, then when the animal touches the lever, and finally only when the animal presses the lever. Most fish learned to press the lever within 2 hours after the onset of training. Seven small goldfish were trained.

The fish were then placed in a "titration" situation. The temperature of the water bath gradually rose and leveled off at 41°C. By pressing the lever for squirts of cold water, the fish could lower the temperature in its container. The temperature was maintained at 41°C for the entire session, once it had reached this level. Thus, a constant temperature stress was provided for the fish.

Two procedures were employed in experimental sessions. In the first procedure, the temperature of the experimental container was raised to 38°C before the fish was permitted access to the lever; the fish was then given access to the lever for 2 hours. The lever-pressing responses and temperature in the container were recorded continuously throughout the 2-hour sessions. As is shown in Fig. 2A, a typical record, the fish almost immediately drove the temperature down from 38°C to approximately 35°C. In almost every 2-hour session, the fish showed a burst of responses when the lever was initially made available. Within a few minutes the temperature was brought down to the level later maintained. The fish very

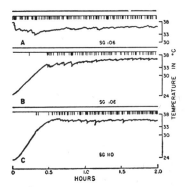

Fig. 2. Typical records of lever-pressing responses of goldfish and temperature in the container. A, Fish drives down environmental temperature. B and C, Fish prevent the temperature from rising above 35° to 36°C.

rarely allowed the temperature to rise above 36.5°C and rarely pushed it down below 33.5°C. The temperature remained with this 3-degree range almost all of the time. The maintained temperature of about 35°C in this experiment is much higher than the value determined by Fry (1) for temperature selected by goldfish in a thermal gradient (27°C for fish adapted at 25°C or more). It is likely that the fish in this experiment were setting the tank at a maximum comfortable temperature. That is, 35°C may be about the highest temperature at which these fish do not get aversive thermal feedback from their environment.

In the second series of experiments, as soon as the lever was made available at the initial temperature of 24.5°C, the water bath was gradually heated to 41°C over a ½-hour period. Sessions lasted 2 hours from the introduction of the lever. In this situation, fish were able to maintain their tank at a given temperature with much less work than under the first procedure. They were not required to bring the temperature down initially to the selected level. If amount of work is an important variable in controlling thermoregulatory behavior, one might predict that the fish would maintain a lower temperature in the second experiment than in the first.

The results of this second experiment, as shown by the examples in Fig. 2, B and C, indicate that there is no difference between the temperatures maintained under the two sets of conditions. Typical records for fish SG 106 under both conditions are shown in the figure.

Fish usually did not press the lever much at temperatures below 33°C in the second experimental series. They usually began pressing consistently at approximately the maintained temperature of 35° to 36°C. Some records (Fig. 2B) show a gradual upward drift in temperature as the session continues. Others show relatively little drift and very close regulation (Fig. 2C).

Control experiments have indicated that the increased activity of the fish at higher temperatures and the slight increase in oxygen tension of the water associated with reinforcement are not important factors controlling thermoregulatory behavior in this situation.

The results of these experiments indicate that the goldfish will regulate its body temperature within certain limits under a constant high-temperature stress. It has been suggested (5) that temperature selection in fish can be accounted for as a direct effect of temperature on the locomotion of fish. This study indicates that other factors are involved in temperature selection, since the goldfish will perform an arbitrary response to change the temperature of its environment (6).

References and Notes

1. F. E. J. Fry, *Publs. Ontario Fisheries Research Lab.* **55**, 5 (1947).
2. B. Weiss and V. G. Laties, *J. Comp. and Physiol. Psychol.* **53**, 603 (1960).
3. F. E. J. Fry, J. R. Brett, G. H. Clawson, *Rev. can. biol.* **1**, 50 (1942).
4. C. B. Ferster and B. F. Skinner, *Schedules of Reinforcement* (Appleton, Century, Crofts, New York, 1957).
5. K. C. Fisher and P. F. Elson, *Physiol. Zoöl.* **23**, 27 (1950).
6. This investigation was part of a thesis submitted in partial fulfillment of the requirements for the Ph.D. degree by one of us (P.N.R.). The research was supported in part by grants from the National Institute for Neurological Disease and Blindness (B-1941) and the Nutrition Foundation.

VOLUNTARY HYPOTHERMIA IN REPTILES

Philip J. Regal

It is commonly assumed that diurnal reptiles become cold and inactive at night because the body cools as the environment cools. In this sense, a lizard's "torpidity" is obligatory and passive. The assumption would seem to follow the fact that few reptiles are able to achieve significant thermal homeostasis with metabolic heat, a python (1) being a notable exception. Rather, some reptiles are known to maintain body temperatures within narrow limits by behavioral exploitation of microclimatic thermal mosaics in their environments. It is the basking lizards of the arid areas of North America which have been most carefully (2) studied, and which form the context for generalizations (3, 4).

The following experiments, however, suggest that low body temperatures are the result of a voluntary and actively initiated process in some lizards. The lizards may thus "prefer" and not simply tolerate low nocturnal temperatures. This would be significant to our conceptualization of the evolution of endothermy in birds and mammals (5).

Casual observations upon *Gerrhonotus* (Anguidae) and *Klauberina* (Xantusiidae) in substrate-heated thermal gradients prompted the experiments. Movement onto the warm surfaces was periodic and usually corresponded to the light periods in the observation room (photoperiod controlled). Inactive individuals would usually become responsive after feeding (6).

For experiments, animals which were judged to be active were placed, two each, in glass terraria (1 by 1 by 2 feet) (0.3 by 0.3 by 0.6 m) maintained in a constant-temperature room (set at 9° to 17°C in various experiments) with a controlled photoperiod (10 hours light, 14 hours dark). The ambient temperature established the cold end of the gradients, and one end of each terrarium was heated by a 250-watt, red-glassed heat lamp. This was suspended from above and provided each gradient with maximum temperatures which would be lethal to the species. Pieces of cardboard were scattered about so that the lizards could retreat

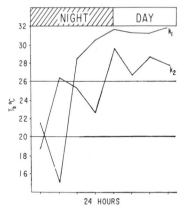

Fig. 1. Each point represents average cloacal temperatures for 3-hour sampling periods (two or three records each period) over 24 hours. K_1 is the larger of two male *Klauberina riversiana* lizards in the same terrarium. K_2 is a male in a terrarium with a female. The available temperatures in the gradient were constant through time, so that the nocturnal lows are achieved by voluntary movements of the lizards.

to shelters over a wide range of temperatures. The room was isolated from all activity and I entered as quietly and briefly as possible. The lizards were not fed, and were not disturbed for 3 days. The cloacal temperatures were then taken with a small-bulb mercury thermometer at intervals no more frequent than 1 hour.

The data revealed that for some period during the night the lizards selected low temperatures at which they became torpid. Orientation to high temperatures is not, therefore, constant. Figure 1 illustrates the results for 24 hours of sampling two *Klauberina*. Both show the drop in cloacal temperatures during the night and the orientation to higher temperatures during the light phase of the photoperiod.

The lizards are torpid at the lower temperatures. They move only with difficulty; they are poorly coordinated and relatively helpless. When replaced, after their temperature is taken, they typically retire rather than bask and move about. The capacity for directed movement is retained, that is, rewarming requires directed activity, and therefore the apparent incapacitation can be only in part due to low body temperatures per se.

The voluntary orientation to low temperatures was unexpected, but on reflection the survival value of such a phenomenon becomes apparent. If these lizards were to "prefer" high temperatures as an invariable aspect of their orientation to the environment, they might face a dilemma at the end of each day. Since at sunset, different parts of the habitat cool at different rates, orientation to isolated warmer areas could lead the lizards away from shelter and make them perhaps more vulnerable to predation. It would seem advantageous for the lizards to possess a mechanism insuring a reversal of high-temperature orientation following appropriate stimuli (exogenous or endogenous), so that they will not direct themselves toward maintenance of warm body temperatures when the environment is cooling and becoming suboptimal for behavioral thermoregulation. It is known that many reptiles take shelter before the natural environment actually becomes unfavorable. It is, however, difficult to distinguish preferences for low temperatures from shelter-seeking from some sort of threat, for example, fright overriding thermal preference. It is also difficult to establish that the shelters are cool at the time the reptiles locate them. In cases where reptiles are found in cool ground on warm days, it is difficult to establish that the animal can sense the more optimal conditions elsewhere. In short, natural history observations suggest that the phenomenon seen in the laboratory may be effective in nature, but field data to support the suggestion that thermal "preferences" change are not yet available.

Little is known of the thermoregulatory responses of "low-temperature" lizards such as *Gerrhonotus* and *Klauberina*, and so the experiments were continued with more familiar heliothermic (basking) lizards: *Phrynosoma cornutum*, *Sceloporus magister*, and *Uma notata*. These forms, in contrast

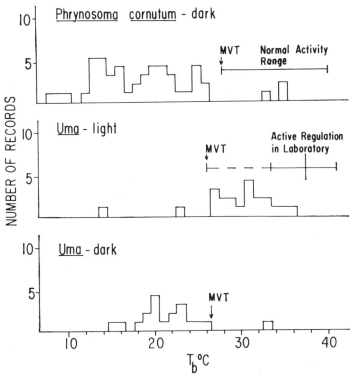

Fig. 2. (Top) Cloacal temperatures of *Phrynosoma cornutum* in sand at night (dark phase of photoperiod). Measurements were made twice a day (once each dark phase) on five individuals over a period of days. Those not in sand at a given time were ignored. *MVT* represents the "minimum voluntary temperature" (3), which is the lower end of the activity range and the imposed temperature at which a diurnally active animal can be forced to retreat from activity. At night, then, the lizards nearly always *voluntarily* select temperatures below the *MVT*. (Middle and bottom) Cloacal temperatures of buried *Uma notata* determined in the same way as for *Phrynosoma*. Like *Phrynosoma*, *Uma* would usually (see text) select temperatures below the *MVT* at night voluntarily, but not in the daytime. Cowles and Bogert's (3) value for the *MVT* is used and Licht's (12) mean and range for active *Uma* (no sand) are illustrated. The lower end of that range bears no essential relationship to the *MVT* which is a temperature eliciting particular types of behavior.

to *Gerrhonotus* and *Klauberina*, burrow in the sand when sleeping at night; *Phrynosoma* and especially *Uma* burrow for considerable lengths of time during the day. Large terraria (6 by 2 by 2 feet) were covered with 3 to 6 cm of fine sand, and two heat lamps established the hot end of the gradients. Lizards would burrow deeper than 1 cm, and since 1 cm of this sand will block all light, lizards buried in different parts of the terrarium presumably experienced only temperature differences. Only buried individuals were measured. Temperatures were taken once during the night and once during the day over several cycles of the photoperiod (10 hours light, 14 hours dark). Food (mealworms) was continually available.

Figure 2 (top) shows that nearly all the buried *Phrynosoma* at night selected temperatures below the so-called "minimum voluntary temperature" (3), which is the lower limit for activity as conventionally determined (3, 4). *Phrynosoma* spent much of the day active, above the surface of the sand, so relatively few data are available. Those that I have indicate day burials above the minimum voluntary temperature and within the basking and activity ranges.

The initial results for *Uma* were variable. While many voluntarily selected low temperatures at night, others were active at night and torpid in the day! Some lizards did not show any indication of the voluntary selection of temperatures below "normothermic" levels on any regular basis. A simple photoperiod appeared to be inadequate to simulate dusk and dawn for this group of individuals under conditions of ad libitum heat.

The procedure was then changed in an effort to reduce the individual variability. One of the two red-glass heat lamps was replaced by a clear-glass lamp and placed in the circuit with the room lights so that when the room lights would go off, the total area comprising the hot end of the gradient would become reduced. Thus, while the temperatures at the extremes of the gradient did not alter, dusk was simulated in that a sharp reduction in light intensity was associated with substrate cooling in some areas and with a reduction in the radiant environment in a part of the gradient.

Under the new conditions nearly all *Uma* selected low temperatures when burrowing at night (Fig. 2) but levels above the minimum voluntary temperature when buried during the day. Parts of the gradient shifted their boundaries at "dusk," but most lizards were found covered where the gradient was stable. A trap effect cannot, then, explain the results.

Only three *Sceloporus* were observed. They would remain active for days at a time, burrowing in cool sand (cloacal temperatures were 14.5° to 18.8°C) and becoming torpid only about once a week. This hypothermia was, of course, voluntary, even though it had no evident relation to photoperiod.

To summarize, the tendency to orient voluntarily to low temperatures has been demonstrated in several species of lizards under conditions that would permit the animals to maintain body temperatures constantly at activity levels. This generally occurs at night, but regular, periodic behavior may require simulation of dusk for certain species. Physiological benefits may well underlie the phenomenon, but the ecological advantage is more obvious at present.

It seems reasonable that the endothermic birds and mammals evolved from behaviorally thermoregulating ectotherms, and that the initial role of metabolic heat production was as an adjunct to an already sophisticated thermoregulatory system (7). These ancestral ectotherms might be assumed to have faced problems similar to those faced by modern behavioral regulators. With daily exposure to suboptimal temperatures, one would expect (using these modern lizards as a model) that they would develop mechanisms to protect them from inopportune thermally directed behavior and consequent exposure to hazards. Given a trend toward endothermy, voluntary hypothermia would gain the added protective function of energy and water conservation. It follows that active, periodic changes in the thermoregulatory system, or "thermostat," were probably part of the heritage of the birds and mammals from ectothermic ancestries and through the transitions to endothermy.

From this point of view it would appear that a hypothermic state such as daily torpidity is primitive, having been modified and retained wherever adaptive. Cade (8) has observed that torpidity appears in the phylogenetically conservative lines of rodents, and on this basis has come to the same conclusion.

The more orthodox view at present is that hypothermic states represent adaptive specializations and are advanced conditions in vertebrate endotherms (9). It is of course possible that thermostatic control became continuous and nonperiodic in the founders of the marsupial and placental lines of mammals, and that the hypothermic states of today represent secondary phenomena. This might have happened if ancestral forms were large at some stage well along toward endothermy. The energy expense of warming a large body every day could have resulted in continuous regulation becoming more practical (10). It is very difficult, of course, to determine if any particular fossil is directly ancestral or, indeed, endothermic.

References and Notes

1. V. H. Hutchison, H. G. Dowling, A. Vinegar, *Science* **151**, 694 (1966).
2. J. E. Heath, *ibid.* **146**, 784 (1964).
3. R. B. Cowles and C. M. Bogert, *Bull. Am. Mus. Nat. Hist.* **85**, 265 (1944).
4. J. E. Heath, *Univ. Calif. Publ. Zool.* **64**, 97 (1965).
5. The semantics of biothermal studies can be confusing. I follow the ecological approach of R. B. Cowles as reviewed in *Science* **135**, 670 (1962). *Ectotherm* includes nearly all reptiles and signifies that thermal homeostasis is dependent upon heat available in the environment. *Endotherm* includes nearly all birds and mammals and signifies the importance of metabolic heat production in heat balance. The term *hypothermia* is unconventional when applied to ectotherms. In speaking of mammals, hypothermia is frequently a purely descriptive term and in clinical studies is even used to denote an induced state. My usage implies a judgment as to the natural operation of the neural control system. My data do not resolve the question of whether a thermostat is "turned off" or "set down," but they do suggest that more than simple inhibition of a thermoregulatory system occurs in lizards, since directed locomotor efforts are required to become cool in these gradients. Use of the word *preference* should not be taken to suggest a mechanism of regulation. Behavioral studies on lizards in nature and in the laboratory indicate, rather, that active lizards maintain body temperatures within a *range* by particular behavior patterns which are elicited toward the extremes of the range (2–4, 11).
6. P. J. Regal, *Copeia* **1966**, 588 (1966).
7. R. B. Cowles, *Science* **103**, 74 (1946); C. M. Bogert, *Evolution* **3**, 195 (1949); K. Johansen, *Physiol. Zool.* **34**, 126 (1961); G. A. Bartholomew and V. A. Tucker, *ibid.* **36**, 199 (1963).
8. T. J. Cade, *Ann. Acad. Sci. Fennicae Ser. A, IV*, **71**, 6, 79 (1964).
9. C. P. Lyman and D. C. Blinks, *J. Cell. Comp. Physiol.* **54**, 53 (1959); N. I. Kalabukhov, *Bull. Mus. Comp. Zool.* **124**, 45 (1960); G. A. Bartholomew and J. W. Hudson, *Physiol. Zool.* **35**, 94 (1962).
10. O. P. Pearson, *Bull. Mus. Comp. Zool.* **124**, 93 (1960).
11. C. M. Bogert, *Sci. Amer.* **200**, No. 4, 105 (1959).
12. P. Licht, *Physiol. Zool.* **38**, 129 (1965).
13. I thank Drs. G. Bartholomew, J. Heath, R. Lasiewski, and P. Licht for reading the manuscript, although their suggestions may not have been adopted. I thank Drs. R. B. Cowles, J. Heath, M. Lloyd, and K. Norris for their encouragement, and particularly Dr. Lloyd for his suggestions on the preparation of the manuscript. The work was completed in part while I was an NIH mental health trainee under PHS 5t1-MH-6415.

A STUDY OF THE NEST-BUILDING ACTIVITY OF THE ALBINO RAT

ELAINE FLITNER KINDER

Psycho-biological Laboratory, Henry Phipps Psychiatric Clinic, Johns Hopkins University, Baltimore, Maryland

TWENTY-FOUR FIGURES

CONTENTS

Introduction	117
Methods	118
Results	125
Section I. The age at which nest-building appears and the average level at which the activity is maintained	125
Section II. The influence of ovulation, pregnancy, and lactation on nest-building	131
Section III. Influence of temperature changes on nest-building	139
Section IV. Relation of nest-building to food-intake	150
Discussion of nest-building as a thermoregulative activity	152
Conclusions	159

INTRODUCTION

The phenomenon of nest-building, like many other important activities, has seldom been subjected to controlled observation and has completely escaped quantitative study. Such data as we have consist chiefly of observations made by naturalists under conditions which often preclude laboratory control and prevent the possibility of evaluating such factors as age, previous experience, food, physical condition, etc. Although there have been a number of studies, especially in birds, of the methods and material used in building nests (studies like those of Herrick, '11), very little is known regarding the factors involved in the production and modification of the nest-building activity itself.[1] It was the purpose

---

[1] Thomson's "Biology of Birds," for instance, in a bibliography of over 200 titles, cites no study of nest-building.

of the present work to study this phenomenon as it appears in the albino rat.

There was little to start from in the way of previous observation. The tendency of the wild Norway rat to build nests has been noted (Lantz, '09), and in laboratory colonies of albino rats provision is usually made for some sort of bedding or nesting material (Small, '00, '01), especially at the time of parturition (Greenman and Duhring, '23). In such colonies, however, the bedding ordinarily provided is sawdust, wood chips, or similar materials not well adapted to the formation of nests. In the rat colony of the Psychobiological Laboratory of the Henry Phipps Psychiatric Clinic, where crêpe-paper waste was used in the stock-room cages, the nest-building activity became especially conspicuous in both non-parturient and parturient animals. Observation of this led Richter, in his study of age variations in the behavior of rats ('22), to include a brief experiment on nest-building—the first effort to bring the nest-building of the albino rat under experimental control and quantitative measurement. Richter used square, cardboard frames, 3 feet on a side and 1 foot high, covered with wire-cloth, and for nesting material he used in each cage 200 strips of absorbent paper toweling distributed over the floor of the cage. Measurement was made by counting the number of strips gathered into the nest during each twelve-hour period. In all, twelve animals of different ages were used, each for three successive days, exclusive of a three-day preliminary period to familiarize the animal with the cage. The present work, in a sense a continuation of this earlier effort, is a more detailed study with larger groups of animals of the activity under conditions of rigorous control and over relatively long periods of time.

METHODS

It has already been mentioned that the significance of the present method lies in the application of quantitative measurement to the study of a phenomenon which had not previously been thus investigated. The method followed throughout was that employed by Richter, who found that a rat will build a new nest each day if the existing nest is removed. Thus, if a rat is presented with a given number of units of nest-building material during successive time intervals, it is possible to measure the nest-building for each period by determining the number of units of material utilized.

With the rat, however, as with other animals whose behavior is varied and complex, a single activity cannot be isolated and studied by itself; it must be considered as one element of a total behavior picture, in which external factors of the environment and varying physiological conditions are so balanced and interrelated that variation in any one factor may materially influence other factors, and may even change the entire behavior. Unless there is a uniform environmental setting, the activity changes of the adult female rat found by Wang ('23) and Slonaker ('24) do not appear clearly, nor is there seen the intimate relation between food-intake, running activity (as measured in revolving drums), and the physiological changes in growth, pregnancy, and lactation shown by Wang ('25). A study of nest-building demands a similar control of environmental conditions, since without it there can be no differentiation between the variations which result from rhythmic physiological alterations and those initiated by changes in the environment. Under 'method,' therefore, we must consider, A) the control of such environmental variables as illumination, food, temperature, noise, and the influence of other animals, as well as, B) the type of the cages and material used and, C) the methods employed in the measurement of the nest-building activity.

A. The environmental setting was similar to that described by Richter ('22, '26) and by Wang ('23, '25). The work was carried on in an inside room of the Psycho-biological Laboratory of the Henry Phipps Psychiatric Clinic of the Johns Hopkins Hospital. The room is equipped with thick walls and double doors which shut out noises from the adjoining rooms and aid in maintaining a constant temperature (70°

to 75°F.). Constant dim illumination, arranged so far as was practicable to light equally all parts of all cages, permitted observation at any time without change of light.

The animals used had been raised in the laboratory colony, the majority of them in cages without nest-building material other than sawdust. That they were in excellent physical condition was shown by their high average body weight. The food used throughout consisted of the standard diet recommended by E. V. McCollum, supplemented by lettuce given once a week.

The daily records included, besides the nest-building data, the amount of food-intake in grams for the preceding twenty-four hours, the readings of a wet and dry bulb hygrometer, and a record of the vaginal-smear findings according to the technique of Long and Evans.[2] On Monday of each week the experiment room and the cages were cleaned and the animals weighed.

B. The cages were constructed of light-weight galvanized iron placed on sheets of the same material and with tops of galvanized wire netting (¾-inch mesh) bound to a rim of heavy wire which fitted securely over the wall. Metal was selected for the cages because it is easily cleaned and because it seemed that a hard, cold surface might stimulate nest-building and contribute to maintaining the activity at its maximum. Though of uniform material, cages were made in three types, A, B, and C. Type-A cages (fig. 1) were 6 inches high and circular, some of them 4 feet and others 6 feet in diameter. Type-B cages were rectangular, 9 inches high, with a floor area 4½ × 6 feet. These cages were built in a four-story tier with a 15-inch space between the cages. Type-C cages (like type-A cages) were circular,[3] but were 20 inches

[2] A small spatula which has been introduced into the vaginal orifice and scraped against the vaginal mucosa carries cells which can be transferred to a microscope slide. Long and Evans have shown that the stages of the oestrous cycle can be determined from these smears, the period of oestrus, immediately preceding ovulation, being characterized by cornified epithelial cells, and the dioestrous interval by leucocytes and nucleated epithelial cells. The intermediate stages can also be recognized by smear (table 2).

[3] Circular cages are very wasteful of floor space, but they have the advantage of presenting a uniform wall.

high with a 2-inch horizontal rim at the top (fig. 2). These cages were 4 feet in diameter for adult animals and 2 feet for young animals. In type-A cages the food and water cups

Fig. 1A Photograph of type-A, 4-foot cage, showing a small nest with the unused strips on the floor of the cage somewhat bunched by the animal's running.

Fig. 1B Photograph of type-A, 4-foot nest-building cage, showing a single very large nest of 2000 strips, built by a twenty-eight-day-old rat. This nest was built during an experiment not discussed in the text in which the nest was not removed for a four-day period, additional papers being offered each day.

were placed in the center of the cage on the floor; in type-B and type-C cages they were attached to the cage wall.

For nest-building material each rat was offered strips of high-grade white crêpe paper which was crinkly and did not flatten out. A few preliminary experiments with different types of material showed that rats will build nests of strips of soft flannel, pieces of twine, soft rope, or paper, but when

Fig. 2A Photograph of type-C cage without nesting material, showing construction of cage with the top raised by knobs to a height of ¾ inch above the rim, thus allowing the strips to slide freely when pulled.

Fig. 2B Photograph of type-C cage with strips in place, but without top, showing the appearance of the inside wall of the cage.

offered a choice they use the lighter and fluffier material. The paper was finally selected because it makes excellent nests, the rats show a distinct liking for it, and it lends itself admirably to a quantitative measurement, a characteristic which excludes such materials as shavings, chips, excelsior, 'wood wool,' and others often used for bedding in rat colonies. Furthermore, the paper strips are light and easily moved so that very young rats can carry them about, they are so tough that they do not tear readily, and, as they are relatively inexpensive, they can be frequently renewed. Fresh strips were distributed at least twice a week and usually more frequently. Soiled or wet strips were replaced daily and papers from one cage, if used again, were returned to the same cage.

Two different methods of presenting the material were used: the first in cages of type A and type B; the second, in cages of type C. The first method consisted of distributing evenly over the floor of the cage strips $\frac{3}{8}$-inch wide and 4 or 5 inches long;[4] the second, of hanging over the horizontal rim at the top of the cage wall strips $\frac{1}{2}$ inch in width and 24 inches long. With the first method of distribution there was greater scattering of the strips, due to the larger floor area. With this method, however, even in the larger cages, the rats would pick up several strips at once, often as many as three or four. They accomplished this by using a scooping movement of the fore paws to assist in carrying the papers to their mouths. In the second method of distribution the ends of the paper hung so that nesting material could be secured only when the animal stood on its hind feet, seized the end of a strip in its mouth, and dragged it down by moving away from the side of the cage. One of these long strips, however, equaled several of the shorter ones, and a nest of even a dozen was sufficient to completely surround a small rat. Figures 2A and 2B are photographs of a type-C cage; A, without strips, showing the construction of the cage, and B, with strips, showing the appearance of the inside wall.

[4] Five hundred 4-inch strips in type-A cages and 600 5-inch strips in type-B cages.

C. The principle of the experimental procedure was to place each rat in a separate cage and, at twenty-four-hour intervals, to distribute a fixed amount of nesting material after the removal of the nest and papers already in the cage. The number of strips gathered together by a rat into nests during each nest-building period was taken as the direct measure of the nest-building activity.[5] Since the size and accessibility of the strips differed with the type of cages used, no direct comparison can be made between the size of nests in one type of cage and those in another. The daily variations in nest-building, however, are comparable, since they are not influenced by the type of cage and can, therefore, be compared throughout this study. It is these fluctuations in the activity, rather than the absolute size of the nest, which are significant for this investigation.

The criteria for determining 'nests' were, 1) papers definitely bunched or gathered together and, 2) evidence that the nest had been used. In larger nests the second criterion was not necessary, but in smaller nests, where there might be some question as to whether the bunching was the accidental result of the animal's running about the cage (fig. 1A), the evidence of use decided whether or not the strips should be counted as a nest. Often an animal would be found asleep on a bunch of papers. This was taken as unquestionable evidence of a nest. Strips touching the bunch, but not definitely a part of it, were not counted.

Although few animals were studied (forty-four), the nests built during the studies numbered nearly 2000, and this seems

[5] Although the size of the nest alone was used in the quantitative study of the data, incidental observations of the position of the nest, its form, and many other items which seemed of interest, were made and included in the daily records. It soon became apparent, however, that the quantitative measurement was to be relied upon for furnishing the most complete insight into the nest-building activity.

For the experiments with the very large cages (type-B) an additional 'weighted' scale of measurement was also used. The 'weighting' was based on the distance through which the animal brought its nesting material. As the use of this scale merely served to accentuate the variations showed by the 'direct' scores, the latter only are given in this paper.

a sufficient basis for the conclusions which have been drawn from the experiments.

[*Editor's Note:* Material has been omitted at this point.]

Section III. Influence of temperature changes on nest-building

The experiments of the preceding sections have had to do with variations in nest-building activity dependent on changes within the organism. The experiments of the present section were undertaken to study the effect on nest-building of changes primarily external to the animal, especially changes in temperature.

The first experiments dealt with the effect of temperature differences on the time of appearance of the nest-building activity in young animals. From a litter of eight animals there were taken two groups of three animals each, the two remaining animals serving as controls. The three rats of the first group were placed in cages out of doors with a temperature varying from 40° to 60°F., the cages (type-C, 2 feet in diameter) being protected by a plaster-board tent; those of the second group were placed in a heated room with a temperature in the neighborhood of 90°F. and the control animals, in a room with a fairly constant, average temperature of about 70°F. Records were taken two, four, eight, twelve, and sixteen hours after the animals were placed in the cage and every six hours thereafter for nine six-hour periods. The nest was removed and fresh papers were added each time the records were taken.

After seventy hours, the outdoor animals were transferred into the heated room, while those previously in the high temperature were placed in the outdoor cages. The control animals were not moved.

Figures 11A, 11B, and 11C give, respectively, the nest-building records for one animal of each of the three groups, the temperature changes being indicated above each curve.

Two of the outdoor animals (the third escaped from its cage) built large nests during the first two-hour period (fig. 11A), and all three built large nests during the second period. Of the animals in the heated room, none built nests during the first two hours and only one during the second two-hour period (fig. 11B). During the first forty hours, the animals in the heated room built fewer nests than either the outdoor animals or the controls, and after the first forty hours they built no nests at all. The total number of nests built by these rats was only fourteen, the average number of strips used by the different rats per six-hour period being 11.5, 11.7, and 6.6, while the outdoor animals during the same time built thirty-eight nests and used, respectively, 44, 52, and 38.6 strips per six-hour period. Such nests as were built in the heated room also differed greatly in appearance from those built out of doors, the former being scattered and loosely made, while the latter, besides being larger, were also very close and compact. It was found, however, that during periods of very low external temperature (below 50°F.), the nest-building activity dropped out almost completely (fig. 11A) and the animals remained inactive, huddled together, and shivering.

When the rats from the outdoor cages (seventy hours after the beginning of the experiment) were placed in the heated room, the nest-building activity ceased almost immediately

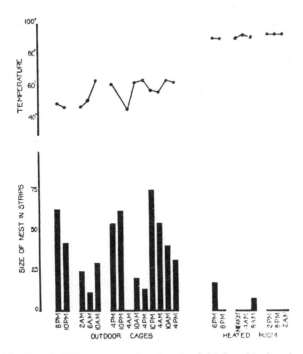

Fig. 11A Nest-building curve for twenty-eight-day-old female rat during seventy hours in an outdoor cage followed by thirty-four hours in a heated room. Curve of temperature variation at top of graph. Breaks in base-line indicate change from two-hour to four-hour, from four-hour to six-hour nest-building periods and removal from cold to warm temperature. First two columns indicate nest-building for two-hour periods; next three columns are nest-building records for four-hour periods; the succeeding records are for six-hour periods. After removal to heated room, there were again two two-hour periods followed first by three four-hour periods and then by three six-hour periods. Type-C, 2-foot cage. One hundred and forty strips of nesting material offered. Ordinates, size of nest in strips of nesting material. Abscissae, successive nest-building periods.

and, with the exception of one six-hour period (fig. 11A), there was no further nest-building by those animals during the remainder of the experiment. The rats from the heated room, on the other hand, used nearly as many strips within two hours after they were placed in the outdoor cages as during the entire seventy hours in the heated room, and they continued to build large and compact nests until the close of the experiment.[8]

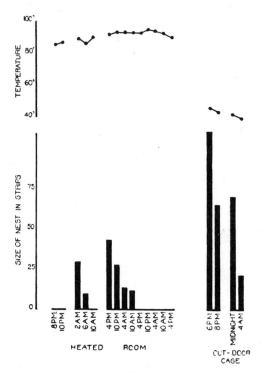

Fig. 11B Nest-building curve for litter mate of the animal whose record is shown in figure 11A, during seventy hours in heated room followed by twelve hours in an outdoor cage. Curve of temperature variation at top of graph. Successive records as in figure 11A. Ordinates, size of nests in strips of nesting material.

The control animals (fig. 11C) built nests during each period of the experiment. Though their activity showed considerable fluctuation, the average level was as high as that of the group in the low temperature (forty-eight strips per six-hour period for one and forty-five strips for the other).

[8] These experiments were terminated after the first few days by a severe storm, so that these observations were somewhat limited.

These rats were placed in their cages four hours later than the other animals, i.e., at the beginning of the four-hour periods.

These experiments, then, demonstrate that nest-building appears more promptly in moderate or low temperatures than in high temperatures and, further, that marked temperature differences not only influence the time of appearance of

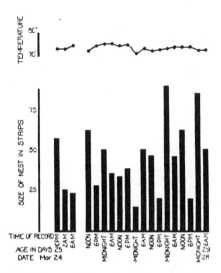

Fig. 11C Nest-building for control animal in normal temperature. Curve of temperature variation at top of graph. Break in base-line indicates change from four-hour nest-building periods to six-hour periods. First two columns indicate nest-building for two four-hour periods; remaining columns, nest-building for successive six-hour periods. Ordinates, size of nest in strips of nesting material. Abscissae, successive nest-building periods.

the activity, but also entirely change a level of activity which has already been established. Thus animals accustomed to building large nests in a low outdoor temperature practically stop nest-building when removed to a heated room, while the nest-building of those acclimated to the high temperature increases enormously when they are placed in outdoor cages.

A series of similar experiments was carried out with adult animals—two males, a normal female, and a pregnant female—all with about ninety days' previous experience in nest-building cages. The experiment was prefaced by a preliminary seven-day period of 'normal' temperature (about 72°F.) to establish a nest-building level. This was followed by three successive nine-day periods; the first, of low temperature,[9] the second, of 'normal' temperature,[10] and the third of high temperature.[11] At the close of the experiment, the temperature was again brought back to the 'normal' level for five days. A thermograph was used for recording temperatures during this experiment, the temperatures indicated on figures 12A, 12B, and 12C being averages of the thermograph records for each twenty-four-hour period.

Of the four animals studied, the nest-building of the males showed most clearly the influence of the temperature variation. Figure 12A, giving the nest-building curve for one of the male rats during the successive periods of the experiment, shows the prompt response of nest-building to temperature variation (i.e., increased nest-building with a drop in temperature and decreased nest-building with a rise in temperature). That this was characteristic of both males is shown in table 3, which gives the average nest-building of each animal for each of the experimental periods.

Figure 12 A shows also that there is a very close correspondence, during the period of low temperatures, between the daily fluctuations of nest-building and the fluctuations of the relative humidity curve, an increase in relative humidity being accompanied by an increase in nest-building and vice versa. Only during the low-temperature period, however, was the influence of relative humidity on nest-building sufficiently clear-cut to suggest this as a significant accessory factor.

[9] The change of temperature for the 'cold' period was effected by admitting outdoor air to the experiment room through an adjoining room. As no regulation was attempted during this period, the room temperature varied with the external temperature from 40° to 68°.

[10] The average temperature during this period was 6°F. lower than during the preliminary 'normal' period. The range of variation was small.

[11] For the high-temperature period electric heaters were used, and a temperature of 80°F. was maintained until near the close of the period, when a severe storm caused a drop of 8°.

The nest-building curve of the adult female rat differs markedly from that of the male in that the influence of the ovulation cycle outweighed the influence of the temperature

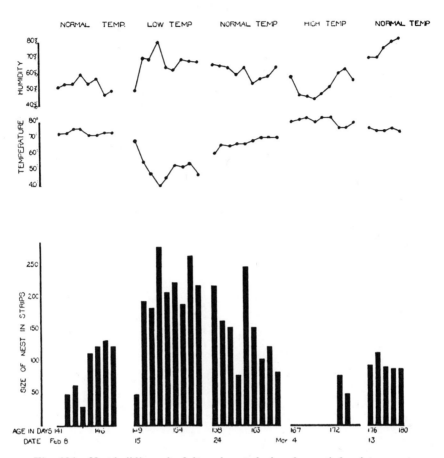

Fig. 12A Nest-building of adult male rat during five periods of temperature changes. Curves of temperature and humidity at top of graph. Breaks in base-line indicate transition from one period to the next. Type-B cage. Six hundred strips of nesting material offered. Ordinates, size of nest in strips of nesting material. Abscissae, successive days of experimentation.

changes. In other words, the peaks and depressions in the curve correspond more closely to the smear record than to the temperature variation. This does not mean, however, that external conditions have no effect at all. The maximum of nest-building activity (fig. 12B) was reached when the

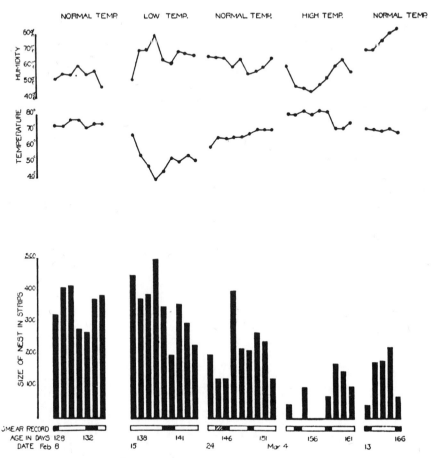

Fig. 12B Nest-building of adult female rat during five successive periods of temperature changes. Curves of temperature and humidity given at top of graph. Breaks in the base-line indicate transition from one period to the next. Smear findings recorded as in figure 8. Type-B cage. Six hundred strips of nesting material offered. Ordinates, size of nests in strips of nesting material. Abscissae, successive days of experimentation.

interoestrous period, a low temperature, and a high humidity were coincident, as on the twelfth day of the experiment. The minimum occurred when oestrus coincided with a high

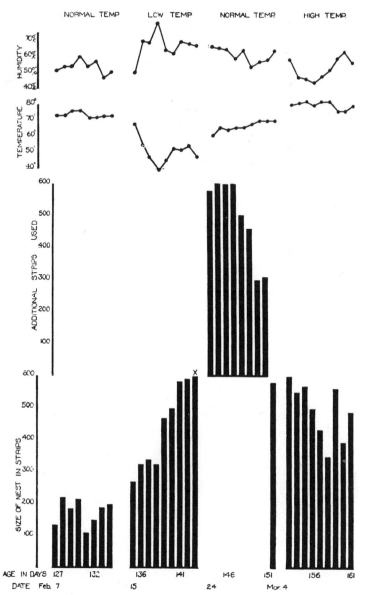

Fig. 12C Nest-building curve for pregnant female rat during successive periods of temperature changes. Curves of temperature and humidity given at top of graph. Litter cast at X. On the day after parturition, this rat was given an additional 600 strips of paper, and for the next eight days 1200 strips were kept in the cage, 600 always being left in the nest. As the record during these eight days represents the number of the additional 600 strips used during each twenty-four-hour period, the base-line of the curve during this time is at 600 instead of at 0. Type-B cage. Ordinates, size of nests in strips of nesting material. Abscissae, successive days of experimentation.

room-temperature. Thus, when the influence of internal physiological factors and the influence of temperature conditions were cooperative, we find extreme fluctuation in the curve; when they were antagonistic, the nest-building shows only slight fluctuations. Figure 8 gives further evidence of this relationship. In other words, we may think of the size of the nest as determined by the resultant of the factors involved. It is quite understandable, then, that in males, where the physiological cyclic variations are lacking, the environmental stimuli are more effective than in females.

The influence of temperature variations upon nest-building was least evident in the case of the pregnant female (fig.

TABLE 3

Nest-building data for periods of different temperatures

RAT NO.	AVERAGE NEST-BUILDING				
	Normal temperature	Low temperature	Normal temperature	High temperature	Normal temperature
69—male............	94	202	145	14	98
70—male............	98	238	111	19	11
72—normal female.....	352	352	215	96	141
73—pregnant female...	173	443	504	495	

12C)—a fact which is not surprising, in view of the extraordinary increase in the nest-building activity found to be normally present at this period. The pregnant animal included in this series showed an increase in nest-building with the onset of the low-temperature period. This was clearly not the preparturition rise, as the latter occurred after the days of lowest temperatures and four days before the litter was cast. Following parturition, which occurred at the close of the low-temperature period, the nest-building remained high, though there was a rise of 15°F. in the temperature of the room. This active nest-building was maintained also throughout the high temperatures (80°F.) of the succeeding period, though with much greater daily variation, for the young animals had by this time become well-furred and active.

In an additional experiment performed to study the influence of high temperatures on the nest-building of pregnancy and lactation, a pregnant animal and three non-pregnant litter mates were observed for a forty-five-day period, during

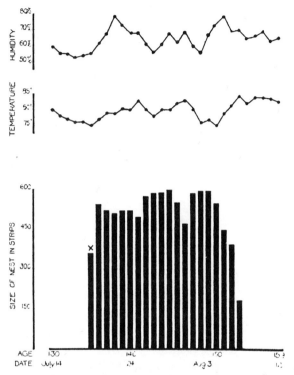

Fig. 13 Nest-building curve for pregnant female rat during high temperatures, five days prior to, and twenty-three days following, parturition. There was no nest built prior to the day the litter was cast (indicated on the graph by X). The temperature variation during the experiment is given at the top of the graph. Type-B cage. Six hundred strips of nest-building material offered. Ordinates, size of nest in strips of nesting material. Abscissae, successive days of the experiment.

which the temperature averaged 80°F. and never dropped below 76°F. The three non-pregnant animals built no nests at all throughout the entire period. The pregnant animal, on the other hand, though she built no nests until the day the litter was cast, built a large nest every day during lactation until the young were well grown. The nest-building curve for this animal, from five days before parturition to twenty-three days thereafter, is given in figure 13. Although all of the nests during the lactation period were large, considerable fluctuation was still present. This corresponded to the temperature variation—the smaller nests being built on days of higher external temperature and larger ones on days of lower temperature.

The nests built by these adult animals during high temperatures were invariably less compact than those built during normal temperatures and were similar in form to the scattered nests built by young animals under similar high-temperature conditions (p. 140). This was observed even of the mother rats, especially as the litter grew older. At times the nest was so scattered as to be merely a ring around the young, though as many strips had been moved in its construction as for the compact, fluffy nests which characterize the activity during normal temperatures.

The experiments of this section, then, demonstrate that these environmental conditions, as well as changes within the organism (studied in the preceding sections), enter into the determination of the quantitative variation in the nest-building activity. When the external temperature was about 80°F., the nest-building activity of all animals was diminished (very young animals built only occasional and scattered nests and adults, except parturient females, stopped nest-building altogether). In low temperatures, on the other hand, the nest-building showed a corresponding increase.

Section IV. Relation of nest-building to food-intake

Throughout the foregoing experiments it was found that the nest-building variations and the food-intake variations paralleled one another very closely. Mention has already been made (p. 135) of the correspondence between food-intake and nest-building during the oestrous cycle, viz., that at the time of oestrus both nest-building and food-intake were low, while dioestrum is accompanied by an increased food-intake and a high level of nest-building.

During the successive periods of temperature variation there was the same relation—low nest-building with a low food-intake during high temperatures and a high nest-building with increased food-intake during low temperatures.

The change in food-intake with a change of temperature is shown in figure 14, which gives, as calories per kilogram of body weight,[12] the food-intake of three rats during two four-week periods: first, when the room temperature averaged 72°F. and the animals were all building nests and, second, when the room temperature averaged 80°F. and no nests were built. This holds equally for a comparison of the food-intake in grams or in calories per kilogram of body weight. Here we find a marked decrease in food-intake accompanying an increase in temperature and, as has already been noted, the nest-building is influenced in the same direction. The average food-intake curve (Wang) for rats of this age is shown by the continuous line.

Incidental observations on a few isolated occasions had shown that absence of food was accompanied by a marked increase in nest-building. (The food cup of one of the animals had been accidentally so placed that the food was inaccessible.) To test this further, several animals were given a four-day period without food following seven days of normal food-intake. Figure 15 gives the nest-building activity of one of these animals (type-C cage) whose average with food was fifty-nine strips per day; without food, ninety-two strips. It will be noted that even on the day of oestrus, when normally there is very little nest-building, the activity of this animal was nearly as high as the maximum of the prestarvation period.

[12] The food-intake in calories per kilogram of body weight was calculated according to the formula used by Wang:

$$\frac{F \times 1000}{\frac{W_1 + W_2}{2}} = C$$

where F is the average daily food-intake for one week in calories (the caloric value of 1 gram of the standard diet being 3.73), and W_1 and W_2 are the body weights of the animal at the beginning and end of the week, respectively.

Thus we find throughout that there is an intimate and consistent relationship between nest-building and food-intake—a variation in one being paralleled by a similar variation (increase or decrease) in the other. This correspondence strongly suggests that we are dealing with two coordinated or parallel responses.

DISCUSSION OF NEST-BUILDING AS A THERMOREGULATIVE ACTIVITY

The results of the present experiments demonstrate that nest-building may be considered a thermoregulative activity. In those experiments we found that during periods of low temperature, when there is a need for conservation of body heat (i.e., protection against heat loss), there was increased nest-building activity, while during periods when such conservation would be a disadvantage to the organism, the activity dropped out. We have seen, further (section IV), that when conservation of body heat is essential, there is an increased food-intake accompanying the increased nest-building, while in temperatures in which the dissipation of body heat is necessary, there is a decrease in food-intake accompanying the decrease in nest-building. Thus, during both low and high temperatures the adjustment of the organism is effected in part by variations in food-intake, which modifies the heat production, and in part by nest-building, which modifies heat loss.

A connection, however, between the thermoregulative function of the nest and the nest-building variations found in lactation, starvation, prepubescence, etc., is not at once evident. Nevertheless, we find in some of these periods a definite need for protection against heat loss. Lactation, for instance, makes a very great demand upon the maternal organism. Furthermore, the heat loss to the young must be considerable, since they become cyanotic if uncovered and away from the mother for even a short time. In starvation and prepubescence also there is need for heat conservation, since in the former the fuel for heat production is diminished, while in the latter we find, in addition to the demands of growth, a relatively large body surface, long periods of inactivity (Richter, '22; Wang, '23, and Slonaker, '24), and a relative instability of the thermoregulative mechanism (Donaldson, '24). All of these suggest a need for accessory protection

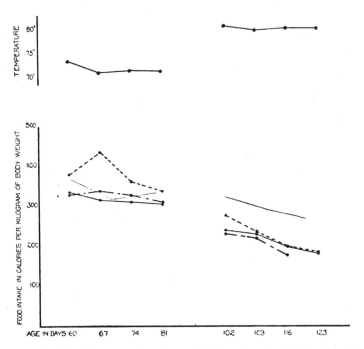

Fig. 14 Food-intake curves for three adult female rats, litter mates, for two four-week periods; first, during normal temperatures when all animals were building nests, and, later, during high temperatures when no nests were built. The unbroken line gives the average food-intake for normal adult female rats during these ages. Temperature variations given at the top of the graph. Type-B cages. Ordinates, weekly average of daily food-intake in calories per kilogram body weight. Abscissae, successive weeks of experimentation.

against heat loss, and in all of these periods we find that both nest-building and food-intake (except in starvation) are increased.

On the other hand, during oestrus, when the running activity is extraordinarily high and heat production is thereby increased, both the nest-building and the food-intake are minimum.[13]

In two periods of high nest-building, however (dioestrum and just preceding parturition), the need for heat conservation is not clear. Both are times of relative inactivity, but in neither does the drop in activity seem sufficient to explain the sudden and very marked nest-building increase. Just preceding parturition, of course, there is the very rapid growth of the foetuses, and undoubtedly a corresponding demand upon the maternal organism. Both periods, also, are probably times of increased sleep which might explain an increased need for heat conservation. In thinking of dioestrum as a recuperative period, however, we would expect the maximal nest-building in the first half of the interval, whereas the peak usually occurs early in the second half.

Dixon and Marshall ('24) have reported that in the bitch, rabbit doe, and marsupial cat the nest-building habits are displayed at the end of pseudopregnancy, and the authors suggest that these processes are "correlated with ovarian changes depending upon the degree of involution of the corpus luteum." In the female rat, also, the excessive nest-building during dioestrum and at parturition is similarly coincident with changes in the corpora lutea of ovulation and pregnancy. This strongly suggests an endocrine influence upon nest-building, but the fact that in both dioestrum and the last days of pregnancy a rise in food-intake accompanies the increased nest-building indicates that some unrecognized need for greater protection against heat loss cannot be excluded.

A summary of the nest-building findings for all of the periods studied, together with the food-intake findings and the running-activity data of Slonaker and Wang, is given in table 4. This shows again that the food-intake fluctuations are at all times synchronous with the nest-building varia-

---

[13] The fact that this period, in contrast to all other periods of the animal's life, is termed 'heat' (a similar connotation is found in other languages) suggests recognition of the body-temperature conditions indicated by these experiments.

tions,[14] and that there is, further, a consistent inverse relationship between these and the running-drum activity, so that

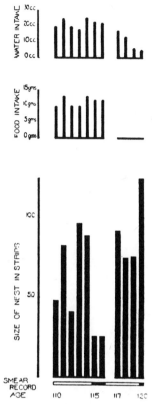

Fig. 15 Nest-building curve for adult female rat during eleven days, the first seven with normal food-intake, the last four without food. Curves at the top of the figure give water and food-intake. Temperature showed only slight variation and averaged 76°. The break in the base-line indicates the transition from the normal to the starvation period. Type-C, 4-foot cage. Two hundred and forty strips of nest-building material offered. Ordinates: for nest-building curve, size of nests in strips of nesting material; for food-intake, grams of food; for water-intake, cubic centimeters of water. Abscissae, successive days of experiment.

---

[14] It has been observed in the rat colony of E. V. McCollum's laboratory (personal communication from Miss Nina Simmonds) that animals fed during pregnancy on a low protein diet fail to show the normal nest-building activity at the time of parturition or to manifest any interest in their young, which, being neglected, soon die.

each period presents a definite behavior picture in which running-drum activity, nest-building, and food-intake are interrelated.

That we deal in nest-building with a behavior pattern definitely adapted to the thermoregulation of the organism is still further seen in the choice of nesting positions. Of over a thousand nests built during this study, every one was built against the cage wall or against the food cup, and of 700 nests

TABLE 4
Summary of nest-building, food-intake, and running-activity findings for various life periods

LIFE PERIODS	GROWTH	RUNNING ACTIVITY	FOOD-INTAKE	NEST-BUILDING	
Prepubescence	High	Low	High	High	Energy needed for growth
Oestrus		High	Low	Low	Increased heat production
Dioestrum		Low	High	High	?
Pregnancy	High	Low	Above average	Above average	Energy needed for growth
Lactation			Low	High	High	Heat loss to young. Energy for milk production
Starvation				High	Heat production low. Need for cutting down heat loss
High temperature			Low	Low	Increased heat production. Need for facilitating heat loss
Low temperature			High	High	Increased heat production. Need for cutting down heat loss

in the rectangular cages, 650 were in corners. In other words, the positions selected were those where the air circulation was at a minimum. Figure 16A shows the nesting positions with the frequency of their use for two males and two females in type-A cages over a forty-day period, while figure 16B gives similar data for four female rats in type-B cages over a period of eighty-five days. These figures give evidence of the preference for a certain side or corner of the cage and of the tendency, common to all animals studied, to select invariably a spot sheltered by the wall of the cage or by a food cup.[15]

[15] In these preferences no consistent relation to illumination or proximity to other animals could be determined.

Though the foregoing experiments indicate that nest-building is a specific adaptive activity, they offer no hint of the neurological or physiological mechanism involved. In the response to changes in external temperature, however, we are reminded of those receptors in the skin of the human being which, when stimulated, result in the perception of heat and cold. Stimulation of these receptors produces reflexly such responses as shivering and sweating (Sherrington, '23; Morgulis, '24) and, according to Morgulis ('24), is apparently

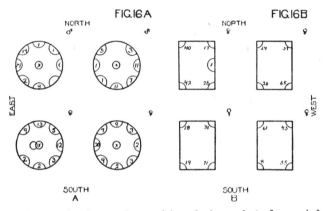

Fig. 16A Figure showing nesting positions during a forty-day period for four rats, two males and two females, in type-A, 4-foot cages. Each circle represents a cage, the small arcs indicating nesting positions, and the numerals, the number of times each position was selected.

Fig. 16B Data similar to that of figure 16A, for four female rats, over an eighty-five-day period in type-B cages.

associated also with changes in tonicity of the vegetative organs. The sensory end organs are also influenced by changes in the peripheral vascular flow (Waller, '93; Ebbecke, '17), and through this mechanism are probably linked up with more general physiological changes such as occur in fever or following vigorous exercise. The feeling of heat or cold in the human organism may, we know, be influenced by both internal and external conditions, and in this it is analogous to the nest-building response of the rat. For, as has already been outlined in this paper, nest-building shows fluctuations which certainly are a resultant of factors (some external to the organism and some internal to the organism), each of which individually influences the activity. These fluctuations, difficult to explain simply in terms of one or another factor, could be more readily understood on the basis of some mechanism influenced simultaneously by both internal and external conditions.

The thermoregulative function of nest-building has been neglected by both the physiologist and the psychologist. The physiologist, though he has studied the chemical and physical basis for heat regulation (metabolism, vasomotor changes, shivering and sweating, etc.; Sherrington, '23) has neglected the complex behavior responses which, through alteration of the animal's relation to its environment (seeking food) or through alteration of the environment itself (nest-building), assist in regulating the body economy. The psychologist, on the other hand, interested primarily in the genesis and development of behavior patterns, has neglected to investigate their physiological significance. A study of nest-building belongs to both fields, since it is a form of behavior which involves the entire organism, and, at the same time, is a thermoregulative activity similar and accessory in function to relatively simple physiological mechanisms.

With the simpler methods for controlling heat production and heat loss we are already familiar. There are the metabolic variations and the specific local changes involved in shivering and sweating, or the more permanent alterations involved in growth of winter fur. More complex are such behavior mechanisms as the modification of food-intake, the transitory search for shelter from wind or storm, or the change of environment by migration, as in birds. Most complex of all is the modification of the environment through the construction of nests or other shelters. The simpler forms of response are more primitive, while complex forms are genetically more recent. Martin ('01) has demonstrated that the physiological mechanisms for regulating heat loss develop much later than those for regulating heat production. The still more complex behavior responses, such as migrating, seeking shelter, building nests, etc., function also in the regulation of heat loss and are accessory to the physiological mechanisms. These require a still greater complexity of organization and are found only in those animals where the relation of the organism to its environment, being relatively unstable, is in need of constant readjustment. We therefore find that the biological basis for the provision of shelter is an organic need. This need is seldom recognized by human beings, who are accustomed to take houses for granted, although, in the absence of some such protection, it assumes an importance similar to the need for food.

CONCLUSIONS

1. The nest-building of the albino rat is a specific and variable form of behavior, contributing to the thermal regulation of the organism, initiated—under certain conditions at least—by the peripheral sensation of cold and inhibited by the peripheral sensation of heat. It is present as early as the twentieth day of age, is relatively independent of learning, appears in animals of both sexes, and is maintained at a fairly constant average level throughout the life of the animal (as far as was here studied—210 days).

2. Nest-building is modified by changes in the environmental temperature. Within normal temperature limits (50°F. to 80°F.) higher temperatures are accompanied by a decrease in nest-building and lower temperatures by an increase. An increase in relative humidity at low temperatures also tends to increase nest-building. The influence of temperature variations is seen most clearly in the male rats. In the adult female the responses to temperature are complicated by cyclic nest-building variations. At temperatures above 80°F. the nest-building drops out for all animals except mothers suckling young. At continued temperatures below 56°F. the nest-building of young animals drops out as the sluggishness from long exposure to the cold increases. The nest-building of adult animals continues through a tempera-

ture drop to 40°F. Lower temperatures were not studied. Nests constructed during low temperatures are compact and closely built, while those built during high temperatures are loose and scattered.

3. The nest-building of the adult female rat shows cyclic variations synchronous with the oestrous cycle, the maximum of nest-building occurring during the dioestrous interval and the minimum at the time of oestrus. The cycle of nest-building activity is thus the reverse of the running activity found by Wang and Slonaker.

4. The nest-building of the adult female rat shows a very great increase at the time of parturition and during lactation. The nest-building activity during this period was many times greater than that of any other period studied and exceeded the quantitative limit of these experiments.

5. An increase in nest-building is accompanied by an increase in food-intake, and a decrease in nest-building by a decrease in food-intake, thus suggesting that these two responses have a similar physiological significance.

6. The thermoregulative function of the nest, seen in the alteration of the nest-building level by variation in external temperature is further shown by, 1) the increased nest-building of prepubescence, lactation, and starvation, 2) the correspondence between nest-building and food-intake, and, 3) by the fact that nests are invariably built in a position where the air circulation is at a minimum, i.e., in a corner or against the wall of the cage.

7. As a thermoregulative activity, nest-building belongs to the field of physiology as well as to the field of psychology. It suggests a biological basis for the shelter provided by the higher animals for whom (because of their more varied environment) the more primitive means of heat regulation are inadequate.

ACKNOWLEDGMENT

To Dr. Adolf Meyer, Dr. Knight Dunlap, and Dr. William H. Howell I am indebted for interest and encouragement in this study. I wish also to make acknowledgment especially to the members of the staff of the Psycho-biological Laboratory; to Dr. C. P. Richter, under whose immediate direction the work was carried on, to Dr. G. H. Wang for the benefit of his experience with the details of experimental procedure, and to Dr. H. C. Syz for his suggestions and assistance both in the preparation of the manuscript and throughout the period of experimental work.

BIBLIOGRAPHY

DIXON AND MARSHALL 1924 J. of Physiol., vol. 59, p. 276.
DONALDSON 1924 The rat. Memoirs of The Wistar Institute of Anatomy and Biology, Philadelphia.
EBBECKE 1917 Pflueger's Archiv., Bd. 169, S. 395.
GREENMAN AND DUHRING 1923 Breeding and care of the albino rat for research purposes. The Wistar Institute of Anatomy and Biology, Philadelphia.
HERRICK 1911 J. of Animal Behavior, vol. 1, pp. 159, 244, 336.
LANTZ 1909 Bulletin no. 33, U. S. Dept. of Agriculture, Biological Survey.
MARTIN 1901–1903 Philosophical Transactions, vol. 195, series B, p. 36.
MORGULIS 1924 Am. J. of Physiol., vol. 71, p. 49.
RICHTER 1922 Comp. Psych. Monograph, vol. 1, no. 2.
——— 1926 Jour. Exp. Zoöl., vol. 44, p. 397.
SHERRINGTON 1923 J. of Physiol., vol. 58, p. 405.
SLONAKER 1924 Am. J. of Physiol., vol. 68, p. 294.
SMALL 1900 Am. J. of Psychology, vol. 11, pp. 80, 133.
——— 1901 Am. J. of Psychology, vol. 12, p. 206.
THOMSON 1923 Biology of birds. Macmillan.
WALLER 1893 J. of Physiol., vol. 15, Proceedings of the Physiological Society, no. 7, p. 25.
WANG 1923 Comp. Psych. Monographs, vol. 2, no. 6.
——— 1924 Am. J. of Physiol., vol. 71, pp. 729, 736.
WANG, RICHTER, AND GUTTMACHER 1925 Am. J. of Physiol., vol. 73, pp. 581–599.

THERMAL BEHAVIOR OF THE SUBNOURISHED AND PANTOTHENIC-ACID-DEPRIVED RAT

BERNARD WEISS[1]

Physiological adjustments to the surrounding temperature are among the most crucial of the activities required to sustain life. This is especially true of homeotherms, such as man, who have to maintain a core temperature within quite narrow limits. Some of the activities that are partly governed by changes in the thermal environment and that help to adjust to it are nest-building, migration, and hoarding. Food seeking and ingestion are also somewhat temperature-bound. Indeed, according to a currently popular theory of food-intake regulation, "animals eat to keep warm and stop eating to prevent hyperthermia" (2, p. 551).

While just the amount of activity directed toward the maintenance of a thermal steady state makes this behavior of interest, it has other attractive features. One is the kinds of variables that may be manipulated in its study. Hardy's (9) scheme is useful in describing these. It views temperature as regulated in two ways: "automatic" regulation, and "servo" control. Automatic regulation (physiological) involves a "... system consisting of two thermal capacities, corresponding to the core and peripheral tissues; three controls, heat production, sweating, and vasomotor activity; two sets of temperature detectors, hypothalamic and cutaneous; and an integrating system, the autonomic centers in the brain and spinal cord" (9, p. 267). With servo control, the external demands are modified by using accessory sources outside the organism: clothing, space heating or cooling, etc.

If an experimenter can modify the processes of automatic regulation, he then has an opportunity to study thermal behavior as a function of both interoceptive and exteroceptive variables. Perhaps one of the simplest ways of varying an important interoceptive factor is to vary the caloric value of the diet. Such a procedure is based on the fact that body temperature is the result of a balance between rate of heat production and rate of heat loss. The increase in heat production due to eating is called the "Specific Dynamic Action" of the food. Under normal conditions, the Specific Dynamic Action must be dissipated to prevent a rise in body temperature. At low environmental temperatures, however, the heat produced helps to avert a fall in body temperature. The point at which this begins is called the "critical temperature." For the rat it is 29° to 30° C. When not enough food is given to maintain thermostasis, heat is produced from the oxidation of body tissues.

Exteroceptive variables may be studied in two ways. One is by regulating the ambient temperature. The second is by providing a means whereby the organism can limit the effect that such a temperature has on it; for example, a thermostat.

In the present study, a situation was used in which the interoceptive conditions induced by a special diet were assessed by the frequency with which S made use of an exteroceptive device that counteracted the effect of a cold stress.

METHOD

Apparatus

The device used was a modified Skinner box, as shown in Fig. 1. Depression of the lever, when reinforcement is made available, turns on the infrared heat lamp for a period of time that can be set by E.[2] In this study, each reinforcement consisted of a 10-sec. activation of a 250-w. lamp. The ambient temperature was 0° ± 2° C. The box measured 10 by 10 by 11 in., and the lamp extended to within 5.5 in. of the floor of the chamber. The occasional reinforcements are not enough to raise the temperature of the chamber, and are therefore in the form of radiant heat. Responses were recorded on a Gerbrands cumulative recorder.

Subjects

The Ss were 17 male albino rats from Sprague-Dawley stock. They were seven months old when

---

[1] This study was conducted in the laboratories of the Department of Experimental Psychology, USAF School of Aviation Medicine.

[2] The first study in this laboratory using heat reinforcement, by R. A. McCleary, G. Rothschild, and R. B. Payne, showed a direct relationship between the ambient temperature and response rate.

started on the special diet. Three groups were formed from these animals, two of 6 Ss each and one of 5 Ss.

Diet

The diet was equivalent to that used in a previous study (12), and was provided by General Biochemicals, Inc. It was deficient in pantothenic acid and was composed of the following ingredients: sucrose, 61 per cent; casein, 24 per cent; salt mixture, 5 per cent; cottonseed oil, 8 per cent; wheat germ oil, 2 per cent. To each kilogram of the diet were added the following synthetic vitamins: thiamine hydrochloride, 10 mg.; riboflavin, 10 mg.; pyridoxine hydrochloride, 10 mg.; nicotinic acid, 60 mg.; biotin, 5 mg.; 2-methyl-naphthoquinone, 5 mg.; folic acid, 10 mg.; p-aminobenzoic acid, 400 mg.; inositol, 800 mg.; B_{12}, 100 mg.; choline chloride, 2 gm.; Vitamin A, 6,000 USP units; Vitamin D, 600 USP units.

Pantothenic acid is a B-vitamin that seems to be involved in the capacity to withstand stress. One of the most prominent effects of pantothenic-acid deficiency is impairment of adrenal function and structural damage to the adrenal gland (3, 6, 11). This is probably because the pantothenic acid in the tissues is bound as coenzyme A (CoA), a biocatalyst essential not only in the intermediary metabolism of lipids and carbohydrates, but in the synthesis of the steroids (10). Deficiencies of pantothenic acid reduce the incidence of survival at low temperatures (7), while supplements have been shown to enhance the capacity to withstand stress (4).

In an earlier study (12), it was found that acute starvation led to a rapid rise of response frequency in the heat-reinforcement situation, and that, when pantothenic-acid deprivation was superimposed, a further significant rise took place. The aim of the present study was to determine the effect of pantothenic-acid deprivation when coupled with inadequate caloric intake (subnutrition).

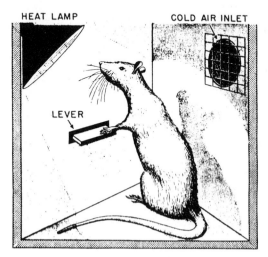

Fig. 1. The heat-reinforcement Skinner box.

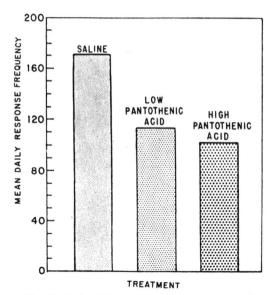

Fig. 2. Mean daily response frequencies during the first seven experimental days after supplements were begun.

Procedure

After a period of preliminary training, the Ss were put on a 5:1 fixed-ratio reinforcement schedule, which was maintained throughout the experiment. This program was used to keep the number of reinforcements under S's control but still produce a fairly even response rate. The records indicate that this was achieved.

All the animals were kept on the pantothenic-acid–deficient diet for the entire length of the experiment. Except for the first week, when they were allowed to feed ad lib., they were given access to food for 2 hr. daily. Because of the daily exposure to cold and, perhaps, the low palatability of the diet, this schedule produced a progressive loss of body weight. Starting with the seventh week, the Ss, which had been divided into three groups, began to receive intraperitoneal injections of the following solutions on Mondays, Wednesdays, and Fridays:[3] Group I, 1 cc. saline; Group II, 0.5 cc/100 gm body weight of calcium pantothenate, 3 µg/cc; Group III, 0.5 cc/100 gm body weight of calcium pantothenate, 3 mg/cc. There were five animals in Group I and six in each of the other two groups.

With occasional exceptions, Ss were run for 35 min. daily, five days per week. Body weights and food intake were measured daily after the injections began. The Ss' coats were kept clipped throughout the experiment.

RESULTS[4]

The main data of this experiment consist of the daily response frequencies of the three

[3] One day was missed on Friday of week 8.
[4] The computations on which this section is based were supervised by Captain E. W. Moore.

groups after the injections were begun. From this time until the first death, which occurred in the nonsupplemented group, there was a period of nine days (seven experimental) during which the pantothenic-acid supplements produced striking results. The data for this period are shown in Fig. 2. The value for a particular group gives the mean number of responses made per day by the Ss in the group. These data are based on adjusted response frequencies obtained from the regression of response frequency during this period on the response frequency of the three preceding experimental days, a procedure that helped to eliminate intergroup differences not resulting from the experimental treatments. The corresponding correlation was .80. The factor of body weight was also examined, but since this showed only a nonsignificant correlation of $-.14$ with response frequency, no adjustments for regression were made. The analysis of variance performed on these data showed significant differences among groups ($p = .02$) and that most of the variance due to groups was accounted for by the difference between Group I and the other two groups. There was increasing divergence between Group I and Groups II and III as the experiment progressed, but its statistical significance cannot rigorously be evaluated because the sample steadily diminished in size due to death.

Another way of looking at the data is to plot mean daily response frequency over comparable periods of life span. This was done by dividing each animal's experimental life span

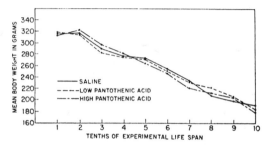

Fig. 4. Mean body weights at comparable stages in the experimental life span for the three groups.

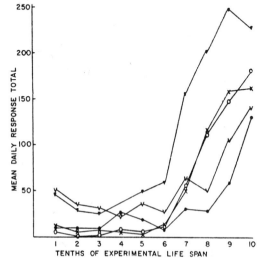

Fig. 5. Individual curves for the saline group showing mean daily response frequencies during each tenth of the experimental life span.

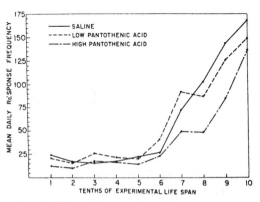

Fig. 3. Mean daily response frequencies at comparable stages in the experimental life span for the three groups.

into tenths, as in learning experiments, starting with the time that the pantothenic-acid-deficient diet was introduced. While such a procedure does not, because the effects of the vitamin supplement are confounded with life span, give very precise information about the efficacy of pantothenic acid, it has the advantage that from it we can make some deductions about how loss of body weight is related to cold sensitivity. The graphs in Fig. 3 show the mean daily response frequency for each of the ten periods.

The most obvious feature of these plots is their positively accelerated shape, indicating that despite a dietary regimen that produced a continual decrease in body weight, as shown in Fig. 4, the rise in response frequency did not take place until roughly 25 per cent of the

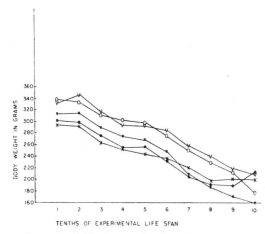

FIG. 6. Individual curves for the saline group showing body weight in grams during each tenth of the experimental life span.

initial weight had been lost. Individual plots for the saline group are shown in Fig. 5 and 6. All display the positively accelerated shape seen in the mean curve. The same is true for the other two groups.

A second aspect of these curves is the differences among the three groups although, as stated above, it is not entirely legitimate to consider these. But even so, they suggest that, for comparable periods during the latter part of the experimental life span, the animals in the high-supplement group were less affected by cold than the animals in the other two groups. This impression was only tentatively supported by the analysis of variance performed on these data, which yielded a Period $\times$ Group interaction p value of .09, and which showed that most of this was due to differences in the first-order components of the trend. While t tests showed significantly fewer responses by Group III for periods 7, 8, and 9, these must also be taken as only suggestive in view of the nonsignificant interaction. The lack of differences among the body weight curves shown in Fig. 4 tends to confirm the interpretation that pantothenic-acid deprivation produces effects that are not directly a result of weight loss.

DISCUSSION

Prolonged subnutrition has thus been shown to lead to a rise in response frequency in the cold when the reinforcement is a short burst of radiant heat. When pantothenic-acid deprivation is coupled with the subnutritive state, a further significant increment in frequency occurs. These results correspond to those found with acute starvation (12). The data also indicate that the thermotactic drive is intensified by pantothenic-acid deprivation in a fashion that is independent of body weight. Since pantothenic acid, as a component of CoA, plays a fundamental role in intermediary metabolism, this should not be too surprising. Ershoff (7) also contends that the heightened susceptibility to cold during pantothenic-acid deprivation is not merely the effect of differential weight loss.

While symptoms related to pantothenic acid deprivation usually take a long time to appear in rats of this age, the previous study (12) indicated that the latency can be shortened by subjecting the animal to starvation. The present study, with subnutrition, also showed a shorter latency of effects related to pantothenic acid than is typically achieved in such animals. This is in accord with the expectation that unideficient diets will result in much fewer symptoms, and take longer to effect them, than multideficient diets (1), with which one can obtain symptoms merely by cutting down on the amount of food ingested.

It was not possible in this experiment to assay directly for pantothenic-acid deficiency. There is, however, indirect evidence, from two sources, of pantothenic-acid insufficiency. One is the survival data. These show that the rats given no supplements (Group I) tended to die significantly earlier than the animals in the other two groups. The other is the degree of adrenal hypertrophy observed in the different groups. Animals in Group I tended to have relatively larger adrenals (a mean of 28.8 relative weight vs. a mean of 24.5 for Group II and of 25.7 for Group III). These values suggest that the combination of stresses (cold and subnutrition) affected the deficient animals more than the others. It is to be noted that throughout the experimental period, there were no significant differences in food intake among the three groups, and that intake fell off only when the animal was close to death.

Some protection seems to be afforded to the rat by a supplement of pantothenic acid as little as 9 μg. per week, which is far below the minimum daily requirement of 75 to 100 μg. given by György and Poling (8), although for a different strain of animals and a diet less fortified with other vitamins. The high dose, however, all things considered, seems to confer the maximum protection. It might be noted, in this connection, that the effect of pantothenic-acid surfeit seems to be a function of the general nutritive state of the animal. With unilaterally adrenalectomized rats, Dumm and Ralli (5) found that large supplements of pantothenic acid increased adrenal cholesterol levels in rats fed a 16-per-cent-protein diet, but not in those fed a 22-per-cent-protein diet. Once more, it is seen that the "daily minimum requirement" of a vitamin is somewhat of a misnomer, in that what is probably wanted is enough of a vitamin in the diet to supply added resistance in cases of unusual stress, especially when widespread dietary deficiencies or abnormalities are superimposed.

SUMMARY

Albino rats exposed to low temperatures were trained to obtain a burst of heat from a heat lamp by pressing a lever. After six weeks on a diet deficient in pantothenic acid, which was made available for short enough daily periods so that the animals lost weight steadily, they were divided into three groups. Beginning with the seventh week, one group was given a large, and one a small, dose of pantothenic acid three times per week. The third was injected with a saline placebo. There were six animals in each of the supplemented groups and five in the control group.

The supplements of pantothenic acid produced significant differences among the three groups in response frequency, the nonsupplemented group responding more often than the other two groups. When each animal's experimental life span was divided into tenths and the response curves for each group plotted, the curves displayed a positively accelerated shape, while corresponding curves for body weight showed a constant loss. Also, the response frequency curve for the high-supplement group was lower than that for the other two groups, suggesting that this group was less affected by the cold during comparable periods of the life span. It was concluded that pantothenic-acid deficiency intensifies the effects of a subnutritive state with respect to cold and that this is not due to differential loss of body weight.

REFERENCES

1. BEZNÁK, A. B. L., & VAN ALPHEN, G. W. H. M. Cardiac automatism and choline acetylation in rats on thiamine and pantothenic-acid-deficient diets. *Canad. J. Biochem. Physiol.*, 1955, **33**, 867–883.
2. BROBECK, J. R. Food intake as a mechanism of temperature regulation. *Yale J. Biol. Med.*, 1948, **20**, 545–552.
3. COWGILL, G. R., WINTERS, R. W., SCHULTZ, R. B., & KREHL, W. A. Pantothenic acid deficiency and the adrenals: some recent experiments and their interpretation. *Int. Z. Vitaminforsch.*, 1952, **23**, 275–298.
4. DUMM, M. E., & RALLI, E. P. Influence of pantothenic acid on response of adrenalectomized and intact rats to stress. *Fed. Proc.*, 1950, **9**, 34–35.
5. DUMM, M. E., & RALLI, E. P. Influence of ascorbic acid, pantothenic acid and protein on the resynthesis of adrenal cholesterol after stress. *Fed. Proc.*, 1954, **13**, 38.
6. ERSHOFF, B. H. Nutrition and the anterior pituitary with special reference to the general adaptation syndrome. In *Vitamins and hormones*, 10. New York: Academic Press, 1952. Pp. 79–140.
7. ERSHOFF, B. H. Comparative effects of pantothenic acid deficiency and inanition on resistance to cold stress in the rat. *J. Nutrition*, 1953, **49**, 373–385.
8. GYÖRGY, P., & POLING, C. E. Pantothenic acid and nutritional achromotrichia in rats. *Science*, 1940, **92**, 202–203.
9. HARDY, J. D. Control of heat loss and heat production in physiologic temperature regulation. In *The Harvey lectures*, Series XLIX. New York: Academic Press, 1955. Pp. 242–270.
10. NOVELLI, G. D. Metabolic functions of pantothenic acid. *Physiol. Rev.*, 1953, **33**, 525–543.
11. RALLI, E. P., & DUMM, M. E. Relation of pantothenic acid to adrenal cortical function. In *Vitamins and hormones*, 9. New York: Academic Press, 1953. Pp. 133–155.
12. WEISS, B. Pantothenic acid deprivation and the thermal behavior of the rat. *Amer. J. clin. Nutrition*, 1957, **5**, 125–128.

Received July 5, 1956.

24

Copyright © 1971 by E & S Livingstone

Reprinted from *Q. J. Exp. Physiol.* 56:231-241 (1971)

EVAPORATIVE COOLING IN THE RAT: INTERACTION WITH HEAT LOSS FROM THE TAIL. By EDWARD M. STRICKER, Department of Psychology, University of Pittsburgh, Pittsburgh, Pennsylvania, and F. REED HAINSWORTH, Department of Biology, Syracuse University, Syracuse, New York.

(*Received for publication* 3rd *May* 1971)

There are two major thermolytic responses of rats to inescapable heat stress: cutaneous vasodilatation and grooming behavior. The former response increases heat loss by radiation, conduction, and convection, primarily from vascularized exposed surfaces such as the tail, and depends upon the positive thermal gradient between the rat and its environment that occurs at ambient temperatures of 32–40°C. The latter response increases heat loss by evaporation of saliva groomed onto body surfaces and becomes increasingly important as ambient temperatures approach 40–41°C and the positive thermal gradient diminishes. These two mechanisms are not independent of one another. For example, rats develop more pronounced hyperthermia following surgical removal of salivary function, thereby increasing the positive thermal gradient and facilitating heat loss. Similarly, rats increase evaporative cooling by salivary grooming following amputation of the tail. Removal of either saliva or the tail impairs the rat's ability to regulate body temperature, while removal of both leaves the rat most vulnerable to heat stress.

There are two major thermolytic responses of rats to inescapable heat stress: grooming behavior, which increases heat loss by the evaporation of saliva [Hainsworth, 1967, 1968], and cutaneous vasodilatation, which increases heat loss by radiation, conduction, and convection, primarily from vascularized exposed surfaces such as the tail [Rand, Burton and Ing, 1965; Thompson and Stevenson, 1965]. Previous reports have described the important contribution of salivary evaporation to body temperature regulation by rats in the heat, and the various factors which influence thermoregulatory grooming (see review by Hainsworth and Stricker, 1970). The present report is concerned with heat loss from the tail, and investigates the effects of tail amputation on evaporative cooling and thermal tolerances of rats during heat stress.

METHODS

Animals

Adult male (280–360 g) and female (230–270 g) rats of the Sprague-Dawley strain were used in all experiments. All animals were housed in individual wire mesh cages in a continuously illuminated temperature-controlled room (21–24°C), and were maintained on Purina Chow pellets and distilled water except during testing.

Operations

To study the contribution of salivary grooming to thermoregulation, rats of both sexes were totally 'desalivated'. That is, salivary flow was abolished, by ligation of the parotid ducts, where they passed along the surface of the masseter muscles, combined with extirpation of the submaxillary and major sublingual glands. An initial disruption of feeding and loss of body weight was observed for several days following surgery, but all rats resumed feeding subsequently and showed the 'prandial drinking' (i.e. frequent interruptions of eating for short drinks of water), polydipsia, and substantial food wastage that is characteristic of a totally desalivated rat [e.g. Epstein et al., 1964; Stricker, 1970]. Experiments began 10–14 days after surgery when body weights had stabilized and rats appeared in good health.

The tails of intact and surgically desalivated rats of both sexes were amputated, to study the contribution of heat loss from the tail and its interaction with evaporative cooling mechanisms. A wire ligature was placed around the base of each tail and amputation occurred caudal to that site, leaving a 1–2 cm stump. The wound was sutured and the wire ligature was removed after several days. Experiments began 3–4 weeks after surgery and, despite cutaneous vasodilatation during thermal stress, bleeding from the stump was never observed.

Ether anaesthesia was used for all operations.

Procedures

Measurement of Body Temperature. Four groups of male rats, i.e. intact and totally desalivated rats with or without tails ($n = 10$ in each group), were exposed in a large incubator to an ambient temperature of 32°C, 36°C, 40°C, or 44°C (± 0.5°C). They were briefly removed from the incubator every 30 min and a thermistor probe was inserted rectally at least 5 cm to measure body temperature (± 0.1°C). Exposures were terminated when body temperatures reached 42.0–43.0°C, or after 4 hr. Each rat was exposed to thermal stress on two separate occasions at least 7 days apart, first at an ambient temperature of 32°C or 36°C and then at 40°C or 44°C.

Comparable procedures were followed with four groups of female rats ($n = 5$ in each group) with the exception that they were exposed to thermal stress only once, at an ambient temperature of 36°C.

Measurement of Evaporative Water Loss. Total evaporative water loss by male rats without tails was measured during steady-state thermal conditions by using an open-flow system described in detail elsewhere [Hainsworth, 1968]. The rats were placed in an air-tight container located in the incubator and were exposed to an ambient temperature between 30°C and 44°C for 37 min. Dried air was passed through the container and then through a series of drying tubes which were connected during the final 15 min of exposure and later were weighed to determine total evaporative water loss. One measurement was made for each rat ($n = 5$, from the above group) every 2–3 days until the rate of evaporation over a range of ambient temperatures (30–44°C) had been assessed.

Data for evaporative water losses from the present experimental rats were compared with values previously obtained from intact male rats using identical procedures and equipment [Hainsworth, 1968]. These values were replicated, with the measurements from intact rats (interspersed with measurements from rats whose tails had been amputated) falling within $\pm Syx$ of the least squares regression line derived previously (see Fig. 3).

Additional Tests

The surgical procedures involving desalivation and amputation of the tail were expected to impair thermoregulation by abolishing salivary evaporation and by eliminating heat loss from the largest noninsulated surface area, respectively. To determine whether rats could regulate body temperature following such procedures if they were again provided with a means of evaporative cooling or with a nude body surface, two additional tests were employed. First, desalivated rats whose tails had been amputated ($n = 5$) were exposed to an ambient temperature of 40°C in the presence of a circular glass dish (5 cm high × 9.5 cm in diameter) filled with water. Previous results from this laboratory have indicated that behavioral thermoregulation by voluntary periodic immersions in a water bath can greatly enhance the thermal tolerance of rats [Stricker, Everett and Porter, 1968; Stricker and Hainsworth, 1970a]. Second, the insulation provided by fur was completely removed with an Oster clipper and the shaved desalivated rats whose tails had been amputated ($n = 5$) then were exposed to an ambient temperature of 36°C until body temperatures reached 42·0–43·0°C. During these two experiments, body temperatures were measured every 30–60 min with a thermistor probe.

RESULTS

Male Rats

Intact Rats. Intact rats had little difficulty in regulating their body temperatures during the 4 hr test period when exposed to an ambient temperature of

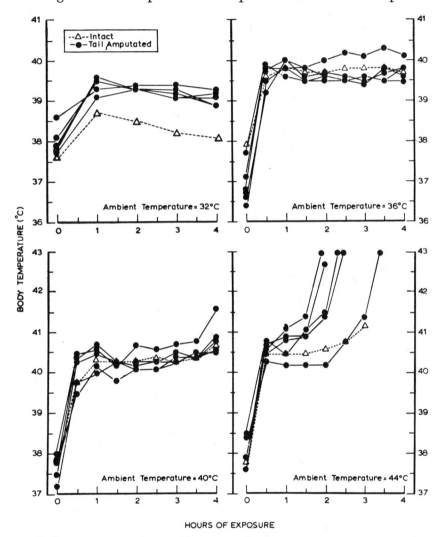

FIG. 1. Body temperatures of rats with amputated tails (●) during exposure to an ambient temperature of 32°C, 36°C, 40°C, or 44°C. Mean responses of rats with intact tails (△) also are presented for purposes of comparison ($n = 5$ for each point.)

32°C, 36°C, or 40°C. All rats showed a controlled hyperthermia, with body temperatures maintained at increasingly higher levels as ambient temperature was increased (Fig. 1). During the severe thermal stress of exposure to an ambient temperature of 44°C, in contrast, body temperatures were maintained only for 2–3 hr before they rose abruptly towards lethal levels. These results confirm previous reports of thermoregulation by intact rats exposed to graded levels of heat stress [Hainsworth, 1967; Hainsworth and Stricker, 1969].

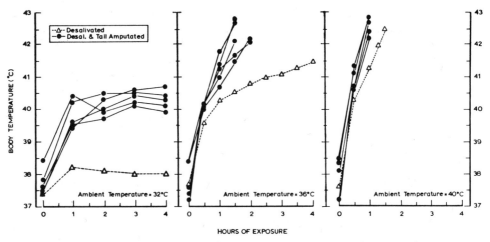

FIG. 2. Body temperatures of desalivated rats with amputated tails (●) during exposure to an ambient temperature of 32°C, 36°C, or 40°C. Mean responses of desalivated rats with intact tails (△) also are presented for purposes of comparison ($n = 5$ for each point).

Desalivated Rats. Thermoregulation in desalivated rats with tails was unaffected at an ambient temperature of 32°C but was markedly impaired at the more elevated ambient temperatures. For example, body temperatures increased slowly but continuously when the ambient temperature was 36°C, and rose rapidly when the ambient temperature was 40°C or 44°C (Fig. 2; Table 1). These results confirm previous reports that desalivated rats are intolerant of heat stress [Hainsworth, 1967; Hainsworth and Stricker, 1969].

TABLE I. *Exposure (min) to elevated ambient temperatures which increased body temperature to 42°C*

Amb Temp	Intact	Desal.	Tail Amput.	Desal. and Tail Amput.
36°C	606.0±95.8 (480–720)	305.0±71.2* (225–390)	456.0±80.4* (390–570)	89.4±22.3* (65–115)
40°C	350.4±35.3 (290–405)	66.6±18.5* (35–80)	320.0±43.7 (250–365)	49.8±4.9* (45–55)
44°C	192.0±40.2 (120–210)	33.0±5.5* (25–35)	130.2±35.1* (100–190)	27.0±6.0* (20–35)

Note: Values are Mean±S D. (range); $n = 5$ in each group; Amb. Temp. = Ambient Temperature.
* $p < 0.05$ in comparison with intact controls exposed to the same ambient temperature.

Rats with Amputated Tails. Rats without tails but with salivary glands showed a pronounced hyperthermia at an ambient temperature of 32°C and regulated their body temperatures 0·5–1·5°C higher than did intact rats. In contrast, rats whose tails had been amputated showed just the same levels of controlled hyperthermia as did intact rats during the first 4 hr of exposure to ambient temperatures of 36°C or 40°C (Fig. 1). More prolonged exposures revealed that rats without tails could not maintain thermoregulation for as long as intact rats at an ambient temperature of 36°C ($P<0.05$; the statistical significance of this and all other results were determined using a two-tailed t test) but showed normal thermoregulation at 40°C (Table I). At an ambient temperature of 44°C, rats without tails again showed somewhat less thermal tolerance than did the intact rats ($P<0.05$; Table I).

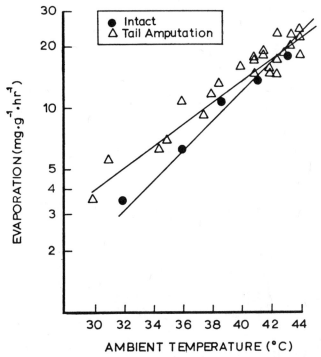

FIG. 3. Evaporation from rats with (●) or without tails (△), as a function of ambient temperature. Each point represents a single measurement. The regression equations are: rats whose tails had been amputated, $\log y = 0.0538x - 1.023$ ($Syx = \pm 0.060$); rats with intact tails (Hainsworth, 1968), $\log y = 0.0748x - 1.941$ ($Syx = \pm 1.16$).

The evaporative water loss of intact rats is known to increase in an exponential fashion as ambient temperatures rise above the thermoneutral range (Fig. 3; Hainsworth, 1968). A comparable relationship also was observed in rats whose tails had been amputated, although the slope of this curve was less steep (Fig. 3; $P<0.001$). Thus, amputation of the tail increased evaporation most clearly at ambient temperatures of 30–36°C and had a smaller and diminishing effect as ambient temperatures were elevated further.

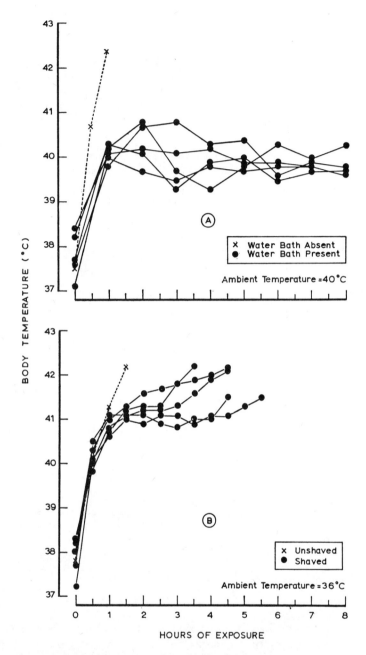

FIG. 4. Body temperatures of desalivated rats whose tails had been amputated (●), (A) in the presence of a water bath or (B) after body fur had been completely removed, during exposures to an ambient temperature of 40°C and 36°C, respectively. Mean responses of unshaved desalivated rats in the absence of a water bath (×) are presented for purposes of comparison ($n = 5$ in each group).

Desalivated Rats with Amputated Tails. Surgical desalivation combined with amputation of the tail produced the most pronounced impairments in the thermal tolerances of rats. Thermoregulation was observed only at an ambient temperature of 32°C, where body temperatures were maintained 1·5–3·0°C above those of desalivated rats with tails (Fig. 2). At more elevated ambient temperatures, heat loss through the fur and from the nude and highly vascularized scrotum and paws apparently was insufficient to support thermoregulation, despite a large thermal gradient to the environment, and body temperatures rose uncontrolled to 42°C (Table I). The most striking effect was at an ambient temperature of 36°C, where rats with tails are known to tolerate exposure for 4–8 hr even in the absence of salivary evaporation (Table I; Hainsworth, Stricker and Epstein, 1968). These results indicate that the tail must represent an important 'thermal window' to the desalivated rat during this heat stress.

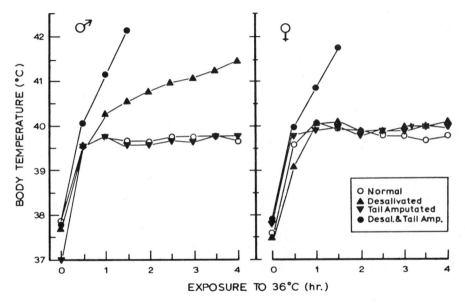

FIG. 5. Mean body temperatures of male and female rats, intact or following surgical desalivation and/or amputation of the tail, during exposure to an ambient temperature of 36°C ($n = 5$ in each group).

Desalivated rats with amputated tails showed excellent behavioral thermoregulation during an 8 hr exposure to severe heat stress in the presence of a water bath (Fig. 4A). Rats exposed to an ambient temperature of 40°C maintained body temperature at approximately the same level of controlled hyperthermia as did intact rats when exogenous water was available for evaporative cooling [Stricker *et al.*, 1968]. In the other test, desalivated rats without tails showed an increased thermal tolerance to an ambient temperature of 36°C after they had been completely shaved (Fig. 4B). It is interesting to note that their body temperatures rose to approximately 41°C within 60–90 min, whereas the body temperatures of unshaved desalivated rats with tails rose more gradually at this ambient temperature (Fig. 2). This difference suggests that the tail may be a greater avenue of heat loss than is the entire denuded body surface.

Female Rats

Whereas desalivated male rats exposed to an ambient temperature of 36°C showed a progressive increase in body temperature, desalivated female rats maintained body temperatures at the same level as intact female rats (Fig. 5). These results are consistent with previous findings that significant increases in salivary evaporation do not occur in female rats until ambient temperatures reach 37·0°C [Hainsworth, 1968], and emphasize the comparatively important role of conductive, convective, and radiative heat loss from the tail for thermoregulation in female rats during mild heat stress [cf. Chevillard, Cadot and Portet, 1962]. Nevertheless, amputation of the tail was without effect on body temperature regulation in female rats with salivary glands, presumably due to increased salivary evaporation (as in male rats at this ambient temperature) since thermoregulation was markedly impaired in desalivated female rats without tails.

DISCUSSION

The thermoneutral range of intact male rats is 28–31°C [Swift and Forbes, 1939]. At an ambient temperature of 32°C, salivary evaporation accounts for only 15–20 per cent of the total heat loss [Hainsworth and Stricker, 1970] and thus desalivation has little effect on thermoregulation (Fig. 2). In contrast, rats lose more than half of their metabolic heat production at this ambient temperature by radiation, conduction, and convection, due to the large temperature gradient between the rats and their environment [Hainsworth, 1968]. The tail appears to be a major avenue of non-evaporative heat loss [Knoppers, 1942; Grant, 1963; Rand *et al.*, 1965]; when it is removed, body insulation increases, the thermoneutral range is lowered, and, consequently, the need for evaporative cooling increases. In other words, an ambient temperature of 32°C represents a more substantial thermal stress to rats after amputation of their tails. Accordingly, rats without tails increase salivary evaporation at this ambient temperature (Fig. 3) and also maintain a greater level of controlled hyperthermia (Fig. 1), which increases the heat loss by radiation, conduction, and convection, and thereby decreases the amount of body fluid which must be lost by salivary evaporation. Nevertheless, note that the increased evaporative water loss by rats without tails probably impairs their survival during prolonged exposure to moderate heat stress (in comparison with intact rats; Table I, ambient temperature = 36°C), due to a more rapid dehydration [Stricker and Hainsworth, 1970b].

Evaporation accounts for more than 60 per cent of the heat lost by intact male rats at an ambient temperature of 36°C [Hainsworth, 1968]. This value increases to an estimated 80–85 per cent following amputation of the tail, as salivary evaporation increases (Fig. 3) to compensate for the reduction of heat loss by radiation, conduction, and convection. The capacity of individual salivary glands to increase their secretion in response to increased thermoregulatory needs for evaporation also has been observed in rats following partial desalivation, but body temperatures in those rats were maintained at slightly higher levels than normal [Hainsworth and Stricker, 1969]. In contrast, it should be emphasized that in the present experiments male rats with and without tails maintained identical body temperatures (Figs. 1 and 5) but had different rates of salivary evaporation at an ambient temperature of 36°C. These results indicate that the regulation of salivary excretion by rats during heat stress is not merely a function of their body temperatures, as has been suggested recently [Malan and Hildwein, 1969; Hainsworth and Stricker, 1971], but may be controlled by a more complex system responsive also to rate of heat gain.

At more elevated ambient temperatures, the positive thermal gradient decreases as body temperatures approach lethal levels and rats become more dependent upon evaporative cooling. For example, evaporation accounts for approximately 90 per cent of the heat lost by intact male rats at an ambient temperature of 40°C [Hainsworth, 1968]. Thus, amputation of the tail has little effect on salivary evaporation or body temperature regulation at this ambient temperature (Figs. 1 and 3) whereas surgical desalivation drastically impairs physiological thermoregulation (Fig. 2). At an ambient temperature of 44°C, heat cannot be lost but is actually gained by radiation and conduction because the ambient temperature is higher than body temperatures. Thus, rats are completely dependent upon evaporative cooling for survival and, in fact, a large exposed area such as the tail may be a liability due to its lack of insulation. This may explain why female rats, with longer tails than male rats of the same body weight [e.g. Chevillard et al., 1962], are less tolerant of severe heat stress than male rats [Hainsworth, 1967]. Nevertheless, male rats with tails seem to tolerate extreme thermal stress somewhat better than rats whose tails have been amputated (Fig. 1), and comparable results also have been reported in mice [Harrison, 1958]. These findings may reflect the facts that male rats and mice groom saliva predominantly on the scrotum and base of the tail [e.g. Herrington, 1940; Hainsworth, 1967; Stricker et al., 1968], and thus amputation of the tail removes an important nude locus for evaporative cooling.

Comparable thermoregulatory impairments to those of desalivated rats without tails have been observed in rats exposed to heat stress after hypothalamic damage or after severe dehydration [Stricker and Hainsworth, 1970a, 1970b]. Both of those treatments abolished salivary evaporation and probably interfered with blood flow adjustments to the tail during hyperthermia as well [e.g. Adolph, 1947; Thompson and Stevenson, 1965]. However, in contrast to the present experimental animals without saliva or tails (Fig. 4A), rats with hypothalamic damage may have suffered additional impairment of blood flow adjustments to the skin since they were not always capable of surviving exposure to an ambient temperature of 40°C in the presence of a water bath [Stricker and Hainsworth, 1970a]. Thus, evaporative cooling from those rats might be meagre even if they were soaked with bath water, and effective thermoregulation might be achieved behaviorally only when they escape from heat stress [cf. Lipton, 1968]. These differences, as well as the controlled hyperthermia in shaved desalivated rats without tails during heat stress (Fig. 4B), emphasize the residual thermoregulatory capabilities of desalivated rats without tails.

ACKNOWLEDGMENTS

The technical assistance of Jen-shew Yen and Richard E. A. Porter is gratefully acknowledged. This study was supported by Research Grant MA-3192 from the Medical Research Council of Canada and by Research Grants GB-12344 and GB-28830 from the National Science Foundation of the U.S.A.

REFERENCES

ADOLPH, E. F. (1947). Tolerance to heat and dehydration in several species of mammals. *American Journal of Physiology*. **151**, 564–575.

CHEVILLARD, L., CADOT, M. and PORTET, R. (1962). Influence de la température d'élevage sur la thermorégulation du rat, croissance de la queue et son rôle dans la thermolyse. *Comptes rendus hebdomadaires des séances et mémoires de la Société de biologie, Paris*. **156**, 1043–1047.

EPSTEIN, A. N., SPECTOR, D., SAMMAN, A. and GOLDBLUM, C. (1964). Exaggerated prandial drinking in the rat without salivary glands. *Nature* **201**, 1342–1343.

GRANT, R. T. (1962). Vasodilatation and body warming in the rat. *Journal of Physiology, London*. **167**, 311–317.

HAINSWORTH, F. R. (1967). Saliva spreading, activity, and body temperature regulation in the rat. *American Journal of Physiology*. **212**, 1288–1292.

HAINSWORTH, F. R. (1968). Evaporative water loss from rats in the heat. *American Journal of Physiology*. **214**, 979–982.

HAINSWORTH, F. R. and STRICKER, E. M. (1969). Evaporative cooling in the rat: Effects of partial desalivation. *American Journal of Physiology*. **217**, 494–497.

HAINSWORTH, F. R. and STRICKER, E. M. (1970). Salivary cooling by rats in the heat. In *Physiological and Behavioral Mechanisms of Temperature Regulation*. Edited by J. D. Hardy, A. P. Gagge and J. A. J. Stolwijk. C. C. Thomas, Springfield, Illinois, pp. 611–626.

HAINSWORTH, F. R. and STRICKER, E. M. (1971). Evaporative cooling in the rat: Differences between salivary glands as thermoregulatory effectors. *Canadian Journal of Physiology and Pharmacology*. **49**, 573–580.

HAINSWORTH, F. R., STRICKER, E. M. and EPSTEIN, A. N. (1968). Water metabolism of rats in the heat: Dehydration and drinking. *American Journal of Physiology*. **214**, 983–989.

HARRISON, G. A. (1958). The adaptability of mice to high environmental temperatures. *Journal of Experimental Biology*. **35**, 892–901.

HERRINGTON, L. P. (1940). The heat regulation of small laboratory animals at various environmental temperatures. *American Journal of Physiology*. **129**, 123–139.

KNOPPERS, A. T. (1942). Le queue du rat, témoin de la régulation thermique. *Archives néerlandaises de physiologie de l'homme et des animaux* **26**, 364–406.

LIPTON, J. M. (1968). Effects of preoptic lesions on heat-escape responding and colonic temperature in the rat. *Physiology and Behavior*. **3**, 165–169.

MALAN, A and HILDWEIN, G. (1969). Thermorégulation en ambiance chaude d'un hibernant le Hamster d'Europe (*Cricetus cricetus*): comparaison avec le rat blanc. *Archives des sciences physiologiques*. **23**, 153–181.

RAND, R. P., BURTON, A. C. and ING, T. (1965). The tail of the rat, in temperature regulation and acclimatization. *Canadian Journal of Physiology and Pharmacology*. **43**, 257–267.

STRICKER, E. M. (1970). Influence of saliva on feeding behaviour in the rat. *Journal of Comparative and Physiological Psychology*. **70**, 103–112.

STRICKER, E. M., EVERETT, J. C. and PORTER, R. E. A. (1968). The regulation of body temperature by rats and mice in the heat: Effects of desalivation and the presence of a water bath. *Communications in Behavioral Biology*. **2**, 113–119.

STRICKER, E. M. and HAINSWORTH, F. R. (1970a). Evaporative cooling in the rat: Effects of hypothalamic lesions and chorda tympani damage. *Canadian Journal of Physiology and Pharmacology.* **48,** 11–17.

STRICKER, E. M. and HAINSWORTH, F. R. (1970b). Evaporative cooling in the rat: Effects of dehydration. *Canadian Journal of Physiology and Pharmacology.* **48,** 18–27.

SWIFT, R. W. and FORBES, R. M. (1939). The heat production of the fasting rat in relation to the environmental temperature. *Journal of Nutrition.* **18,** 307–318.

THOMPSON, G. E. and STEVENSON, J. A. F. (1965). The temperature response of the male rat to treadmill exercise, and the effect of anterior hypothalamic lesions *Canadian Journal of Physiology and Pharmacology* **43,** 279–287

Part VI
EVOLUTION

Editor's Comments
on Papers 25 and 26

25 HEATH
The Origins of Thermoregulation

26 COWLES
Possible Origin of Dermal Temperature Regulation

Most, if not all, thermoregulatory responses evolved out of systems that were originally used for other purposes. The final two papers in this volume offer provocative and intriguing speculations about what those other systems might be for two of the most important mammalian thermoregulatory responses. James E. Heath (Paper 25) argues that shivering and nonshivering thermogenesis, the major methods of heat production, originally evolved because the change from a reptilian to a mammalian posture caused increased muscle tension and an accidental increase in heat production. Raymond B. Cowles (Paper 26) discusses the heat-conserving or heat-dissipating functions of peripheral vasomotor tone in mammals and argues that this basic thermoregulatory mechanism first served as a supplemental respiratory organ in amphibia. Thus these two papers illustrate that a mechanism evolved for one purpose has as a side benefit an adaptive value in an entirely different system.

Given that evolution takes a great deal of time and that all the thermoregulatory mechanisms available to endotherms could not have evolved concurrently, we have here a plausible, though of course untestable, evolutionary underpinning for the multiple thermostat model of temperature regulation proposed in Part III. Suppose we start with an organism that senses temperature and uses some form of behavior to cool down when it is too hot outside or warm up when it is too cold. That would be one complete integrating system. Later on in changing its posture, it accidentally develops a mechanism for producing heat internally. When the temperature sensors gain control over this new form of heat production, we have another complete integrating system. It would be advantageous to be able to lose some of that heat more quickly. The animal already has a good vasomotor system for respiratory purposes, and it would also be useful for losing heat if it were under thermal control. When temperature sensors come to control peripheral blood flow, however long it takes, we have another thermoregulatory integrating system. And so on. Thus it appears that multiple thermostats might have arisen because of design constraints on the evolution of endothermy (Satinoff 1978).

REFERENCE

Satinoff, E., 1978. Neural organization and evolution of thermal regulation in mammals. *Science* **201**:16-22.

25

Copyright © 1968 by Yale University Press

Reprinted by permission of Yale University Press from pages 259–278 of *Evolution and Environment*, E. T. Drake, ed., Yale University Press, 1968, 470 pp.

THE ORIGINS OF THERMOREGULATION
James Edward Heath

Some years ago C. Ladd Prosser observed that the real marvel of living things is that they are so similar (1960). Much of the excitement of modern molecular biology derives from this fact. Yet the generality that is so clearly seen at the molecular or even cellular level often becomes obscured at higher levels of integration. In spite of similarity, delineation of distinctions between a veneer of adaptations, either general or special, and the more profound pattern is often difficult and hazardous. The study of thermoregulation proves to be no exception.

Over 2,000 years ago Aristotle proposed an organization of the mineral and living world which included warm versus cold blood as an important distinction between higher and lower animals. His arrangement of living forms was flexible, and it admitted of some fine gradations between levels of complexity.

A recent study of behavioral versus physiological regulators (Yamamoto, 1965) reveals that this ancient Greek dichotomy is as lively as ever. However, the accumulation of knowledge has changed greatly the definition of the two categories. Perhaps equally important is a new viewpoint developed by systems engineers and lately applied to biological problems (Ashby, 1956; Hardy, 1961). This approach is concerned less with the nature of regulatory mechanisms than with the way a mechanism operates to accomplish regulation. It discards superficial differences and promises to expose the essential fabric of regulatory mechanisms.

Yamamoto (1965) has proposed that behavioral or poikilostatic responses can be distinguished from homeostatic ones by only two basic properties. First, poikilostasis does not require internal receptors since it can depend upon external senses. It is therefore not necessarily error-actuated. Second, poikilostasis does not alter "sources and sinks" (in a narrow context, heat production and heat loss) but acts by altering the "class of random processes" the animal is exposed to (behavioral responses). I wish to examine these properties critically, and then, with a clearer vision of the nature of thermoregulation among terrestrial organisms, to suggest trends in the evolution of thermoregulation.

THE RECEPTORS

The location of receptors can be examined in two ways. First, are all physiological responses to temperature necessarily cued internally? Second, are some behavioral responses independent of external sensation?

Among mammals, current research shows that deep internal thermal receptors, especially the hypothalamus, can no longer be considered the absolute control point for body temperature. Often responses to heat and cold are initiated by peripheral receptors. A few examples will make this clear.

One component of mammalian and avian thermoregulation involves changes with temperature in the rate of perfusion of peripheral tissues. These vasomotor changes influence the flux of heat from the body. Experimental heating and cooling of the hypothalamic region cause major changes in the perfusion of superficial structures. For example, the pinna of the ear may act alternatively as an insulator or heat exchanger (Hammel, 1965). On the other hand, insulative changes due to vasomotion occur more generally as "anticipatory" responses to stress (Hardy, 1961, 1965). A sudden exposure to cold ambient conditions causes constriction of superficial vessels and an increase in insulation before there is a detectable change in hypothalamic temperature. The hypothalamic temperature may even increase temporarily (Hardy, 1961). Presumably, receptors located at the extreme periphery sense a sudden temperature change and initiate vasomotor adjustments. The amount of vasomotor change depends on the rate of change of temperature at peripheral receptor sites. Among reptiles, and perhaps other vertebrates, vasomotor response to temperature is present (Cowles, 1958b; Bartholomew and Tucker, 1963). It is largely a peripherally sensed, rate-control process. In lizards, vasomotion favors increased heat flux between a cold animal and hot surroundings, and decreased flux between a hot animal and cold surroundings (Bartholomew and Tucker, 1963). However, warming and cooling of the reptilian hypothalamus produces vasomotor adjustments in turtles similar to those of mammals (Rodbard, 1948; Rodbard et al., 1950; Heath et al., MS).

Panting among reptiles, birds, and mammals begins when the internal temperature exceeds some set-point value (Hardy, 1961; Heath, 1965; King and Farner, 1961). Evaporative hyperventilation occurs even in insects in response to high temperature of the interior of the body (Edney and Barrass, 1962; Adams and Heath, 1964a). Among sheep, hyperventilation and evaporative cooling depend more upon the temperature in the scrotum than upon hypothalamic temperature (Waites, 1961, 1962; Waites and Voglmayr, 1963).

Even the more internalized mechanisms of temperature control such as heat production are subject to modification by external temperature receptors. Table 1 relates heat production to hypothalamic temperature in a man. Both the critical temperature for the initiation of increased heat production and the magnitude of the response are sensitive to external temperature conditions (Benzinger et al., 1963). Clearly, even large mammals under nearly ideal conditions rely upon peripheral receptors to trigger adjustments in heat production.

Among higher taxa there are no purely poikilostatic or homeostatic animals. Externally sensed, non-error-actuated behavioral responses to temperature are found among all the groups. There are also internally sensed error-actuated behavioral responses in each group (Satinoff, 1964; Heath, 1964a). Yamamoto's criteria based upon receptor site fail in all ways to distinguish a homeostat from a poikilostat.

Table 1. Set Points of Human Temperature Regulation
From Benzinger et al., 1963.

Ambient temperature (°C)	Critical temperature, cranial (°C)	Heat production (cranial temp. 36.4°C) (cal/sec)
30	36.8	~40
28	37.0	~52
26	37.1	~64
20	37.1	~88

CONTROL OF SOURCES AND SINKS

The second specific set of criteria relates to effector mechanisms. Physiological regulators control production and loss of heat whereas behavioral regulators alter their location and postures. The previous discussion of panting and vasomotor control applies equally to the effector side of the argument. For example, whether one chooses reptiles (Warburg, 1965), birds (King and Farner, 1961), or mammals or insects (Edney and Barrass, 1962), the net result of panting and hyperventilation is a decrease in body temperature, provided random processes of ambient temperature and humidity are favorable. At high temperatures all of these animals are homeostatic.

Ignoring some special mechanisms, such as sweating in mammals or special evaporative devices of moths (Adams and Heath, 1964a), the distinction between homeostat and poikilostat reduces to a single criterion, the production of heat to resist cooling. Birds and mammals adjust their heat production against heat loss to arrive at a balance that keeps their internal temperature constant. The rate of heat loss depends upon the amount of insulation and the area of the interface between the organism and the environment (Scholander et al., 1950). Mammals have arrived at a thermal balance in three ways. The arctic fox is an insulative specialist. Its heat production is low over a wide range of temperatures. At ambient temperatures above a critical level it must lose heat largely by vascular, evaporative, and behavioral changes, or suffer hyperthermia. An elephant is a surface-to-volume expert. Simply by being large, its rate of heat loss is low over a wide range of temperatures (Benedict, 1936). The shrew is a heat-production expert (Morrison et al., 1959). Its small size precludes a long, dense pile. However, a large scope of metabolism permits this animal to maintain its body temperature in very hostile situations. All other mammals and also birds fall within the extreme limits represented by the fox, elephant, and shrew.

These three animals, and in fact most birds and many mammals, maintain their body temperatures at high and relatively constant levels from birth to death. However, some mammals and birds have compromised their thermal homeostasis to some extent either seasonally or even daily (Pearson, 1960). Hibernators, aestivators, and animals which undergo regular daily torpor restrict thermal homeostasis in the ordinary sense to critical periods in their lives. During these periods they carry out the prime functions of existence. Similarly, reptiles and insects may restrict their activity to critical times of day or season when ambient conditions are favorable. The line between poikilostat and homeostat becomes tenuous in this context.

Among vertebrates, only birds and mammals have developed external insulative materials—fur and feathers. However, some insects have an external pile. Beneath the "furry" surface of a large moth, for example, lies one of the most energetic tissues in the entire animal kingdom—flight muscle. The machinery for heat production and an insulative material to decrease heat loss are both present. A thermocouple implanted in the thorax reveals that a moth can produce sufficient heat to warm many degrees centigrade above ambient temperature (Dotterweich, 1928; Oosthuizen, 1939; Krogh and Zeuthen, 1941; Dorsett, 1962). Periods of an hour or more of high heat production may occur regularly and spontaneously in these insects (Adams and Heath, 1964b) and over a wide range of ambient temperatures (Heath and Adams, 1965). Figure 1 shows a spontaneous period of activity measured from sphingid, saturniid, and arctiid moths, and a hummingbird. During active periods the moths feed and seek mates. The initiation of a period of activity begins by a special "warmup" behavior (Dotterweich, 1928; Dorsett, 1962). In all these moths the warmup is not an accidental result of flight. Indeed, one of the traces shown is from *Celerio lineata* (Sphingidae), and this moth cannot fly at all below 28°C. Except for the duration of activity and the range and apparent set points of temperatures, there is little to distinguish these traces from those of warm-blooded vertebrates. The extent and importance of heat production among insects is unknown. Insulative materials are found among Lepidoptera (Church, 1960), Hymenoptera and Coleoptera (Krogh and Zeuthen, 1941), and Neuroptera (Adams, personal communications). Church (1960) discovered that some dragonflies have a series of air sacs, lying just beneath the surface of the thorax, which restrict the outward flow of heat. Increased thoracic temperature resulting from heat production in the flight muscles has been described for only a few species of each order. Until more is learned of the thermostat and integrative mechanisms involved in insect heat regulation, it is perhaps best to call their responses behavioral thermogenesis.

Even the so-called cold-blooded vertebrates occasionally generate noticeable heat. Some large varanid lizards produce an internal heat excess of 2°C during artificially stimulated activity (Bartholomew and Tucker, 1964). However, the most significant case is that of brooding pythons. Female pythons not only warm their clutch of eggs by violent contraction of their trunk muscles but also regulate the temperature of the clutch to 31–32°C at ambient temperatures from 26–30°C (Hutchinson et al., 1966).

Cold-blooded vertebrates have also had their bulk specialists. Undoubtedly, the giant Mesozoic saurians encountered only small fluctuations in body temperature from day to day (Colbert et al., 1946). Given a small heat production and the vasomotor mechanisms of modern reptiles, it is conceivable that these giants were as thermally stable as any bird or mammal.

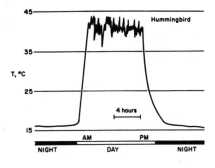
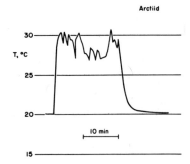
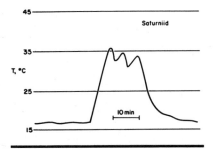
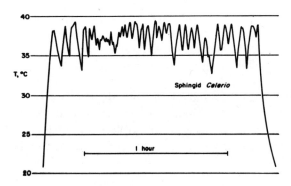

Fig. 1. Heat production during activity by moths and a bird. The arctiid weighed 0.13 gm; the other moths averaged 1–2 gm. The hummingbird curve was synthesized from data given by Pearson (1950), Bartholomew et al., (1957), and Lasiewski (1964). Note the difference in time scale.

In summary, all properties of warm-bloodedness are found at least incipiently in reptiles and insects. Peripheral temperature receptors are more important in mammalian thermoregulation than was realized a few years ago. On the other hand, the cold-blooded terrestrial animals may rely largely on internal receptors. Most of the effectors of mammalian and avian thermoregulation are present among reptiles. However, sustained and controlled heat production is perfected only by birds and mammals and to a lesser extent by insects.

Cowles (1962) recognized this and suggested that the temperature responses of terrestrial organisms be divided into two broad categories. Animals which rely upon internal heat production to regulate temperature he called endotherms. Those that rely upon external heat sources he called ectotherms. This classification disposes of confusion generated by misleading terminology and focuses upon the essential difference, and it will be adopted here. Part of my task is to account for the development of heat-producing mechanisms in pre-avian and pre-mammalian reptiles and to account for a paralellism by some insects. However, this will require some background information on behavioral regulation of temperature.

BEHAVIORAL REGULATION

Just as there are similarities between endotherms and ectotherms, there are also some important differences. Most terrestrial organisms regulate internal temperature by more or less elaborate systems of motor patterns. If one allows for the diversity of morphology and adaptation between groups such as butterflies and lizards, some general properties of behavioral regulators emerge. Temperature is maintained not to a single level but to a range of temperatures. The limits of the range are often sharp, and each limit is associated with a set of distinctive behavior patterns. The execution of one of these patterns results in the return of body temperature to the prescribed range. Within the regulated range the organism need do little to alter heat flux. Table 2 compares ranges, set points, and motor patterns for several animals. The range is usually 5°C or more wide. Under natural conditions the organism often reaches and stays within the range for long periods in thermal equilibrium and without need for regulatory adjustment (Heath, 1964, a and b; Heath, 1967). A regulated range rather than a single set point means the animal need not alter repeatedly the arena of its activity in order to regulate temperature. One can hardly imagine a more commodious arrangement.

Each limit of the regulatory range is to some extent independently determined by the demand of the habitat. Thus, the desert horned lizard *Phrynosoma m'calli* has a set point of 40.4°C for the shade-seeking motor pattern, whereas the coastal species *Phrynosoma coronatum* exhibits the same pattern at 37.7°C. Both species leave the shade between 34° and 35°C. This relationship is repeated among several other thermoregulatory patterns (Table 3).

The set points of the regulatory range are largely insensitive to the rate of heat flux, and they remain constant in spite of varying thermal conditions. Horned lizards change the frequency of shuttling movements between sunlit areas and shade with time of day (Heath, 1965). Even when the average duration of forays into direct sunlight averages as little as two minutes, the animals still retreat to shade at about 37.7°C. This is strong evidence that the pattern is elicited by a central receptor sensing deep-body temperature. On the other hand, peripheral

Table 2. Set Points of Behavioral Temperature Regulation

Animal	Range (°C)	Set point (°C)	Motor pattern
Ectotherms			
Iguanidae			
Phrynosoma coronatum	3.5	37.7	Shade-seeking
		34.2	Leave shade
Phrynosoma m'calli	5.5	40.4	Shade-seeking
		34.9	Leave shade
Cicadadae			
Magicicada cassini	6.8	31.8	Shade-seeking
		25.0	Leave shade
Endotherms			
Sphingidae			
Celerio lineata	2.9 – 3.2	38.0	Cessation of shivering
		37.7	Shade-seeking
		34.8	Resumption of shivering
Saturniidae			
Rothschildia jacobae	~4	36	Cessation of shivering
		~32	Resumption of shivering

Table 3. Set Points of Behavioral Temperature Regulation in Two Closely Related Ectotherms

Animal (habitat)	Panting (°C)	High temp. burrowing (°C)	Shade seeking (°C)	Area change (°C)	Leave shade (°C)	Head-up burrowing (°C)
Phrynosoma coronatum (coastal grassland)	43.2	40.5	37.7	37.4	34.2 NS*	35.3 NS
P. platyrhinos (desert)	44.1	41.0	39.1	38.9	35.7 NS	36.9 NS

* NS = P > 0.5

reception seems relatively unimportant in controlling regulatory behavior. These patterns also show the same detector-to-effector relationships, on-off, proportional, and rate controls, that are found in mammals (Heath, 1965).

Among tropical reptiles such as the green iguana (Bogert, 1959) or shade-selecting anoles (Ruibal, 1961) responses to prevent overheating are well developed. Because of the warmth of their habitat they do not need elaborate responses to cold. Very likely, the early terrestrial vertebrates also possessed only evaporative cooling, vasomotion, and a few behavioral patterns to regulate temperature (Cowles, 1958, a and b).

One major line in the evolution of thermoregulation has been an elaboration of behavioral mechanisms as adaptation to severe and more "temperate" habitats. Figure 2 demonstrates the complexity achieved by some modern reptiles.

Although the receptor sites for temperature and many of the effector mechanisms of temperature regulation are common between ectothermic and endothermic vertebrates (Hammel et al., 1967), the regulated internal temperature is rather constant in birds and mammals; but it regularly fluctuates in reptiles through a 3–5°C range. Is the endothermic integrative system derivable from that of ectotherms?

ORIGIN OF THERMAL REGULATION IN MAMMALS

The origins of mammals from reptilian stock are well known in their general aspects. Paleontologists, familiar with the mammal-like appearance of therapsid remains from Triassic beds, tend to favor the idea that these animals were progressing rapidly toward homeothermy. By convention, the loss of the reptilian jaw suspension and making of the dentary–squamosal joint separate the therapsids from true mammals (Simpson, 1960). However, many mammalian characteristics may have developed either before or after the acquisition of the mammalian jaw suspension (Brink, 1955). Indeed, Crompton (1958) described an intermediate animal, *Diarthrognathus,* which had a double articulation involving both the reptilian and the mammalian jaw suspension.

Brink (1955) listed several mammalian characteristics of therapsids. He found evidence of a diaphragm and development of ethmoturbinal bones for increased olfactory surface or warming and moistening inspired air. He also discussed the possible presence of vibrissae and sweat glands on the snout. The dentition of therapsids was differentiated, although it was mammal-like only in general aspects. At the same time the secondary palate developed, possibly to permit simultaneous mastication and respiration. These data led Brink to propose that some advanced therapsids were homeothermic; later (1959) he described a specimen of *Thrinaxodon liorhinus* discovered in a peculiarly mammal-like position. He suggested the position may have served to reduce heat loss. The animal was buried by a flood without disturbing its posture. It may have died during a hibernation period. This is a tempting explanation. Van Valen (1960) agreed that therapsids had some degree of homeothermy. Although all the mammal-like

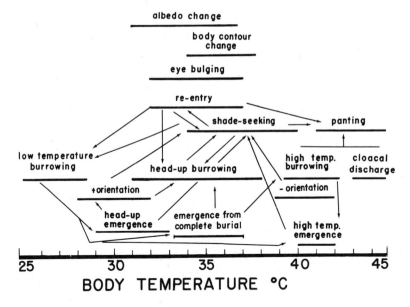

Fig. 2. Thermoregulatory ethogram of *Phrynosoma coronatum* (after Heath, 1965). The bars represent the range; arrows represent transitions. The sum of these activities results in regulation of body temperature to a prescribed range.

characteristics could be explained away—for example, the position of *Thrinaxodon* was accidental, or crocodilians have longer secondary palates than most mammals —the occurrence among therapsids of these mammalian facies when viewed in concert is impressive.

Therapsids possibly behaved more like mammals than like reptiles (Colbert, 1958). For example, the differentiated dentition of therapsids indicates that their feeding behavior was mammalian. On the other hand, the therapsid ear structure was still quite reptilian. Behavior associated with hearing was presumably also reptilian. Reed (1960) nevertheless proposed that therapsids be included in the class Mammalia. He argued that therapsids were homeothermic and that this was in part responsible for the changed posture and activity of therapsids.

I believe that most of the mammalian characteristics of therapsids can be traced not to "warm-bloodedness" but to a shift from the sprawling gait of pelycosaurs to the limb-supported posture of therapsids (Fig. 3). A diaphragm, secondary palate, differentiated teeth, even a moist nose, and expanded ethmoturbinal bones could be modifications to permit rapid respiration during high activity and rapid replacement of depleted energy stores. I believe that the shift in posture led not only to the development of mammalian temperature regulation but also to basic changes in the organization of the central nervous system.

During periods of rapid locomotion, therapsids must have experienced an increase in internal temperature above that of the surroundings, much like modern varanids. However, lizards produce insignificant quantities of heat at rest.

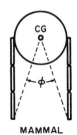

Fig. 3. Resting stance of a reptile and a mammal. The weight of the mammal is borne by the limbs; the reptile rests directly on the ground. CG, center of gravity; ø, angle of stability.

Very likely therapsids differed in this respect. The skeletal remains of therapsids suggest that they adopted the fixed pillar stance of modern mammals. This stance requires a great deal of activity in the muscles about the fixed joints, in the antigravity muscles of the back, and in the muscles of the pectoral girdle. Further, this posture demands more than increased muscular activity and tucking the limbs under the body. It requires sweeping changes in muscular organization from the action of muscle masses. (Vaughn, 1956) to the activity of the postural motor unit.

Among amphibians and to a lesser extent living reptiles, the muscles receive a double innervation—large nerve fibers with few endings on the muscle and smaller fibers with many endings. Stimulation of the large fibers produces an all-or-none twitch in the muscle. Stimulation of small fibers produces a graded local contraction within some motor units (Kuffler and Gerard, 1947). Muscular tone in the frog is produced by summing many small graded contractions. This is largely the case in reptiles as well (Zhukov, 1965). Mammals do not possess the graded contractile mechanism in skeletal postural muscles at all. Rather the postural motor unit responds in an all-or-none fashion. A steady pull upon a muscle's tendon is produced by summing the random twitch contractions of many motor units. The mammalian mechanism is simpler and perhaps offers greater control than the amphibian system of the magnitude of tension produced about the joint (Ruch and Patton, 1965). The tension in a mammalian muscle can change rapidly and with less delay than can frog or tortoise muscle (Zhukov, 1965). As a result of a shift in posture, then, mammal-like reptiles probably acquired *accidentally* a system to provide heat even at rest.

Among modern mammals, the tonic mechanism provides about 30 per cent of the heat production at basal levels (Jansky, 1962). The contractile mechanism accounts for 60 per cent of the response of rats to cold (Davis and Mayer, 1955b). Jansky (1962) obtained similar figures but included the basal metabolism of muscles to arrive at 73 per cent of the response to cold. Davis and Mayer (1955a) found a smooth transition from tonic activity in rats to full shivering. This smooth transition suggests that shivering is closely related to maintenance of tone and may have evolved from the tonic mechanism.

Noncontractile thermogenesis was probably elaborated later in mammalian evolution. Maher (1965) found increased metabolism after thyroid treatment of lizards. This response is sensitive to temperature and occurred only when the body temperature of the lizard was high (over 30°C). Increased sensitivity to thyroid hormones and a high body temperature may account for part of the difference between the basal metabolic rates of mammals and reptiles.

A high internal heat production requires an effective transport system to provide adequate nutrients and remove wastes. A four-chambered heart and reorganized vascular tree could have developed in response to a high oxygen demand resulting from exercise in therapsids. Tucker (1966) suggested that differences in the physiological properties of mammalian and reptilian circulation can be related in part to the high respiratory demand of mammals during exercise.

Adoption of a new posture required changes in the central nervous system to handle increased sensory input and to produce finer muscular control. There is a progressive increase in the size of the cranial vault among therapsids (Van Valen, 1960) which probably reflects the increasing role of the forebrain in sensory analysis and integration. Although endocranial casts of Jurassic and early Cretaceous mammals show only conspicuous increases in the size of the olfactory bulbs, by late Cretaceous, major alterations in the structure of the tectum indicate the culmination of a long trend toward reorganization of acoustic and visual input (Edinger, 1964). Both lines of modern mammals have somatotopically organized sensory projections into the forebrain (Lende, 1964), while most modern reptiles (Kruger and Berkowitz, 1960) and birds (Northcutt, 1966) have only diffuse projection. The parallelism between monotremes, and marsupial and placental mammals, suggests that these trends began with therapsid reptiles. The therapsid forebrain must have had localized and specific areas for sensory projection, which implies that the neocortex had begun to differentiate from the general cortex of earlier reptiles. Although the olfactory sense may have been accentuated at first (Edinger, 1964), the primary projection of olfaction is not to neocortex in any case. Further, modern reptiles which rely upon olfactory senses have not extended the size or importance of general cortex (Ariens-Kappers et al., 1936). More likely, a greater significance of visual input to motor coordination in rapidly moving therapsids compared to slowly moving reptiles increased tectocerebral interaction. In a more general way, Westoll (1963) suggested that the shift in posture from the reptilian to the mammalian stance presented therapsids with more information about the environment and began a trend toward greater intelligence.

There is evidence of changes in motor coordination as well. Amphibians possess the beginnings of the cortically originated extrapyramidal motor system (COEPS) of mammals (Herrick, 1948). In modern mammals the extrapyramidal system is concerned with gross postural control (Patton, 1965). It is connected through basal and midbrain nuclei to lower motor centers. It is at least partially somatotopically organized in placental mammals, but its condition in monotremes is not known. The COEPS is absent in lizards, crocodilians, and probably birds (Northcutt, 1966). It probably underwent modification during the initial changes in therapsid posture. However, the extrapyramidal system must have proved inadequate to control locomotion in therapsids.

Mammals have a second motor system absent in all other living vertebrates. This system projects directly from neocortex to both the red nucleus of the pons and to spinal motor centers. It is concerned with fine motor control. The pyramidal system originates in somatotopically organized cortex in marsupials, placentals, and monotremes (Lende, 1964). This suggests it originated with therapsids. The pyramidal system is evident in the gross anatomy of the brain and spinal cord, where it shows as cerebral peduncles, pons, and pyramids. Edinger (1962) doubted that any of these structures existed in therapsids. However, Olson (1944) described a pons in several therapsids. The existence of a pons correlates with an increase in cerebellar size, which signaled the beginning of the neocerebellum of mammals. Therapsids may have possessed the short feedback circuit of mammals which operates in locomotor integration. The appearance of a pons may also indicate changes in respiratory control of therapsids, perhaps involving a diaphragm. The origin of all mammalian brain specializations may have begun in response to the locomotor shift of therapsids.

I now hope to piece together the information available to outline a possible history of mammalian thermal regulation. Although Reed (1960) suggested that sphenacodonts might have been endothermic, there seems little to support his view. Therapsids must have possessed some vasomotor control of temperature and effective panting. Their small, reptile-like forebrain suggests that their behavior was highly stereotyped. Support of their bodies above the ground implies some basic shifts in the postural mechanism, especially a more active tonic mechanism intermediate between reptiles and mammals. This change in muscle activity proceeded simultaneously with the development of a diaphragm to provide ventilation of the lungs largely in response to exercise. The first therapsids probably possessed only white or fast muscle; "slow" or red postural muscles developed later. The red muscle system would grant greater stability. Its development might parallel the reduction in tail size among late therapsids.

Body temperature probably depended upon a combination of physiological mechanisms and selection of suitable environment. The central nervous integration of temperature remained reptilian. Therapsids probably experienced fluctuations in body temperature with activity but restricted the fluctuations to within a regulated range of temperature. The position of *Thrinaxodon*, referred to earlier, might suggest that it could arouse from low body temperature by increased tone in the skeletal muscle. However, the undisturbed repose of this animal suggests its arousal mechanism was limited. Therapsids, in general, probably lacked a complete shivering mechanism. Large therapsids may have had consistently high, but variable, body temperatures because of poor dissipation of heat.

For several million years therapsids remained the largest terrestrial carnivores. Their rapid locomotion might have forced contemporary sauropsids to develop more rapid running ability. By late Triassic, sauropsids developed a bipedal gait and possibly as great a speed as therapsids, at lower metabolic cost. In any case the therapsids disappeared, leaving the large carnivore niche to bipedal dinosaurs. The limb-supported posture and short tail of Cretaceous ceratopsians suggest a parallel with the mammal-like reptiles of several tens of millions of years before.

The successors to therapsids were much smaller animals. The brain of these early mammals continues the trend toward greater utilization of peripheral sensing. If the Jurassic and Cretaceous mammals resulted from capitalization upon internal heat production to exploit small size, then certain limitations of small "warm-blooded" animals may suggest something of their biology. These small mammals may not have maintained a high body temperature during inactive periods. Perhaps they conserved energy by undergoing daily torpor like modern bats or murine opossums (Morrison and McNab, 1962). The full exploitation of small size required development of a shivering mechanism in order to permit rapid warming from torpor to activity independent of surrounding temperature. Such activities also require effective insulation. If these small mammals underwent frequent torpor, there would be little advantage in refinement of the regulatory control inherited from therapsids.

The final stage in the development of mammalian temperature regulation began in late Cretaceous and Eocene times with elaboration of neocortex and full utilization of peripheral sensing. Mammals in these periods became larger and could no longer afford major fluctuations in internal temperature (Johansen, 1962). Peripheral receptors gradually modified the role of hypothalamic receptors of temperature in the control circuits of thermal regulation. The hypothalamic centers of modern mammals still require large changes ($\sim 2°C$) in brain temperature to stimulate thermoregulatory responses. However, such large changes do not occur frequently in nature. Instead, there seem to be two independent set points, one for cold and one for warm responses, just as in reptiles (Table 4). Each set point is variable and at least partially determined by ambient conditions (Hammel, 1965). Although the set points may completely overlap in modern mammals, under fixed ambient conditions, there is a surprisingly reptilian aspect to the temperature sensitivity of the hypothalamus. The modern mammalian thermoregulatory system anticipates changes in heat flux of the body by peripheral receptors and thereby avoids fluctuation of body temperature.

The chronology set forth is hypothetical. It does correspond to changes in size and adaptive trends in the phylogenetic lines leading to modern mammals.

Table 4. *Set Points of Temperature Regulation in the Dog*
From Hammel, 1965.

Ambient temperature	Hypothalamic Temperature		
	Thermogenesis	*Panting*	*Range*
(°C)	(°C)	(°C)	
13.5	38.8	41	2.2
23.0	36.8	38.8	2.0

THE ORIGIN OF THE THERMOREGULATION OF BIRDS

The development of thermoregulation in birds followed a different course than in mammals. Although the general aspects of avian thermoregulation have been well studied, the details of the mechanisms have not received the intense inspection given to mammals. In addition, the fossil record of birds is fragmentary. However, the close morphological relationship of birds to reptiles and their analogy with insects give clues to the development of avian thermoregulation.

The impressions of *Archeopteryx* in the Jurassic beds clearly show that this bird-like reptile possessed aerodynamic feathers. A long, feathered tail means these animals belong to the class of stable fliers. They could glide from perch to perch, but they had not developed the musculature for flapping flight. Cowles (1946) suggested that the evolution of insulative materials must precede large heat production. The development of feathers prior to the acquisition of flight musculature in birds supports this contention.

The next fossils are found in Cretaceous strata, but by that time birds were fully developed. Clearly, the fossil record is inadequate to detail the early evolution of birds. However, if heat production is the primary difference between reptilian and avian thermoregulation, we can trace, by analogy from insects, some of the events leading to avian homeothermy.

Powered flight is an energetic process. Since muscle is not a perfect machine, a great deal of heat is produced incidental to flight. For example, Church (1960) found that the desert locust *Schistocerca gregaria* produces a heat excess of 6°C during flight. An excess of 4°C is found among periodical cicadas (Heath, unpublished). Whether a heat excess is ecologically important in insects can be documented only by field studies. In June 1965, while studying cicadas in Ohio, I had such a rare opportunity. One morning began with a heavy overcast. Male cicadas, *Magicicada septendecim*, sat about on prominent twigs in basking positions waiting for the sun to break through the overcast. Body temperatures of these animals were all within a fraction of 19°C. Although they can fly at this temperature, they will not do so voluntarily. Yet a few individuals were flying and actively singing. Body temperatures of these animals were 22–24°C. Since the flying cicadas could not have reached those temperatures by basking, they must have begun flying for one reason or another, and had warmed sufficiently to reach temperatures of normal activity. I verified this by simply knocking quiet males from their perches. The disturbed animals spread their wings and flew about awkwardly. Within a few moments, they began alternately flying and singing in cicada fashion. They were then captured and their body temperatures measured. They had reached 22–24°C. Meanwhile, inactive basking animals remained at 19°C. The few active individuals gained two to three hours of activity that morning over their quiescent comrades. Even small and incidental increases in body temperature may occasionally extend the activity period significantly. The cicadas probably do not encounter conditions like those I described often enough to develop behavioral patterns to capitalize on this heat production. Further, *Magicicada* is gregarious, and mating is not dependent upon the performance of any individual male. Among solitary species, for example *Tibicen canicularis*, more active males might have an advantage. The example illustrates what could be called incipient behavioral thermogenesis.

Among other insect groups, selection has operated differently so that much larger temperature excesses are produced, and body temperature is maintained to a narrow temperature range.

There is little doubt that heat production and regulation among birds evolved along similar lines. "Warm-bloodedness" in birds and in moths resulted from capitalization upon the incidental heat produced in flight. Many of the refinements in avian respiratory and circulatory anatomy and physiology probably evolved, as with mammals, in response to the metabolic demands of exercise.

Similarly, pterosaurs were endothermic to some extent (Romer, 1945). These flying reptiles were not strictly gliders. At least one group of pterosaurs, Pterodactyloids, developed unstable, powered flight. Further, some pterodactyls possessed hair-like structures (Romer, 1945). However, these may have served only tactile functions.

Further discussion of the evolution of avian temperature regulation is hampered by insufficient data. The general response of birds to cold is shivering. Nonshivering thermogenesis may exist in some birds (Freeman, 1966), but it may still involve the contractile tissue. What is the nature of tonus in birds? Is there a graded tonic response like that of reptiles, or have birds independently developed the mammalian, all-or-none motor unit? If there was a shift to twitch tonus, did this begin with the shift to bipedal posture in ancient reptiles? Since crocodilians are near the stem of thecodont reptiles, similar questions might be posed of these animals.

Birds have largely eliminated the cerebral cortex. Extrapyramidal motor control probably does not exist in birds (Northcutt, 1967). What then, is the nature of the central nervous tonic circuitry? Since the bird forebrain continues a trend begun in reptiles to striatal development and suppression of cortex, does this indicate that birds have retained an essentially reptilian perception of the thermal environment?

Birds normally undergo large fluctuations of body temperature especially in response to activity. Such fluctuations might mean the hypothalamus of birds still functions as the absolute sensor of body temperature and that it utilizes a minimum of peripheral sensory information. Are there anticipatory mechanisms other than vasomotion in birds? Dawson (1962) suggested that the central nervous integration required for thermoregulation of birds was probably present in ancestral reptiles. Is the bipartite behavioral temperature control of reptiles still discernible in the responses of birds? Without answers to these and many other questions, the problem of the origin of avian thermoregulation must remain very sketchy. However, when the details of avian thermal reception and integration are worked out, they need not mirror the mammalian system. Rather, they may reflect the independent development of thermoregulation by birds.

SUGGESTIONS FOR FURTHER WORK

The speculations made here cannot be fully explored without comparative investigations of the muscular mechanisms of tonus and its central nervous circuitry among higher vertebrates. Much more must be known of the control of temperature in reptiles, birds, and mammals. However, the crucial evidence of evolution always lies in the fossil record.

We need information on the angle of stability of the bodies of fossil reptiles (Fig. 3), on the distribution of weight upon the limbs, and on the size of postural muscles. With these data we can calculate the stresses and forces about supporting joints. The heat production of muscles supporting the joints and maintaining posture can then be estimated. We can also estimate the temperature excesses produced by activity. We would then be in a better position to speculate upon the time of appearance and the extent of endothermy among fossil reptiles. We can continue to hope to discover imprints of hair associated with the bones and trackways of fossils as a further clue.

Progressive refinement of locomotion and increased receptivity to sensation are associated with changes in the size and external features of the brain. Our interpretation of these changes depends upon a better knowledge of the neuroanatomy of living forms. With this knowledge in hand, we will be better able to interpret the endocranial casts made of fossil forms.

CONCLUSIONS

Thermoregulation developed in higher vertebrates and insects probably to withstand the rapid fluctuations of temperature in the terrestrial environment. The differences in temperature responses among each of the groups, although superficially very striking, are few. The most significant is the independent development of high internal heat production by birds, mammals, and some insects to oppose cold stress.

Endotherms appear to regulate body temperature to a single set point, whereas ectotherms have both high and low points separated by several degrees centigrade. Endothermic insects have clearly retained the bipartite regulatory mechanism, while mammals and perhaps birds have obscured the two ectothermic set points by increasing the role of peripheral temperature receptors. Ectotherms commonly undergo fluctuations in temperature and can rely upon a rigid internal receptor or receptors. Endotherms cannot afford the large losses or gains in heat content required to produce deviations of internal temperature. They rely heavily upon externally sensed "anticipatory" responses to thermal stress. This hypothesis opposes the generally held view, but it is consistent with the trend among mammals toward increased peripheral sensitivity.

Endothermy probably developed several times among Mesozoic animals. Early birds, several groups of insects, and possibly pterodactyls capitalized upon the high energy production in flight to arrive at varying degrees of endothermy. Among mammal-like reptiles, a shift to limb-supported posture required refinements in the musculature which ultimately led to control of internal heat for temperature regulation. The development of endothermy in each group accompanied or proceeded from experimentation with new locomotor patterns.

Acknowledgments. I thank J. S. Willis and R. G. Northcutt for critically reading the manuscript and for their penetrating comments which helped immeasurably in the formulation of the ideas presented here. Many refinements of my speculations resulted from conversations with B. C. Abbott, P. A. Adams, R. B. Cowles, W. R. Dawson, M. S. Gordon, H. S. McDonald, and C. L. Prosser. This work was supported by National Science Foundation Grant GB-3702.

REFERENCES

Adams, P. A., and J. E. Heath, 1964a. An evaporative cooling mechanism in *Pholus achemon* (Sphingidae). *J. Res. Lepidoptera,* 3(2): 69–72.

———, 1964b. Temperature regulation in the sphinx moth, *Celerio lineata. Nature,* 201: 20–22.

Ariens-Kappers, C. U., B. C. Huber, and E. C. Crosby, 1936. *The Comparative Anatomy of the Nervous System of Vertebrates Including Man,* reprinted 1960. Hafner, New York.

Ashby, W. R., 1956. *An Introduction to Cybernetics.* Chapman & Hall, London.

Bartholomew, G. A., and V. A. Tucker, 1963. Control of changes in body temperature, metabolism, and circulation by the agamid lizard. *Amphibolurus barbatus. Physiol. Zool.,* 36(3): 199–218.

———, 1964. Size, body temperature, thermal conductance, oxygen consumption and heart rate in Australian varanid lizards. *Physiol. Zool.,* 37(4): 341–54.

Bartholomew, G. A., T. R. Howell, and T. J. Cade, 1957. Torpidity in the White-throated Swift, Anna Hummingbird and Poor-will. *Condor,* 59: 145–55.

Benedict, F. G., 1936. *Physiology of the Elephant.* Rept. No. 474, Carnegie Inst., Washington.

Benzinger, T. H., C. Kitzinger, and A. W. Pratt, 1963. The human thermostat. In *Temperature—its measurement and control in science and industry,* J. D. Hardy, ed., pt. 3, Biology and Medicine. Reinhold, New York, p. 637–65.

Bogert, C. M., 1959. How reptiles regulate their body temperature. *Sci. American, 200:* 105–20.

Brink, A. S., 1955. Speculations on some advanced mammalian characteristics in the higher mammal-like reptiles. *Palaeontologia Africana, 4:* 77–96.

———, 1959. Note on a new skeleton of *Thrinaxodon liorhinus*, *Palaeontologia Africana, 6:* 15–22.

Church, N. S., 1960. Heat loss and the body temperature of flying insects. Parts I and II. *J. Exp. Biol., 37(1):* 171–212.

Colbert, E. H., 1958. Morphology and behavior. In *Behavior and Evolution*, A. Roe and G. G. Simpson, eds. Yale Univ. Press, New Haven, p. 27–47.

Colbert, E. H., R. B. Cowles, and C. M. Bogert, 1946. Temperature tolerances in the American alligator and their bearing on the habits, evolution and extinction of the dinosaurs. *Bull. Am. Mus. Nat. Hist., 86:* 327–74.

Cowles, R. B., 1946. Fur or feathers; A result of high temperatures? *Science, 103:* 74–75.

———, 1958a. Additional notes on the origin of the tetrapods. *Evolution, 12:* 419–21.

———, 1958b. Possible origin of dermal temperature regulation. *Evolution, 12(3):* 347–57.

———, 1962. Semantics in biothermal studies. *Science, 135:* 270.

Crompton, A. W., 1958. The cranial morphology of a new genus and species of ictidosaurian. *Proc. Zool. Soc. London, 130:* 183–216.

Davis, T. R. A., and S. Mayer, 1955a. Nature of the physiological stimulus for shivering. *Am. J. Physiol., 181:* 669–74.

———, 1955b. Demonstration and quantitative determination of the contributions of physical and chemical thermogenesis on acute exposure to cold. *Am. J. Physiol., 181:* 675–78.

Dawson, W. R., 1962. The evolution of avian temperature regulation. In *Comparative Physiology of Temperature Regulation*, J. P. Hannon and E. Viereck, eds. Arctic Aeromedical Lab., Ft. Wainwright, Alaska, p. 45–72.

Dorsett, D. A., 1962. Preparation for flight by hawk-moths. *J. Exp. Biol., 39:* 579–88.

Dotterweich, H., 1928. Beiträge zur Nervenphysiologie der Insekten. *Zool. Jahrb. Physiol., 44:* 399–450.

Edinger, T., 1964. Midbrain exposure and overlap in mammals. *Am. Zool., 4:* 5–20.

Edney, E. B., and R. Barrass, 1962. The body temperature of the tsetse fly *Glossina morsitans* Westwood (Diptera, Muscidae). *J. Insect. Physiol., 8:* 469–81.

Freeman, B. M., 1966. The effect of cold, noradrenaline and adrenaline upon the oxygen consumption and carbohydrate metabolism of the young fowl (*Gallus domesticus*). *Comp. Biochem. Physiol., 18:* 369–82.

Hammel, H. T., 1965. Neurons and temperature regulation. In *Physiological Controls and Regulations*, W. S. Yamamoto and J. R. Brobeck, eds. Saunders, Philadelphia, p. 71–98.

Hammel, H. T., F. T. Caldwell, and R. M. Abrams, 1967. Regulation of body temperature in the blue-tongued lizard. *Science, 156:* 1260–62.

Hardy, J. D., 1961. Physiology of temperature regulation. *Physiol. Rev., 41(3):* 521–606.

———, 1965. The "set-point" concept in physiological temperature regulation. In *Physiological Controls and Regulations*, W. S. Yamamoto and J. R. Brobeck, eds. Saunders, Philadelphia, p. 98–116.

Heath, J. E., 1964a. Head-body temperature differences in horned lizards. *Physiol. Zool., 37(3):* 273–79.

———, 1964b. Reptilian thermoregulation: An evaluation of field studies. *Science, 146:* 784–85.

———, 1965. Temperature regulation and diurnal activity in horned lizards. *Univ. California Publ. Zool., 64*(3): 97–136.

———, 1967. Temperature responses of the periodical "17 year" cicada, *Magicicada cassini* (Homoptera, Cicadidae). *Am. Midland Naturalist, 77* (1): 64–76.

Heath, J. E., and P. A. Adams, 1965. Temperature regulation in the sphinx moth during flight. *Nature, 205:* 309–10.

Herrick, C. J., 1948. *The Brain of the Tiger Salamander.* Univ. Chicago Press, 409 p.

Hutchinson, V. H., H. G. Dowling, and A. Vinegar, 1966. Thermoregulation in a brooding female Indian python, *Python moluris bivittatus. Science 151:* 694–96.

Jansky, L., 1962. Maximal metabolism and organ thermogenesis in mammals. In *Comparative Physiology of Temperature Regulation,* J. P. Hannon and E. Viereck, eds. Arctic Aeromedical Lab., Ft. Wainwright, Alaska, p. 133–74.

Johansen, K., 1962. The evolution of mammalian temperature regulation. In *Comparative Physiology of Temperature Regulation.* J. P. Hannon and E. Viereck, eds. Arctic Aeromedical Lab. Ft. Wainwright, Alaska, p. 73–131.

King, J. R., and S. D. Farner, 1961. Energy metabolism, thermoregulation, and body temperature. In *Biology and Comparative Physiology of Birds,* v. 2, A. S. Marshall, ed. Academic Press, New York.

Krogh, A., and E. Zeuthen, 1941. The mechanism of flight preparation in some insects. *J. Exp. Biol. 18:* 1–10.

Kruger, L., and E. C. Berkowitz, 1960. The main afferent connections of the reptilian telencephalon as determined by degeneration and electrophysiological methods. *J. Comp. Neurol. 115:* 125–41.

Kuffler, S. W., and R. W. Gerard, 1947. The small-nerve motor system to skeletal muscle. *J. Neurophysiol. 10:* 383–94.

Lasiewski, R. C., 1964. Body temperatures, heart and breathing rate, and evaporative water loss in hummingbirds. *Physiol. Zool. 37:* 212–23.

Lende, R. A., 1964. Representation in the cerebral cortex of a primitive mammal. Sensori-motor, visual, and auditory fields in the Echidna. *J. Neurophysiol., 27:* 37–48.

Maher, M. S., 1965. The role of the thyroid gland in the oxygen consumption of lizards. *Gen. Comp. Endocr., 5*(3): 320–25.

Morrison, P., and B. K. McNab, 1962. Daily torpor in a Brazilian murine opossum (*Marmosa*). *Comp. Biochem. Physiol., 6:* 57–68.

Morrison, P. R., F. A. Ryser, and A. R. Dawe, 1959. Studies on the physiology of the masked shrew *Sorex cinereus. Physiol. Zool., 32:* 256–71.

Northcutt, R. G., 1966. Analysis of reptilian cortical structure. *Nature, 210:* 848–50.

———, 1967. Architectonic studies of the telencephalon of *Iguana iguana. J. Comp. Neurol., 130:* 109–48.

Olson, E. C., 1944. Origin of mammals based upon cranial morphology of the therapsid suborders. *Geol. Soc. Am. Spec. Paper 55:* 1–136.

Oosthuizen, M. J., 1939. The body temperature of *Samia cecropia* (Lepidoptera, Saturniidae) as influenced by muscular activity. *J. Entomol. Soc. S. Africa, 2:* 63–73.

Patton, H. D., 1965. Reflex regulation of movement and posture. In *Physiology and*

Biophysics, T. C. Ruch and H. D. Patton, eds. Saunders, Philadelphia, p. 181–206.

Pearson, O. P., 1950. The metabolism of humming birds. *Condor*, 52: 145–52.

———, 1960. Torpidity in birds. In *Mammalian Hibernation*, C. P. Lyman and A. R. Dawe, eds. *Bull. Mus. Comp. Zool.*, 124: 93–105.

Prosser, C. L., 1960. Comparative physiology in relation to evolutionary theory. In *Evolution after Darwin*, v. 1, S. Tax, ed. Univ. Chicago Press, p. 569–94.

Reed, C. A., 1960. Polyphyletic or monophyletic ancestry of mammals, or: What is a class? *Evolution*, 14: 314–22.

Rodbard, S., 1948. Body temperature, blood pressure, and hypothalamus. *Science*, 108: 413–15.

Rodbard, S., F. Sampson, and D. Furguson, 1950. Thermosensitivity of the turtle brain as manifested by blood pressure changes. *Am. J. Physiol.*, 160: 402–09.

Romer, A. S., 1945. *Vertebrate Paleontology*. Univ. Chicago Press, 687 p.

Ruch, T. C., and H. D. Patton, 1965. *Physiology and Biophysics*. Saunders, Philadelphia. 1242 p.

Ruibal, R., 1961. Thermal relations of five species of tropical lizards. *Evolution*, 15: 98–111.

Satinoff, E., 1964. Behavioral thermoregulation in response to local cooling of the rat brain. *Am. J. Physiol.*, 206(6): 1389–94.

Scholander, P. F., R. Hock, V. Walters, F. Johnson, and L. Irving, 1950. Heat regulation in some arctic and tropical mammals and birds. *Biol. Bull.* 99: 237–58.

Simpson, G. G., 1960. Diagnosis of the classes Reptilia and Mammalia. *Evolution*, 14: 388–92.

Tucker, V. A., 1966. Oxygen transport by the circulatory system of the green iguana (*Iguana iguana*) at different body temperatures. *J. Exp. Biol.*, 44: 77–92.

Van Valen, L., 1960. Therapsids as mammals. *Evolution*, 14: 304–13.

Vaughn, P. P., 1956. The phylogenetic migrations of the ambiens muscle. *J. Elisha Mitchell Scient. Soc.*, 72: 243–62.

Waites, G. M. H., 1961. Polypnoea evoked by heating the scrotum of the ram. *Nature*, 190: 172–73.

———, 1962. The effect of heating the scrotum of the ram on respiration and body temperature. *Quart. J. Exp. Physiol.*, 47(4): 314–23.

Waites, G. M. H., and J. K. Voglmayr, 1963. The functional activity and control of the apocrine sweat glands of the scrotum of the ram. *Australian J. Agric. Res.*, 14(6): 839–51.

Warburg, M. R., 1965. The influence of ambient temperature and humidity on body temperature and water loss from two Australian lizards, *Tiliqua rugosa* and *Amphibolurus barbatus*. *Australian J. Zool.*, 13(2): 331–50.

Westoll, T. S., 1963. The functional approach to paleontological problems. *Proc. XVI Internat. Cong. of Zool.*, 3: 273–77.

Yamamoto, W. S., 1965. Homeostasis, continuity, and feedback. In *Physiological Controls and Regulations*, W. S. Yamamoto, and J. R. Brobeck. Saunders, Philadelphia, p. 14–32.

Zhukov, Ye. K., 1965. Evolution of physiological mechanisms of tonus. In *Essays on Physiological Evolution*, T. M. Turpayev, ed. Pergamon Press, London, p. 339–49.

POSSIBLE ORIGIN OF DERMAL TEMPERATURE REGULATION [1]

Raymond B. Cowles [2]

University of California, Los Angeles

The importance of blood shunting from superficial to deep tissues as a means of heat conservation in warmblooded vertebrates has been amply reviewed by Bazett (1948), Hammond (1954) and Sholander (1955). From the evidence it is clear that the mechanisms for changes in superficial blood flow including temperature-induced vasoconstriction and vasodilatation are adaptations that serve a vital need in these warm-blooded, endotherm, animals for a heat regulating mechanism that in warm-to-hot environments or during violent physical effort, will prevent a) thermal injury or even death from overheating and b) under cold environmental conditions, heat depletion and lowering of the body temperature requisite for their well being or even for their survival. In either case, the tissues of the surface of the body as well as those in the extremities may survive thermal limits that would be paralyzing or lethal to the organism as a whole.

Heat Regulation

Both the necessity and the effectiveness of heat control mechanisms normally depend on the fact that for organisms in which heat is an endogenous product with only very rare environmental exceptions (notably for diurnal desert animals) the thermal gradient between the organism and its environment is almost always favorable for heat dissipation by convection, conduction, and radiation. In other words, but for the heat conserving devices such as insulation and vascular control, body heat would flow unimpeded and more or less continuously to the organism's surroundings. When heat flow to the environment is prevented by fur or feathers, panting and evaporation in the buccal cavity and in a few mammals by sweat evaporation supplement the normal discharge of excess heat.

It is generally tacitly assumed that these heat regulating phenomena are concomitants of the evolution of warmbloodedness, i.e. the evolution of effective internal heat production, and that the need for producing heat and the devices for conserving it must therefore have been in some way associated with the existence of a cool or cooling environment. However, instead of being peculiarly endotherm adaptations for temperature regulation, it seems probable that at least the basic mechanisms for changing blood flow in the surface may have originated at a very early time in history, and that the adjusting vascular changes already may have been developed by the amphibia and subsequently passed on to the reptiles from whom this mechanism was then inherited by the warmbloods.

In the amphibia, dermal vascularity serves and probably must have served in the remote past as a necessary adjunct to their notable dependence on supplemental dermal respiration. In keeping with the general principle of size trends and specialization we can surmise that these earliest semiterrestrial and terrestrial tetrapods were probably small and unarmored, characters that would increase the effectiveness of accessory dermal respiration.

Transitional stages in the development of the reptiles to the fully evolved rep-

---

[1] These investigations were supported jointly by a research grant from the Richfield Oil Corporation, and from the University of California, Los Angeles.

[2] I wish to acknowledge here the excellent assistance given by Mr. Robert Phelan.

tilian type with wholly dry skin and pulmonary respiration would profit by the employment of these same mechanisms for dermo-vascular heat absorption because of the reptile's dependence on external sources of heat and their need for regulation of body temperatures.

The reason for noting the dermal-vascular-respirational complex in the amphibia is based on the apparently universal dependence of these animals on accessory dermal respiration through what of necessity must be a moist skin. Because of amphibian dependence on this supplemental gaseous exchange at the skin's surface, this accessory respiratory organ must be maintained in a moist condition and hence it must be supplied with blood at all times in order to maintain this essential moisture under conditions of varying atmospheric humidity. Additionally, higher environmental temperatures should be normally accompanied by a compensating increase in blood circulation at the respiratory surface since higher body temperatures produce increased metabolic rates and hence rising demands for O_2 and the dissipation of CO_2 in all known terrestrial organisms, furthermore rising temperatures are normally accompanied by increased evaporative rates.

Evidence that these compensatory dermal changes actually take place may be deduced from the well known fact that the temperature of amphibians is essentially that of the wet bulb thermometer and that a rapid increase of moisture to the skin can be effected only by a compensating vascular flow. Crude confirmation also may be obtained by observations of changes in the capillary flow in the feet and webs of amphibians, although precise direct observations should be obtained by experimentation on terrestrial albino amphibia where vascular changes in the integument should be easily observable and could be recorded by use of infrared photographs.

Adaptations in the form of supplemental respiration employing vascular changes in accordance with the respiratory needs in the ancient amphibia from among whom arose progressively the dryer-skinned lung breathing reptilia, would presumably set the stage for dermal heat-regulation by basking, in successive stages of reptilian evolution and ultimately the transitional stages of ectotherm to endotherm heat metabolism.

VALIDITY OF EXPERIMENTAL ANIMAL

In order to test for the existence of reptilian thermovascular mechanisms in the skin it is necessary to employ modern and highly evolved reptiles which raises the question of the advisability of employing them as criteria for paleophysiological phenomena. It is of course impossible to state the degree of reliability of physiological-ecological concepts pertaining to extinct animals when these concepts must be based on evidence derived from any living animal whether mammal, bird or the amphibians and reptiles, all of them creatures which presumably have accumulated numerous adaptations necessary for existence under present conditions, nevertheless these organisms are the only remaining samples of any kind by means of which we may hope to throw light on the living conditions of now extinct forms of animals.

With this possible evidential defect in mind we may nonetheless assume the probability that where only physical and chemical laws are involved, studies of modern forms may throw at least a little light on past biological events when such unchanging phenomena as respiratory exchanges of gases, or surface volume relationships in rate of temperature change, or the influence of temperatures on chemical reactions are employed. These physical and chemical phenomena certainly have not been subject to change during biotic time.

Although modern reptilia, including our experimental animal *Dipsosaurus dorsalis*, may be only scarcely related to the archaic reptiles they are nonetheless subject to the limitations of physical laws that would be involved in the heating and cooling of creatures that depend on external sources of heat. On this basis we may assume that these animals approximate the physical and some physiological relationships to heating and cooling that would be found in archaic forms of ectothermic organisms such as most if not all of the evolving reptiles as well as incipient or semiwarm bloods such as the heterotherms.

Temperature: Ectotherm versus Endotherms

Because of the still commonly employed terms that provide a basis for uncritical thinking it is asserted that cold animals live "at the temperature of their invironment." This gross oversimplification in terminology should be abandoned because, in fact, reptile temperatures may differ very greatly under a complex of environmental conditions but most notably in those forms that for their body temperatures are dependent on basking in the sunlight.

Unlike the warmblooded animals with a normal heat flow that passes from their warm body to a normally cooler environment, cold blooded terrestrial creatures depend on apparently primitive reverse functioning of this process and they draw their heat from their environment which for many (possibly all) at one time included more or less direct solar radiation. Except when they moved into an environment cooler than their bodies the necessary flow of heat for the ectotherms must have been from their environment into their bodies as compared with the usual outward flow of heat in the endotherms, i.e. from body to the environment.

This role of the body surface as a source of necessary ectotherm heat led to the speculations (Cowles, 1946) that fur and feathers or other insulating external structures would have been deleterious to all externally warmed organisms except under conditions requiring protection from too great amounts of external heat, most notably from excessive solar radiation. Under cool or cooling conditions the interposition of any heat occluding barrier between a cold blooded animal and its source of heat would still further deprive the organism of the advantage of warmth. Conversely the internal generation of heat by a naked-surfaced animal under such cooling or cold conditions would impose the necessity for very great and possibly unattainable increases in their requirements for food and for rapid digestion, and the attainment of the latter would be extremely doubtful if the enzymes of both the present day ectotherms and the warm bloods resembled those of their ancestral forms.

Insulation in an Ectotherm

The effects of covering a present day cold blooded animal with insulating fur (a crude "wrap-around" of mink!)[3] are shown in figures 1 and 2, and they illustrate the workings of a principle that could be calculated equally well but apparently less convincingly by simple mathematics. More effective insulation could be achieved by employing a duckling or a gosling down "wrapper" and the more efficient the insulation employed the greater would be the retardation of both warming and cooling. From these graphs it can be seen that the animals might tend to profit from insulation by their retention of heat when once warmed, but this advantage would in turn be cancelled by the retardation in warming rates. For diurnal heliothermic animals there might be some extension of activity during the latter part of each day, but a delayed or late start each morning would incur a penalty. For nocturnal animals there would seem to be similar cancellations of advantages and disadvantages, the disadvantages being aggravated by evaporative cooling if the insulated covering had absorbed either dew or rain.

Because of these purely physical rela-

---

[3] Class assignment experiment by Mr. Richard C. Grossman.

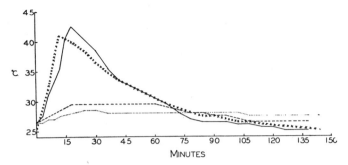

Fig. 1. Differential heating and cooling rates in fur-insulated and uninsulated lizards. Exposure to heat source 12 minutes for *Sauromalus* and 18 minutes for *Dipsosaurus*.

- - - - - - - - Fur-insulated *Dipsosaurus dorsalis*.

─────── Uninsulated *Dipsosaurus dorsalis*.

. . . . Fur-insulated *Sauromalus obesus*.

x x x x x x Uninsulated *Sauromalus obesus*.

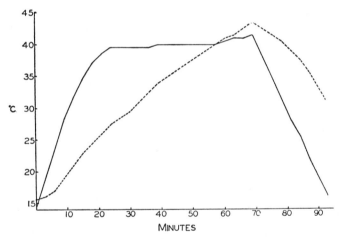

Fig. 2. Pre-chilled to approximately 15° C., exposed to heat capable of increasing temperature to + or − 43° C. and returned to chilling at 69 minutes.

- - - - - - - Fur-insulated *Dipsosaurus dorsalis*.

─────── Uninsulated *Dipsosaurus dorsalis*.

tionships to heating and cooling it is reasonably safe to assume that, since the dermal layers in cold-blooded animals served as heat accumulators rather than heat dissipators as in the endotherms, this surface heat would reach the deeper tissues and viscera by the very slow process of simple tissue conduction unless it was accelerated by an effective extension of a vascular system that in a perfected state should promote heat distribution from the dermal collector to the muscular and visceral areas where it would be needed. To serve this purpose the skin on being warmed should be flushed with the cooler arterial blood in order that

TABLE 1

	Bleb	Cloacal temp.	Cold side	Hot side	Elapsed time to absorption	
					Cold	Hot
1	0.5 cc	19.5° C.	16° C.	36.5° C.	60 min.	40 min.
2	0.2 cc	30°	15	37.5	17	10
3	0.3 cc	28°	8	44	23	10
					Av. 33	20

the deeper tissues should be warmed by the returning venous flow. Furthermore, in order to promote maximum efficiency under marginal conditions, the cool or cold, i.e. the shaded non-irradiated ground-contacting ventral surfaces of the skin should remain with constricted arterioles, and virtually emptied capillaries and venules. This is essentially the same type of mechanism that functions for heat conservation in the endotherms, and if this vascular functioning can be demonstrated in modern reptiles and postulated for reptiles ancestral to the warm bloods, we may consider vascular shunting adaptations from amphibians through the reptiles but especially in the latter, as pre-adaptations that were needed in order to set the stage for the successful evolution of warmbloodedness.

Changes in Dermovascular Flow

Because of its heliothermic (basking) nature and its excellent eurythermal characteristics, and its availability and size, *Dipsosaurus dorsalis,* an iguanid lizard of the southwestern United States, was selected for experiments in testing differential dermal blood flow under different localized skin temperatures. The loose flabby surface of *Sauromalus,* the chuckawalla, prevented their use in these experiments.

In a preliminary series of experiments 0.2 cc, 0.3 cc and 0.5 cc of water were hypodermically injected into the subject so as to form small but conspicuous blebs or swellings.[4] The injection sites were dorsolateral, on opposite sides of the 7th dorsal light spot counting anteriorly from the center of the pelvis, and far enough from the center line so that one bleb could be heated by localized infrared illumination while the other was being cooled by applications of ice. Throughout the experiment each of three animals was tied by its legs to a small operating board and flexions of the body were compensated for by manual movement of the ice or radiant heat source. The rate of dissipation of the subdermal water globules proceeded as illustrated in table 1.

It is significant that despite the permutations of hot and cold lateral temperatures, the size of the bleb, and the over-all body temperatures as represented by cloacal readings, in all instances the rates of absorption (interpreted as indicating the corresponding rates and volume of vascular circulation), were markedly most rapid on the heated side.

In order to "load" the experiment against the theory, fractionally more water was injected on the warm than on the cold side.

In a second series of experiments designed to test for the existence of vascular responses and the heat transporting adaptations to thermal differences, an ice pack was adpressed to one side of the animal and radiant heat from an infrared lamp was then directed against the opposite

[4] The experimental animals were furnished through the courtesy of Charles Shaw, curator of reptiles, San Diego Zoological Park, San Diego.

side, and the thermal gradient through the skin was obtained by means of constantin copper thermocouples and a recording potentiometer.

These experiments were conducted on the working hypothesis that if there are differences in vascular flow in accordance with changes in skin temperature then heat should be rapidly dispersed throughout the body by means of either or both an enhanced volume or rate of blood flow, and that if there is such a mechanism the temperature immediately beneath the heated skin would more closely correspond to the temperature of the body as measured in the cloaca than to its heat receiving surface. Conversely if blood flow in a cold or chilled skin is retarded the chilled area and its adjacent tissues should demonstrate localized cooling and hence the temperatures should deviate markedly from temperatures noted in the cloaca to which heat must be transported from the surface. In effect, absorption and dispersal of heat throughout the body should supplement the information obtained by the dispersal rates of the water blebs.

The results of these simple experiments which were designed to reveal the degree of

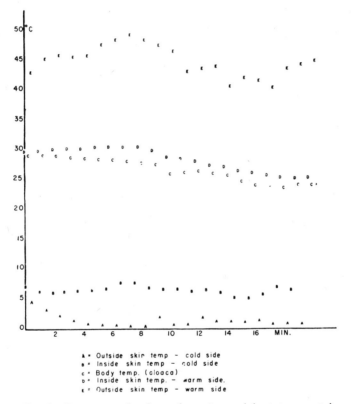

A = Outside skin temp - cold side
B = Inside skin temp - cold side
C = Body temp. (cloaca)
D = Inside skin temp. - warm side
E = Outside skin temp - warm side

FIG. 3. Dermo-vascular heat absorption and heat transport in *Dipsosaurus dorsalis*. The approximately 16° C. difference between the skin surface E and immediately below the skin D is indicative of the heightened efficiency of vascular heat transport under a warmed skin. The much reduced blood flow-heat movement on the cooled side is indicated by the mere 4° C. difference between A, the skin surface, and B, its undersurface, as well as by the great disparity between B and the cloaca C when compared with D and C.

conformity of changing dermal circulation with the theoretical construct are plotted in the accompanying graphs showing the temperature gradients across the skin of animals that were simultaneously cooled on one side of the body and heated on the other.

Figure 3 demonstrates that in accordance with the theory, the thermal gradient across the skin is notably steeper on the hot E to D than on the cold side, A to B. That there should be so little thermal conformity between the heated outer surface of the skin and the underlying area less than a millimeter distant, is indicative of the efficient transportation of heat by means of a large vascular flow in the intervening tissues, whereas on the contrary the lack of such a steep gradient on the cold side indicates that there is little vascular intervention in the depletion of local heat. On the cold side, external and subdermal temperatures are more nearly alike because the reduced flow of blood and resultant retardation of its movement from the deeper tissues fails to modify greatly the locally chilled surface area. The near conformity of subdermal temperature D and cloacal C indicates that heat from the skin dominates the body temperature and that through reduced flow the influence of the cold side is blocked.

Figures 4 and 5 present the results of differences in thermal gradients employed as a check on the validity of the assumptions drawn from observations presented in figure 3.

In two cross checks against the previous experiments an animal was first heated and then chilled (at about minute 20), the result being plotted in figure 4. The test was then reversed and the animal was started at a low temperature and warmed as shown in figure 5.

In the first check (fig. 4) when subjected to heating the temperature of the skin surface and the tissues immediately adjacent to its inner surface showed a wide difference even though the thermocouples were less than a millimeter apart, a clear indication of rapid transportation of heat from the skin by the passage of cooler blood from deeper levels and its uptake of heat. On chilling (see at min. 20) there was a precipitous drop at the surface of the skin followed by a much retarded fall in subsurface temperature and ultimately by an almost identical skin surface-skin subsurface temperature that was accompanied by barely perceptible thermal change in the cloaca. These observations indicate a rapid diminution in the surface circulation and a conductively rather than a circulatively penetrating reduction in the local temperature.

In the second check (fig. 5) radiation was directed against the skin surface of a cooled animal. A sharp rise in both the outer and inner skin surface temperatures resulted, but this was soon followed by an increasing divergence as circulation increased and the incoming heat was carried from the skin and circulated deeper into the body until an equilibrium was approached. Body temperature as represented by the cloaca levelled off as a result of heat loss from the ventral surface to the substratum and heat gain from the radiation.

These phenomena can be accounted for only by changes in rate or volume, or both in the dermal and subdermal circulation effecting transport and a blending of all blood heat.

It is possible that the increased flow of blood that accompanies dermal heating and the resultant transport of heat from the surface may also serve in protecting the skin from uncomfortably high or damaging insolation. This function might be of considerable importance to animals existing under arid or semiarid conditions, especially in those that lacked a heavily cornified or thick-scaled covering.

The discontinuities in the graphs occurred whenever the animal gulped air and thus bloated the thoracic and abdominal areas, an act that apparently resulted in restricting the flow of blood in the skin and its vascular network.

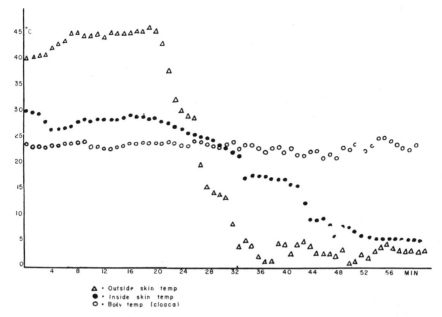

Fig. 4. Effects of surface cooling on dermo-vascular flow in *Dipsosaurus dorsalis*.

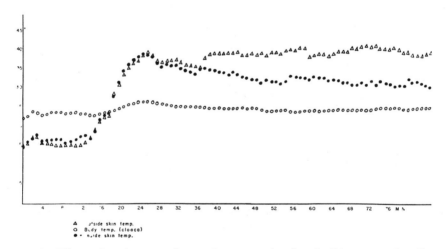

Fig. 5. Effects of surface warming on dermo-vascular flow in *Dipsosaurus dorsalis*.

Albedo in Relation to Heat Regulation

The role of an uninsulated skin as a heat absorbing surface in ectothermic terrestrial vertebrates revives the still unsettled arguments as to the utility or "purpose" of animal pigmentation. Gloger's rule (Gloger, 1833) is adequately descriptive but is not explanatory and opinion is still divided as to whether the pigmental gradient of dark to light from areas of high humidity to those that are arid, expresses physiological needs or represents color adaptations for purposes of concealment.

Density of vegetation is also correlated with amounts of precipitation and humidity, and where vegetation is abundant shadows comprise a larger percentage of darkness than where dryer conditions prevail. Lessened intensity of the cloud-filtered sunlight and the over-all darkening of the earth's natural surface that results from increasingly large amounts of moist and decomposing detritus additionally enhance darkening. Combining these factors simply reinforces the concept of an environment that itself approximates Gloger's rule, with the animal life conforming to the general albedo of the environment. The situation perfectly fits both the requirements for concealing coloration in animals, and the color gradient described by Gloger's rule.

That there are many adherents to explanations based on the functioning of dark pigment in absorbing heat is not surprising. Increased rainfall and humidity are almost invariably accompanied by a downward trend in environmental temperatures and it is therefore not surprising that so much emphasis has been placed on this physiological aspect of animal albedo. It is entirely possible that both metabolic requirements and those for concealment of the animals are mutually supportive and thus produce as perfect a conformity to Gloger's rule as could be expected by the proponents of either viewpoint.

In order to clarify the situation insofar as vertebrates alone may serve the purpose, from what has been shown it would seem as though the ectotherms would be more responsive to the thermal factor than would be the endotherms and that because of their slowness and general helplessness, that they should be equally susceptible to the dictates of albedo as a dominant factor in concealing coloration. However, their very susceptibility to heating requirements in cooler environments, which are notably prone to be areas of high humidity and precipitation, should make them conformists to the rule to an even greater degree than the more heat independent endotherms. And toward increasing humidities they do conform with no more exceptions to the rule than in the endotherms. In this direction it is certainly possible that the need for both heat and concealment have a synergistic effect on compliance with the Rule.

The answer to physiology versus concealment may be found in the ectotherms of desert regions where during active portions of their lives these extremely heat sensitive organisms are almost daily exposed to the grave dangers of overheating (Cowles and Bogert, 1944). Under these conditions the variable albedo of most diurnal desert reptiles (Atsatt, 1939) should serve the physiological requirements of heat protection and it probably does afford an advantage to the animal although preliminary experiments (Cowles, unpublished) strongly suggest that even slight differences in body habitus may and often do cancel out the theoretical significance of the change in albedo.

Under these circumstances it is necessary to turn to yet another source of information, namely the variations in color or hue of desert ectotherms on areas of differently colored substrata but under identical climatic conditions. When viewed from this standpoint these extremely heat-responsive and heat-sensitive animals do not conform to Gloger's rule. With very few exceptions desert reptiles are prone to match very closely the substratum on which they are found. this conformity has been most closely measured by K. S. Norris (ms in press) employing a photocolorimeter and populations of the sand dune lizard *Uma*, but even without such precise measurements as he reports, no experienced desert collector is unaware of the fact that it is possible to seek out areas of some specific hue and obtain closely matching *Phrynosoma platyrhinos* (a lizard) and other species of reptiles. In this *Phrynosoma*, and in *Crotalus cerastes* (a sand dwelling rattlesnake) as in the populations of the pink Algodones sand dunes and the more pallid dunes of the Coachella valley, as well as in numerous other species of snakes and lizards, the imprint of color-matching is clearly impressed on local populations.

Probably the most informative color-matching is to be found in the reptilian residents of extensive black lava beds when their pigmentation is compared with those of closely adjacent regions. This is notable in the Amboy-Pisgah crater lava flows of the Mojave desert where the populations of rock-dwelling *Sauromalus* and *Uta* are permanently very black whereas those from nearby areas of normal soil color have what might be called normal desert coloration. The unchangeable color of these animals indicates that the color is not a transient visual response but is innate and probably due to genetic factors.

In all of these cases climatic conditions are identical as far as casual temperature studies reveal except that in the black lava beds conditions may be even hotter than on the more highly reflective deserts a few feet distant.

I believe that there can be no doubt but that Gloger's rule is expressed in desert reptiles only because they must match the soil color or be eliminated due to their conspicuousness. Even the conspicuously changing albedo of such species as *Dipsosaurus dorsalis* and *Callisaurus ventralis* for example is accompanied by even more intense changes of diurnal illumination and glaring reflection from the soil surface, and hence their change may be more important for camouflage than for heat regulation.

The great temperature changes that many reptiles can achieve through flattening their body on a hot substratum or elevating it out of such contact or even by mere movement from sun to shade may be more effective in heat control than degree of albedo.

In order to live, all vertebrate ectotherms are certainly dependent upon accumulating adequate amounts of heat from their surroundings and, of equal importance, avoiding the deleterious or lethal effects of overheating. They exist in a precarious thermal balance and are subject to many thermal vagaries in environment. Nonetheless they are also susceptible to such a high degree of predation that despite the possible benefits that might accrue from albedos adjusted solely to their thermal needs, predation appears to force a compromise that is largely balanced in favor of concealment.

Conclusion

The dermal surface and its contained vascular systems in the cold bloods, the ectotherms, served first as a supplemental respiratory organ in the amphibia, then as a heat collector and disperser in the reptiles and finally as an essential temperature regulator for the endotherms. In each of the shifts of function the thermal reaction expressed in terms of changed blood flow was the same, namely contraction under cold and expansion under heat.

Since the skin and its contained vascular mechanisms is essential to reptiles as a heat collector and disburser the interposition of barriers to heat absorption, i.e. fur, feather, or hypodermal adipose tissue would be harmful except where surplus heat is a problem, and the reduction in effective heat collecting and the heat distributing blood flow would retard body warming. To demonstrate some aspects of thermal adaptations three simple experiments were devised: 1) insulation of the body skin by a fur "coat," 2) absorption of water blebs beneath the skin, and 3) analysis of the thermal gradient through the skin.

In each case the results conformed to the theory and the conclusion was drawn that vascular blood shunting originated in the amphibia to serve respiratory needs, was then employed in the reptilia to absorb heat, and in the warm bloods to regulate but, except for cold-environment species, more particularly to dissipate excess heat.

If the evidence and the arguments prove acceptable, we may assume the probability that these preadaptations in the dermo-vascular system began to set the stage for vertebrate warmbloodedness in the tetrapods as long ago as possibly Devonian times. Because the skin is a heat absorbing mechanism for the ectotherms, Gloger's rule should be particularly applicable to them, but where conformity of their color to their backgrounds in desert regions where vegetation and vegetational debris do not masque the varicolored ground surface they show no evidence of the rule nor do they support the concept that color is a physiologic advantage rather than a concealing mechanism.

Literature Cited

Atsatt, S. R. 1939. Color changes as controlled by temperature and light in the lizards of the desert regions of Southern California. Univ. Cal. Press. U.C.L.A. Series, **1**: 11, 237–276.

Bazet, H. C. (and co-authors). Temperature changes in blood flowing in arteries and veins in man. Jour. Physiol., **1**, 1, 1–19.

Cowles, R. B. 1946. Fur and feathers; a result of high temperatures? Science **103**; 2664, 74–75.

Cowles, R. B., and Bogert, C. M. 1944. A preliminary study of the thermal requirements of desert reptiles. Bul. Am. Mus. Nat. Hist., **38**: art 5, 265–296.

Gloger, C. L. 1833. Das abwandern der vogel durch einfluss des klimas. Breslau.

Sholander, P. F. 1955. Evolution of climatic adaptation in homeotherms. Evolution, **9**: 1, 15–26.

AUTHOR CITATION INDEX

Abrams, R. M., 204, 258, 353
Adair, E. R., 192, 218
Adams, P. A., 352, 354
Adolph, E. F., 330
Ailver, J., 193
Allison, F., Jr., 255
Alpers, B. J., 156, 157
Alpher, G. W. H. M. van, 320
Altman, P. L., 204
Andersen, H. T., 163
Andersson, B., 79, 163, 192
Andral, 235
Ariens-Kappers, C. U., 352
Ashby, W. R., 352
Atkins, E., 254, 255, 258
Atsatt, S. R., 366
Aub, 43

Baldwin, B. A., 217
Barbour, H. G., 6, 116, 156
Bard, P., 156, 157, 192
Barenne, J. G. Dusser de, 116
Barker, J. L., 204
Barrass, R., 353
Barrett, J. A., Jr., 255
Bartholomew, G. A., 71, 72, 204, 284, 352
Baxter, D. W., 218
Bazett, H. C., 116, 156, 170, 192, 366
Beaton, L. E., 6, 163
Beeson, P. B., 226, 254, 255
Benedict, 43
Benedict, F. G., 352
Bennett, I. L., Jr., 226, 254, 255, 256
Benzinger, T. H., 352
Bergquist, E. H., 204
Berkowitz, E. C., 354
Bernard, C., 49, 229, 243
Bernheim, H. A., 260
Berry, C., 6, 163
Berthold, A. H., 265, 266
Berthrong, M., 255
Beznák, A. B. L., 320

Billroth, 231, 241
Birzis, L., 192, 193
Bleier, R., 192
Blatteis, C. M., 226, 260
Bligh, J., 6, 204
Blinks, D. C., 284
Bodel, P., 258
Bogert, C. M., 6, 72, 284, 353, 366
Botkin, 243
Bourneville, 242
Boyce, R., 117
Braude, A. I., 255
Brett, J. R., 281
Bretz, W. L., 205
Bright, E. M., 43
Brink, A. S., 353
Britton, S. W., 43
Brobeck, J. R., 6, 79, 157, 163, 168, 192, 204, 320
Brodie, B. C., 241
Brouwer, B., 117
Brück, K., 194, 217
Burton, A. C., 156, 204, 330
Buzzard, E. F., 117

Cabanac, M., 217, 218, 226, 258
Cade, T. J., 71, 284, 352
Cadot, M., 330
Cajal, S. R., 117
Caldwell, F. T., Jr., 258, 353
Calvert, D. T., 217
Cannon, W. B., 43, 49, 156
Carey, F. J., 255
Carlisle, H. J., 168, 192, 217
Carlton, P. L., 6
Carpenter, 43
Carpenter, D. O., 204
Carpenter, R. E., 72
Casby, J. U., 192
Casterlin, M. E., 226, 260
Chai, C. Y., 192, 204
Chambers, W. W., 192, 193

Author Citation Index

Charcot, J. M., 243
Chatonnet, J., 218
Chevillard, L., 330
Chew, R. M., 72
Church, N. S., 353
Clark, G., 79, 156, 157, 192
Clark, S. L., 255
Clawson, G. H., 281
Cluff, L. E., 255
Colbert, E. H., 353
Confavreux, J., 218
Conner, J. D., 192
Constatin, L. L., 192
Cooper, K. E., 260
Corbit, J. D., 204, 217
Cornew, R. W., 176
Covert, J. B., 226, 260
Cowgill, G. R., 320
Cowles, R. B., 6, 72, 284, 353, 366
Cramer, 43
Crawford, I. L., 192
Crompton, A. W., 353
Crosby, E. C., 352
Crowe, S. J., 117
Cushing, H., 117
Cushman, A. J., 204
Cytawa, J., 79

d'Aquili, E., 192
Davis, L., 194
Davis, T. R. A., 353
Davison, C., 156
Dawe, A. R., 354
Dawson, H., 204
Dawson, W. R., 71, 72, 353
Dearborn, G., 218
DeWitt, C. B., 258, 260
Dittmer, D. S., 204
Dixon, 315
Donaldson, 315
Dorsett, D. A., 353
Dotterweich, H., 353
Dougherty, M., 193
Dowling, H. G., 284, 354
Dubois, E. F., 157
Duhring, 315
Dumm, M. E., 320
Dworkin, S., 43, 157, 193

Ebbecke, 315
Edinger, H. M., 204
Edinger, T., 353
Edney, E. B., 353
Eisenman, J. S., 6, 79, 204, 226
Eliasson, S., 163
Elkan, E., 258

Elson, P. F., 281
Epstein, A. N., 330
Erb, 243
Erb, W. H., 156
Erickson, T. C., 157
Ershoff, B. H., 320
Everett, J. C., 330
Ewald, C. A., 243

Farner, S. D., 354
Ferster, C. B., 281
Findlay, J. D., 217
Fisher, C., 157, 163
Fisher, K. C., 281
Forbes, A., 117
Forbes, R. M., 205, 331
Forgrace, P., 193
Forman, 43
Francou, M., 218
Frazier, C. H., 157
Freeman, B. M., 353
Freeman, W. J., 194
Frings, H., 72
Frings, M., 72
Frumin, M. J., 194
Fry, F. E. J., 281
Furguson, D., 355
Fusco, M. M., 163

Gale, C. C., 79, 163, 192, 193
Gamgee, A., 117
Gavarret, 235
Gelineo, S., 204
Gerard, R. W., 354
Girard, P. F., 218
Glaubach, S., 157
Gloger, C. L., 366
Goldblum, C., 330
Goodfield, G. J., 6
Gordon, M. S., 72
Grant, 276
Grant, R., 163, 255, 330
Green, P. P., 157
Greenman, 315
Groot, J. de, 204
Guieu, J. D., 193
Guttmacher, A. F., 315
Guttman, L., 193
György, P., 320

Hahn, L. J., 157
Hainsworth, F. R., 204, 330, 331
Haldane, J. S., 117
Hales, J. R. S., 193
Hamilton, C., 163, 168
Hammel, H. T., 6, 79, 163, 176, 204, 258, 353

Author Citation Index

Hammouda, M., 157
Hardy, J. D., 6, 79, 163, 193, 204, 258, 320, 353
Hare, W. K., 193
Harrison, F., 6, 79, 157, 163, 204
Harrison, G. A., 330
Hartman, 43
Hasama, B., 157, 172
Hashimoto, M., 157
Hatschek, 117
Hausmann, 265
Heath, J. E., 205, 258, 260, 284, 352, 353, 354
Hellon, R. F., 79, 260
Hemingway, A., 163, 192, 193, 194, 204
Hendersen, R., 226
Herdman, S., 168
Herrick, C. J., 315, 354
Herrington, L. P., 330
Heymans, J. F., 117
Hildwein, G., 330
Hill, 43
Hirtz, H., 235, 243
Hock, R., 73, 355
Hokfelt, B., 79, 192
Holmes, G., 117
Holmes, R. L., 204
Holmquist, B., 193
Homans, J., 117
Houpt, T. R., 73
Howell, T. R., 72, 352
Huber, B. C., 352
Hudson, J. W., 71, 72, 204, 284
Hutchison, V. H., 284, 354

Ing, T., 204, 330
Ingram, D. L., 217
Ingram, W. R., 157, 163, 193
Iriki, M., 169, 176, 193, 217
Irving, L., 73, 355
Isenschmid, R., 43, 117, 157, 170

Jackson, J. H., 6
Jacobson, R. H., 79
Jansky, L., 354
Jarnum, S. A., 73
Jeddi, E., 260
Jessen, C., 169, 193, 217
Jochmann, 235
Johansen, K., 284, 354
Johnson, F., 355
Juch, 266

Kahn, R. H., 157
Kalabukhov, N. I., 284
Karplus, J., 117
Keene, W. R., 255, 256
Keller, A. D., 157, 168, 170, 193

Kim, Y. B., 226
Kinder, E. F., 49
King, J. R., 354
King, M. K., 255, 256
Kirby, G. P., 255
Kitzinger, C., 352
Kluger, M. J., 258, 260
Klussman, F. W., 169, 217
Knoppers, A. T., 330
Koenig, H., 192
Kohler, Cl., 218
Kosaka, M., 168, 169, 193
Krehl, L., 117
Krehl, W. A., 320
Kreidl, A., 117
Krogh, A., 354
Kruger, L., 354
Kuffler, S. W., 354

Lantz, 315
Larsson, B., 79, 192
Larsson, S., 163
Lasiewski, R. C., 354
Lassar, O., 235
Laties, V. G., 163, 281
Lator, R., 235
Lee, D. H. K., 205
Le Fèvre, J., 117, 226
Legge, K. F., 217
Leitner, P., 72
Lende, R. A., 354
Leschke, E., 117
Lewy, F. H., 157
Leyden, 235
Licht, P., 284
Liebermeister, C., 226, 235, 243
Lilienthal, J. L., Jr., 79, 157
Lin, M. T., 192
Lipton, J. M., 169, 193, 204, 330
Liu, C. N., 192, 193
Loder, 43
Loeb, 43
Longet, F. A., 117
Ludwig, O., 193, 217
Lundberg, A., 193
Lusk, 43
Lyman, C. P., 284

McClaskey, E. B., 168
McCordock, 43
McCouch, G. P., 192
McCrum, W. R., 79
McEwen, G. N., Jr., 260
Macht, M. B., 192
McIver, 43
McKinley, W. A., 6, 163

Author Citation Index

McLean, J. A., 217
Macleod, J. R., 117
MacMillen, R. E., 72
McNab, B. K., 72, 354
Macnab, R., 243
Magoun, H. W., 6, 79, 156, 157, 163, 204, 258
Maher, M. S., 354
Malan, A., 330
Marey, 235
Marks, R. A., 6
Marshall, 315
Martin, 315
Martin, J. R., 169, 204
Mathews, M., 193
Mayer, E. T., 193, 217
Mayer, S., 353
Means, L. W., 205
Menkin, 43
Meurer, K. A., 169, 193
Meyer, H. H., 6, 79
Minaire, Y., 218
Monakow, C. v., 117
Monnet, P., 218
Mooney, R. D., 204
Moore, 43
Moore, L. M., 157
Moorhouse, V. H. K., 157
Morgulis, 315
Morrison, P., 72, 354
Mount, L. E., 260
Mu, J. Y., 192, 204
Murgatroyd, D., 204
Murie, M., 73
Myhre, K., 226

Nakayama, T., 6, 79, 204
Naunyn, 243
Nelson, J. E., 72
Neumann, 243
Newman, P. P., 204
Newton, 43
Ngai, S. H., 194
Norris, K. S., 73
Northcutt, R. G., 354
Novelli, G. D., 320

Ogilvie, D. M., 260
Olson, E. C., 354
Olszewski, J., 218
O'Neill, E. J., 71
Oosthuzien, M. J., 354
Otenasek, F. J., 79, 157
Owen, 276

Pasquier, J., 218
Patton, H. D., 354, 355

Pearson, O. P., 284, 355
Pellegrino, L. J., 204
Penfield, W. G., 156, 170, 192
Peter, 243
Peters, G., 157
Petersdorf, R. G., 254, 255, 256
Pick, E. P., 157
Pierau, F. K., 217
Pinkston, J. O., 157
Pittman, Q. J., 260
Poling, C. E., 320
Pollock, L. J., 194
Porter, R. E. A., 330
Portet, R., 330
Pratt, A. W., 352
Prince, A. L., 157
Probst, 117
Prosser, C. L., 355

Querido, A., 43
Quincke, 243

Ralli, E. P., 320
Rand, R. P., 204, 330
Ranson, S. W., 6, 79, 156, 157, 163, 204, 255, 258
Rasmussen, A. T., 163, 204
Rasmussen, T. W., 163, 204
Rautenberg, W., 169, 176, 217
Reed, C. A., 355
Regal, P. J., 284
Reichenbach-Klinke, H., 258
Rengger, 266
Reynolds, W. W., 226, 260
Richet, C., 79, 117, 194
Richter, C. P., 49, 315
Riddoch, G., 117
Rioch, D. McK., 156, 157
Roaf, H. E., 117
Roberts, W. W., 169, 204, 205
Robinson, K. W., 205
Robinson, T. C., 204, 205
Rodbard, S., 355
Rogers, F. T., 117
Romer, A. S., 355
Rothmann, M., 117, 118
Ruch, T. C., 194, 355
Ruibal, R., 355
Rutstein, J., 79, 169, 194
Ryser, F. A., 354

Sachs, E., 157
Samman, A., 330
Sampson, F., 355
Satinoff, E., 6, 79, 169, 194, 226, 335, 355
Schmidt-Nielsen, B., 73

Schmidt-Nielsen, K., 73, 205
Schnitzler, W., 117, 157, 170
Schober, W., 205
Scholander, P. F., 73, 355, 366
Schultz, R. B., 320
Sée, 243
Selby, N. E., 156
Selle, W. A., 256
Senator, H., 235, 243
Shea, S., 193
Sheriff, W., Jr., 163
Sherrington, C. S., 79, 117, 118, 157, 194, 315
Simpson, G. G., 355
Simon, E., 168, 169, 176, 193, 217
Simon, J., 231
Skinner, B. F., 281
Slonaker, 315
Small, 315
Smith, M. R., 255
Spector, D., 330
Squires, R. D., 79
Steinberg, M. L., 205
Stevenson, J. A. F., 205, 331
Stinson, R., 260
Stolwijk, J. A., 192
Stricker, E. M., 169, 330, 331
Strom, G., 163
Strømme, S., 176
Stuart, D. G., 194
Sundsten, J. W., 163
Sutherland, K., 79
Swift, R. W., 205, 331

Tang, P. C., 194
Tarr, R. S., 258
Taylor, C. R., 205
Teague, R. S., 157
Teitlebaum, P., 79
Templeton, J. R., 72
Thauer, R., 6, 157, 168, 169, 176, 193, 194, 217
Thomson, 315
Thompson, G. E., 205, 331
Traube, 235
Trevan, J. W., 118
Tucker, V. A., 72, 73, 284, 352, 355

Uotila, U. U., 157

Vallin, 243
Van Valen, L., 355
VanZoeren, J. G., 169
Vaughn, L. K., 260
Vaughn, P. P., 355
Veale, W. L., 260
Vinegar, A., 284, 354
Voglmayr, J. K., 355

Wade, G. N., 205
Waites, G. M. H., 355
Waller, 315
Walters, V., 73, 355
Walther, O., 193
Wang, G. H., 315
Wang, S. C., 192, 194, 226
Warburg, M. R., 355
Warrier-Jines, P. C., 118
Watson, D. W., 226
Weber, E., 231, 235, 243
Weiss, B., 163, 281, 320
Werner, L., 205
Westoll, T. S., 355
Whalen, W. J., 255
Wikoff, H., 163, 204
Williams, B. A., 205
Wilson, S. A. K., 118
Windle, W. F., 192
Winters, R. W., 320
Wit, A., 226
Wolstencroft, J. H., 204
Wood, W. B., Jr., 254, 255, 256
Woodhead, G. S., 118
Woods, J. W., 192
Woodworth, C. H., 205
Woodworth, R., 118
Wunnenberg, W., 194, 217
Wunscher, W., 205
Wyndham, C. H., 193

Yamamoto, W. S., 355
Young, J., 193

Zalesky, M., 255
Zeuthen, E., 354
Zhukov, Ye. K., 355
Zimmermann, 231
Zucker, I., 205

SUBJECT INDEX

Adrenal glands, 9, 23, 27, 28, 30, 45
  and cold exposure, 39, 119
  secretion of adrenin in, 27–28, 30, 35, 36
Animal heat, 2–3
Autonomic nervous system, 27, 234
  sympathetic division, 27, 40, 155, 231

Bar pressing, for ambient temperature change
  in pigs, 206–217
  in rats, 159–163, 203, 263, 316–320
Bees
  individual body temperature of, 271–272
  nest temperature of, 269–270
Behavioral thermoregulation, 5, 9, 40–41, 168
  during brain heating in rats, 195–205
  in fish, 5, 262, 280–281
  during hypothalamic cooling in rats, 78, 158–163
  in infants, 225–226, 259–260
  in insects, 5, 269–272, 339, 350
  in mammals, 5, 45–48, 55, 327
  in pantothenic-acid deprived rats, 316–320
  in reptiles, 5, 58–59, 225, 257–258, 282–284, 341–343
  during thermal stimulation of hypothalamus and spinal cord in pigs, 167, 206–216
Birds, origin of thermoregulation in, 349–351
Body temperature regulation, 4, 21
  in insects, 264, 271–272
  maintenance of, in decerebrates, 94–105, 127
  means of, 119
  in spinal animals, 240–242

Calorimeter, 82–85, 231
Carbonic acid, 146, 232–233, 237–238
Circadian rhythms, 203, 263

Cold-blooded animals. *See* Ectotherms
Corpus striatum, 4, 85, 86, 120

Decerebrate cats
  panting in, 101–103, 113, 152
  rigidity in, 90, 99, 113–116
Desert
  diurnal mammals in, 56–58
  nocturnal rodents in, 55
  reptiles in, 58–61
Determinism, 3, 9
Diathermy. *See* Radio-frequency heating

Ectotherms, 20, 265, 339, 358
  desert species of, 58–61, 341–343
  fever in, 225, 257–258
Electrical stimulation, of the brain, 197–198
Endotherms, 341, 358
Evolution, 334–335
  of chemical heat production, 345–346
  of dermal temperature regulation, 356–366
  of endothermy, 283, 336–355
  of fever, 258
  of shivering, 346

Fever, 17, 19, 81–82, 224–226
  induced behaviorally, 225–226
  in infants, 225–226, 259–260
  in lizards, 257–258
  in rabbits, 244–254, 257–258
  symptoms of, 228–229

Goldfish, 262, 280–281
Grooming, thermoregulatory, in rats, 168, 197, 202, 263, 321–331

Heat loss
  evaporative, in rats, 321–331
  mechanisms of, 4–5, 24, 158, 263
  morphological adaptations for, in birds, 65

Subject Index

neural center of, 77, 82, 121–122, 145, 172
Heat production
  after hypothalamic cooling, 158
  during fever, 224, 230–235
  mechanisms of, 5
  neural center of, 77, 82, 172
Heat reinforcement
  in pigs, 206–216
  in rats, 160–163
Homeostasis, 1, 4, 9, 23, 44, 50, 52, 54, 63, 70, 338
  maintained by behavioral mechanisms, 45, 47–49, 55, 59
Hyperthermia
  in cats with medullary transections, 171
  in cats with preoptic/anterior hypothalamic lesions, 144, 152
  in desert animals, 62–63, 70–71
  in goats during hypothalamic cooling, 158
  in monkeys with preoptic/anterior hypothalamic lesions, 150, 152
Hypothalamus
  lesions of, 4, 126–145, 171
  posterior, 77, 151, 153, 198
  preoptic/anterior, 4, 77–78, 151, 158, 198
  thermal stimulation of, 121–126, 158–163, 198, 206–216
  as thermostat, 4–5, 41, 76–78, 154, 177, 202
Hypothermia, in reptiles, 282–284

Insects, 262, 264–279, 339–341
Internal environment
  Bernard's theory of, 3, 8, 10–11
  constancy of, 1, 8, 19, 44

Life, forms of
  constant, 8, 10–18
  latent, 8, 18
  oscillating, 8, 13, 18
Locomotion
  as heat-escape response, 168
  during local brain heating, 197, 202

Mammals
  diurnal, 56–57
  origins of thermoregulation in, 343–348
Medulla, 81, 171, 198
  thermosensitivity of, 178, 190, 203
  transections through, 171, 187–188
Metabolic rate, 22, 35–38, 56
  in desert species, 65–68
Midbrain
  anatomy of, 107–108
  development of, 106
  thermosensitivity of, 202

transections of, 171–172, 178–179, 188
Nerves, vasomotor, 9, 14, 234, 239–241
Nervous system, 12, 15
  lesions of, 4, 78
  levels of control in temperature regulation, 111–112, 167
  thermal stimulation of, 4, 78, 120–125
  transections of, 4, 41, 171
Nest building, 5, 9, 46, 168, 263, 285–315
  in hypophysectomized rats, 46–47
  in thyroidectomized rats, 46–47
Nutrition, 17
  and behavioral thermoregulation, 316–320

Operant behavior. See Bar pressing
Oxygen, 14, 23
  consumption, 22

Pain, indifference to, 218–221
Panting, 26, 337, 347
  in birds, 62, 64
  during hypothalamic heating, 121–126
  after hypothalamic lesions, 133–134, 136, 140
  in midbrain-transected animals, 166, 172
  nervous control of, 41
Pantothenic-acid deprivation, 316–320
Piloerection, 26–27, 30, 155
  in decerebrates, 178, 189
  during spinal cord cooling, 181–183, 189
Poikilotherms. See Ectotherms
Pons, 77, 81, 113, 171–172, 187–188, 347
Posture, 203, 347
  changes in, during evolution, 346
  cold-defense, in pigs, 206
Prone body extension. See Sprawling, in heat
Pyrexia. See Fever
Pyrogen, 224
  action on CNS, 225, 244–254
  exogenous or bacterial, 225, 244–254, 257, 259–260
  leucocytic, 225, 244–254

Radio-frequency heating, 4, 78, 195–205
Respiration
  in bees, 264, 271, 275–277
  in decerebrates, 101–103, 113
  in fever, 232
  as generator of animal heat, 2
  after hypothalamic lesions, 136, 140
  increased (panting), 4–5
  as mechanism for water loss, 12, 25
  in midbrain preparations, 172

Rodents, nocturnal, in desert, 55

Setpoint, 202, 337, 341, 348, 351
  in fever, 224
Shivering, 5, 9, 35–40, 96, 119, 191
  in decerebrates, absence of, 104–105, 112, 114, 116, 171, 181
  in decerebrates, presence of, 178, 181–183, 185–187, 189
  during hypothalamic cooling, 161–162
  after hypothalamic lesions, 142–143
  nervous control of, 41
  in spinal animals, 167, 189
  during spinal cooling, 173–176, 187
Spinal cord, 5, 347
  behavior during thermal stimulation of, in pigs, 206–216
  shivering during cooling of, in cats, 173–176, 181, 183, 185–187, 189
  transections of, 77, 80–81, 166–167, 177–178, 240–242
Sprawling, in heat, 168, 203
  during localized brain heating, 197, 202, 203
Sunstroke, 224, 236
Sweating, 5, 12, 24–26, 119, 356
  in cats with hypothalamic lesions, 142–143
  nervous control of, 41, 155

Thalamus, 77
  role in temperature regulation, 146–149
Thermal comfort, in humans, 218–221

Thermogenesis, nonshivering, 5, 9, 119
Thermoregulation. *See also* Behavioral thermoregulation; Body temperature regulation
  centers, 5, 76–77, 151–153
  dermal, 356–366
  physiological, 40–42
  and water economy, 55, 57
Thermostat, 41, 76, 78, 284
  dual control of, 42, 76
Thyroid, 9, 22–23, 39, 47, 119, 346
Torpidity, 68–69, 338, 348

Units, temperature sensitive, 4, 78–79, 225, 348
  in spinal cord, 167, 216
Urea, production during fever, 233, 237

Vasoconstriction, peripheral, 5, 9, 26, 119, 155, 356
  in decerebrates, 178, 189
  during spinal cord or medullary cooling, 189
  in spinal preparations, 184, 187, 189
Vasodilation, peripheral, 5, 9, 24, 119, 321–331, 356
  during localized brain heating, 196, 198
  in low-level decerebrate cats, 182
  in midbrain-transected cats, 172
  during spinal cord or medullary heating, 189
  in spinal preparations, 184

Wasps, nest temperature of, 278–279

About the Editor

EVELYN SATINOFF is professor of psychology and physiology and biophysics at the University of Illinois at Urbana-Champaign. She received her B.S. from Brooklyn College in 1958 and her Ph.D. from the University of Pennsylvania in 1963. She works on the central neural regulation of thermoregulation in mammals and is the author of numerous research articles and review chapters in this area. She has served on the editorial boards of the *Journal of Comparative and Physiological Psychology*, the *American Journal of Physiology*, and the *Journal of Applied Physiology*. She is currently a member of the Thermal Physiology Commission of the International Union of Physiological Sciences.

ST. MARY'S COLLEGE OF MARYLAND LIBRARY
    ST. MARY'S CITY, MARYLAND